Elementary Computer-Assisted Statistics

REVISED EDITION

FRANK SCALZO
QUEENSBOROUGH COMMUNITY COLLEGE
CITY UNIVERSITY OF NEW YORK

ROWLAND HUGHES
FORDHAM UNIVERSITY
LINCOLN CENTER, NEW YORK CITY

VNR VAN NOSTRAND REINHOLD COMPANY
NEW YORK CINCINNATI ATLANTA DALLAS SAN FRANCISCO
LONDON TORONTO MELBOURNE

Van Nostrand Reinhold Company Regional Offices:
New York Cincinnati Atlanta Dallas San Francisco

Van Nostrand Reinhold Company International Offices:
London Toronto Melbourne

Library of Congress Catalog Card Number: 78-4209
ISBN: 0-442-80316-8

Manufactured in the United States of America

Published by Van Nostrand Reinhold Company
135 West 50th Street, New York, N.Y. 10020

Published simultaneously in Canada by Van Nostrand Reinhold Ltd.

15 14 13 12 11 10 9 8 7 6 5 4 3 2 1

Library of Congress Cataloging in Publication Data
Scalzo, Frank, 1941–
Elementary computer-assisted statistics.

Includes index.
1. Mathematical statistics—Data processing.
2. Probabilities—Data processing. I. Hughes, Rowland, 1935– joint author. II. Title.
QA276.4.S27 1978 519.5′028′54 78-4209
ISBN 0-442-80316-8

TO ANGELO AND ANNE

Contents

PREPACKAGED COMPUTER PROGRAMS

Acknowledgments

The authors gratefully acknowledge permission from the following companies for the photographs used in this book: Teletype Corporation, Skokie, Illinois, for photographs of a computer terminal, its keyboard, and its paper tape unit; and Anderson Jacobson, Inc., Sunnyvale, California, for the photograph of an acoustic coupler.

The authors wish to thank Professors Anthony Behr and Michael Brozinsky of Queensborough Community College, City University of New York, for their careful reading of the text and many helpful suggestions; and Dr. Anne Hughes of St. John's University, Jamaica, New York, for her constructive criticism.

Our appreciation and gratitude are extended also to Mel Lutwak for his drawing of diagrams and flowcharts, and his perceptive comments.

Especially, we thank Malcolm Mills, whose expertise in typing, suggestions concerning format and text clarity, and patience proved invaluable.

FRANK SCALZO
ROWLAND HUGHES

Suggestions to the Instructor

During the past two decades we have become a computer-oriented society. In the 1950s colleges and universities used the computer primarily for research and administrative purposes, but the computer was soon to be assimilated into education as a tool of instruction. Within the relatively short span of twenty years, nearly all institutions of higher learning have come to provide computer services for both classroom instruction and research.

Computers are now being used by educators in the instruction of mathematics, physics, chemistry, biology, medicine, business, environmental health, electrical technology, mechanical technology, and social science. The close relationship of computer science and mathematics suggests that an introductory course in elementary statistics would be made more relevant if the use of computers were an integral part of the course.

The purpose of the text is to illustrate the use of prepackaged computer programs as practical tools to solve statistical problems. More specifically, the authors have attempted (1) to introduce students to the BASIC programming language; (2) to guide students in the use and operation of computer terminals; (3) to illustrate the use of prepackaged computer programs to solve statistical problems; and (4) to provide a greater understanding of statistical concepts via the use of an intuitive-algorithmic approach to the instruction of elementary statistics.

The instructor will notice that at the end of each important statistical concept there will be a problem set for the students to solve manually. Immediately following this problem set there will be a prepackaged computer program. Each prepackaged computer program consists of an instruction sheet, a flowchart, coding in the BASIC programming language, and an additional problem set to be solved using the prepackaged program as a tool. The instruction sheet gives the program name, a brief description of purpose, and a sample run of the prepackaged program. To use a prepackaged computer program the student simply must load the program into his user's space (by program name); enter the data to solve a specific problem; command the computer to run the program with the given data; and interpret the computer output. It is important that the student fully understands the mathematical procedure before he or she uses a prepackaged computer program. The student must be able to read a mathematical problem, extract the given data from the problem, decide which prepackaged program to use, and follow the instructions for using the program.

If Chapter 1 is covered with the emphasis placed on the use of computer terminals and prepackaged computer programs, the students will be able to utilize these prepackaged materials as practical tools of problem solving. Before the semester begins the instructor should have each of the fourteen prepackaged programs loaded into the computer's memory by the appropriate name, such as STAT1, STAT2, STAT13, and BINOM.

If the BASIC programming language in Chapter 1 is covered in detail, the students will be able to comprehend the logic and coding of the prepackaged programs and perhaps revise some of them or develop their own programs.

Preface

This textbook is designed to present to the student the use of computers and prepackaged computer programs as tools to solve problems in elementary statistics. Completion of two years of high school algebra will adequately serve as a prerequisite for the student's study of both the elementary statistics and the computer applications included in this textbook.

The authors wish to suggest some possible ways to use the material presented herein. If the instructor wishes to disregard Chapter 1 and the prepackaged programs given at the end of specific topics, he will find the remaining material adequately constitutes a basic one-semester course in elementary statistics. If Chapter 1 is covered with emphasis placed on the use of computer terminals and prepackaged computer programs, the students will be able to utilize these prepackaged materials as practical tools of problem solving. Finally, if the BASIC programming language in Chapter 1 is covered in detail, the students will be able to comprehend the logic and coding of the prepackaged programs and perhaps develop their own programs.

Detailed flowcharts used in the formulation of each of the prepackaged statistical programs are included for further possible discussion. The authors have chosen to place problem sets at the end of presentations of important statistical concepts so that students can gradually build their statistical competence through solving applicable problems.

1

Understanding the Use of Computers

1.1 INTRODUCTION

During the past two decades we have become a computer-oriented society. The introduction of programming languages such as Fortran and Basic has made the computer more readily adaptable to the work of education, economics, finance, and medicine, as well as mathematics and science. The use of the computer has opened the door for ready accomplishment of complicated statistical studies not previously feasible. Programming instructions are presented throughout this text in the Basic language because it is easily learned and conveniently used in time sharing on computer terminals.

1.2 STEPS IN COMPUTER PROBLEM SOLVING

How does a person use a computer to solve problems? The answer to this question need not remain a mystery to the general public. The computer is not a genius; it is an idiot waiting to be told what to do. Let us look at the steps one should follow in order to use a computer to solve problems.

FLOWCHARTING

After the problem has been clearly defined, the programmer should construct a flowchart. A flowchart is a diagram of the step-by-step procedure the computer must follow to solve the problem. That is, what do you want the computer to do first, second, ... , last?

Coding

The second step is to translate this flowchart into a machine language or a compiler language program—a series of statements which instructs or directs the computer to perform a specific task. The only language the computer can understand is the *binary* (machine) language. In the binary language, all numbers, instructions, and commands must be written using only two symbols, 0 and 1. Writing in the binary language requires the utmost precision. Other programming languages such as Basic, Fortran, and Cobol are constructed to resemble and adapt to the languages used in certain sciences or professions. For example, the computer language used throughout this text is Basic (Beginner's All-Purpose Symbolic Instruction Code), which is easily learned and applied to the solution of problems in mathematics and the physical sciences. Any programming language other than the binary language is called a *compiler* language. When we feed our Basic language program to the computer, a binary language program called the *compiler*, which is stored in the computer's memory, will translate our Basic language program into a binary language program. Then the computer can execute our program.

Inputting

The third step is to put our program on some device so as to feed it to the machine. These devices will be discussed when computer terminals are introduced later in this chapter.

Running

The fourth step is to have the computer run through the steps of our program and hopefully give us the results we are looking for.

Debugging

Very few original programs will run smoothly the first time. Therefore, the fifth and final step is to find the errors in our program, correct the errors, and run the program again. A person should not get discouraged when a program does not run. It should be a challenge to find and correct any errors.

1.3 FLOWCHARTING

There are many standard symbols used in flowcharting. Only a few of them are necessary for the topics covered in this textbook. They are:

INPUT

In order to show that we want the computer to accept some numerical or alphabetic data, we use the following symbol:

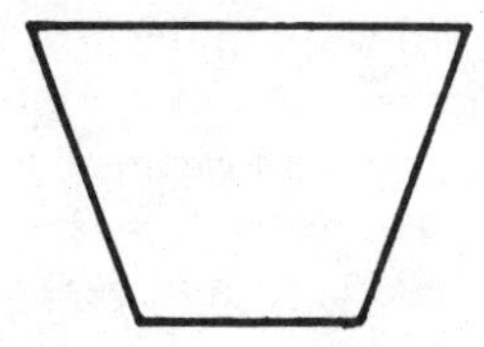

Suppose we wanted to show that the computer must read a value for A, B, and C. This would be illustrated in your flowchart as follows:

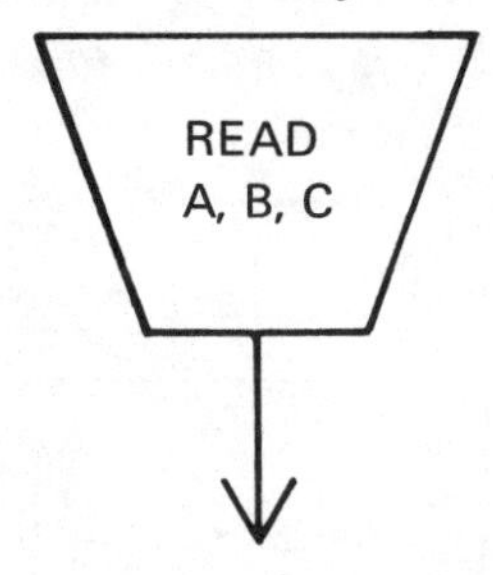

ORDER OF STEPS

The arrow used in the symbol represents the flow of the program. That is, in the illustration above, if we want to know what the computer must do next after reading values for A, B, and C, we follow the arrow.

ARITHMETIC STATEMENTS

In order to show that we want to compute the value of some variable (on the left side of the equation), given the computer already knows the values of any variables on the right side of the equation, we will use the following symbol:

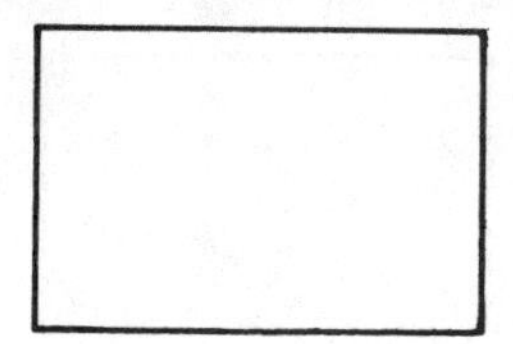

Suppose, after reading values for A, B, and C, we want the machine to compute 3A + 2B and store it in a place called X, then compute $3X + A^2$ and store it in a place called Y, then compute A + Y and store it in the place reserved for A. This is illustrated in Fig. 1.1.

Note: An arithmetic statement like A = A + Y may appear in a flowchart. Moreover, this may appear when $Y \neq 0$. Use of such a statement leads us to the discussion of the = sign in computer programming. When the computer knows the value of A and B and we tell it to compute X = 3A + 2B, it will store the value of 3A + 2B in a place called X. However, if the computer has the value of 2 stored in a place called A and the value of 3 stored in a place called Y and we tell it to compute A = A + Y, the = sign means "is replaced by." That is, the old A = 2 is replaced by the new A = 2 + 3 = 5. Therefore, the 2 stored in a place called A is erased and replaced with 5.

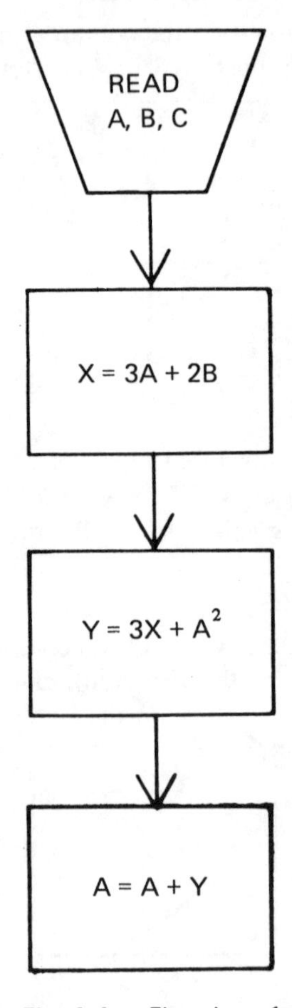

Fig. 1.1. Flowchart 1

The memory of a computer might be compared to the work of a tape recorder. What happens if a song is recorded on a certain length of tape and you play it back at a later date? The song is repeated. Now, what happens if you record another song on the same length of tape? As you know, the first song is erased and the new song takes its place.

Decisions

In order to show that we want the computer to make a decision or test the value of some arithmetic expression, we use the following symbol:

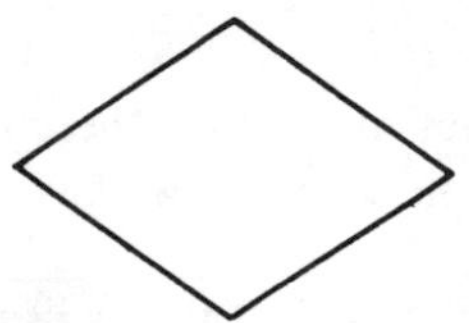

For example, consider the previous illustration where we want the machine to read values of A, B and C, and then compute values for $X = 3A + 2B$, $Y = 3X + A^2$, and replace A with $A + Y$. Now suppose what we do next depends on whether A is less than zero. Our flowchart is illustrated in Fig. 1.2. As you can see, this symbol allows you to express a desire to branch to two different locations in your program after testing to find whether A is less than zero.

Output

The symbol for showing that we want specific results given to us is the same as the symbol for input.

If we wanted the computer to print out the values stored in A, B, C, X, and Y we would write:

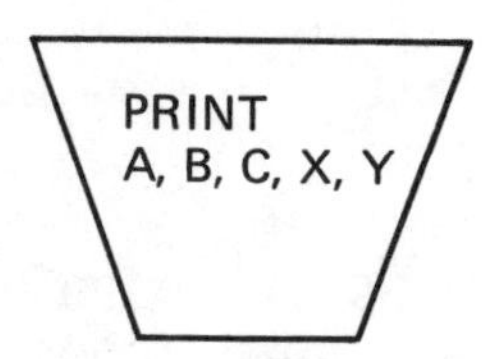

Unconditional Branch (Connector)

If the programmer desires to break the natural order of steps in a flowchart, then he will use the symbol ⓐ, where "a" is *any letter of the alphabet.*

The symbol ⓐ will appear somewhere else in the flowchart to express a desire to branch unconditionally ahead to another step in the program or to return unconditionally to a previous step in the program.

TERMINAL

In order to show that we want the computer to stop executing our program, we write:

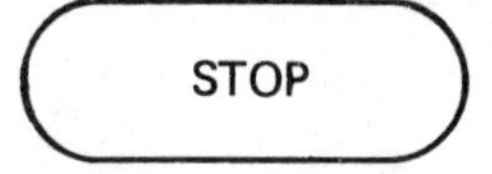

1.4 SAMPLE FLOWCHARTS

Given: A,B,C,D;A,B,C,D; ... , a collection of groups of four numbers, one group of four at a time, where all the D's are zero except the last, D = 1. For example, suppose we have 95, 52, 27, 0; 77, 82, 51, 0; 75, 83, 91, 0; 0, 0, 0, 1.

Wanted: A flowchart (Fig. 1.4) that will process four numbers at a time and either

(1) Compute and print out MEAN = (A+B+C)/3 if D = 0, *or*

(2) Print out your name and today's date if D = 1. That is, the value of D will tell the computer whether to compute the MEAN of A,B, and C, *or* stop executing the program.

(b) Construct a flowchart (Fig. 1.5) to compute and print out SUM = 1 + 3 + 5 + 7 + ... + 10001. That is, have the machine compute and print out the sum of these odd counting numbers. The computer will generate the final sum as follows:

SUM	X
0	1
1	3
1 + 3	5
(1 + 3) + 5	7
.	.
.	.
.	.
(1 + 3 + 5 + ... + 9997) + 9999	10001
(1 + 3 + 5 + ... + 9999) + 10001	

It is important to note that flowcharts can be constructed in more than one way to solve the same problem. To illustrate this possibility, consider the alternate flowchart (Fig. 1.6) for this same example, SUM = 1 + 3 + 5 + ... + 10001.

In Fig. 1.5 the computer tested for X = 10001 to decide when to print the SUM, because the incrementing of X by 2 is done *after* the

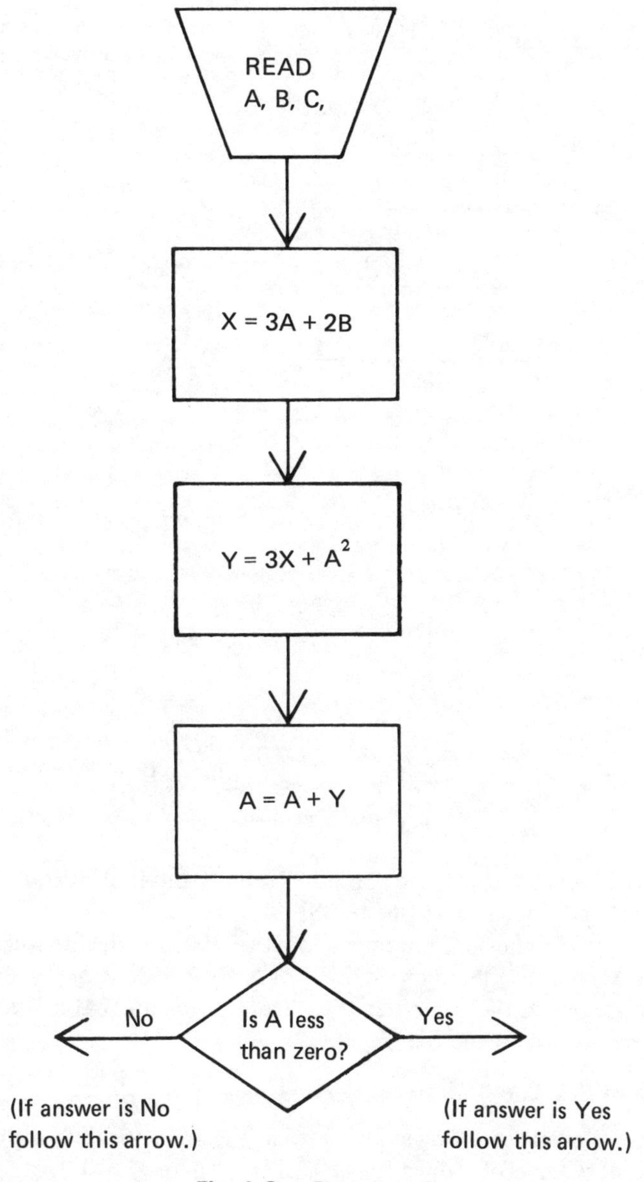

Fig. 1.2. Flowchart 2

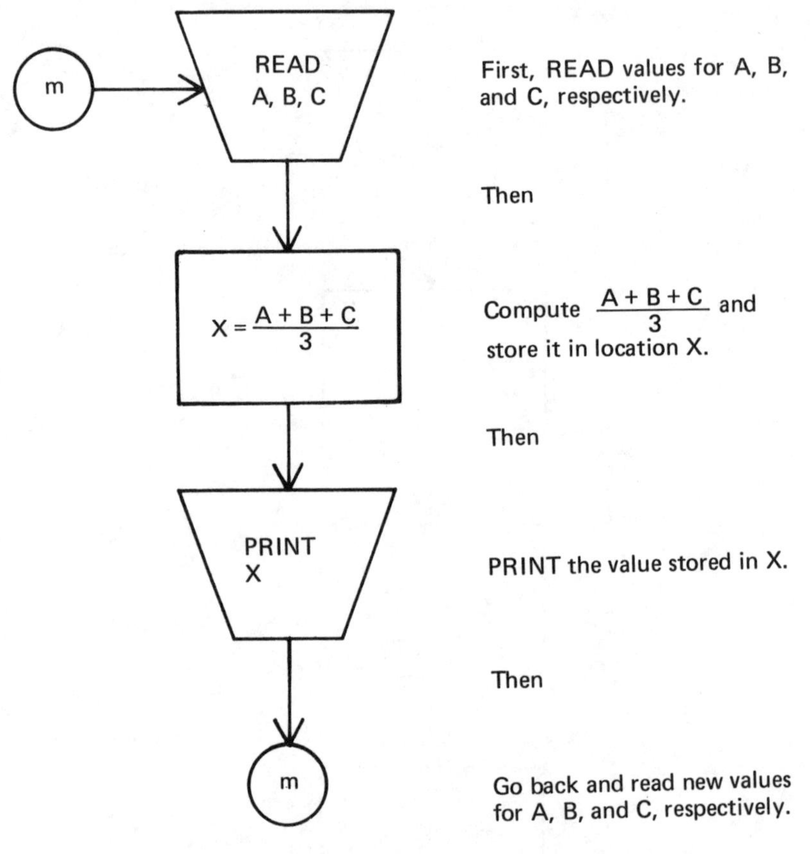

Fig. 1.3. Flowchart 3

testing. Therefore, when the computer tests X and finds it equal to 10001, it has already been added to the SUM.

In Fig. 1.6 the computer tests for X = 10003 to decide when to print the SUM, because the incrementing of X by 2 is done *before* the testing. Therefore, when the computer tests X and finds it equal to 10001, it has not been added to the SUM yet.

Problem Set 1.1

1. Construct a flowchart that can be used to write a program to print out a table of values for X and X^2, where X = 1,2,3, . . ., 10.

2. Given: NAME, AGE, NAME, AGE, NAME, AGE, . . . NAME, AGE. Wanted: A flowchart that can be used to write a program to read each NAME and AGE and print out a list of names and ages for people younger than 30.

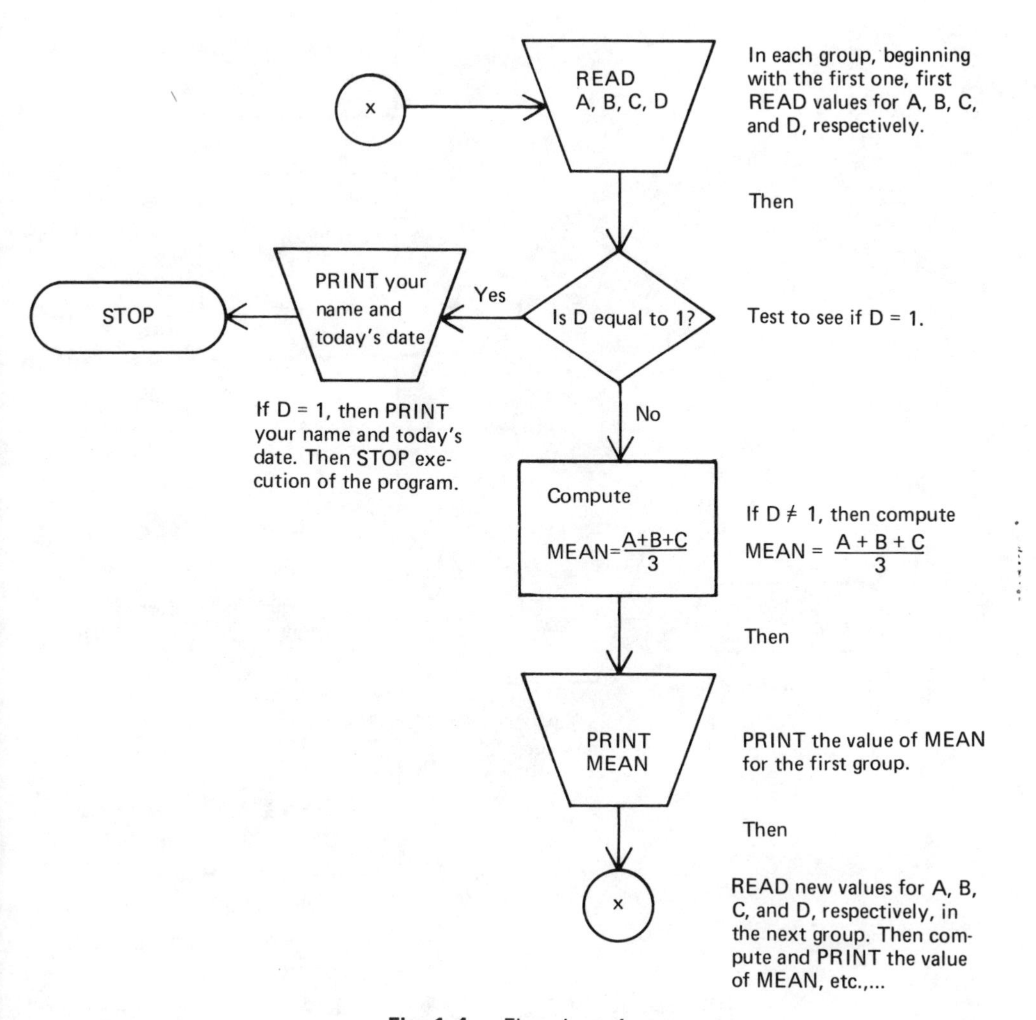

Fig. 1.4. Flowchart 4

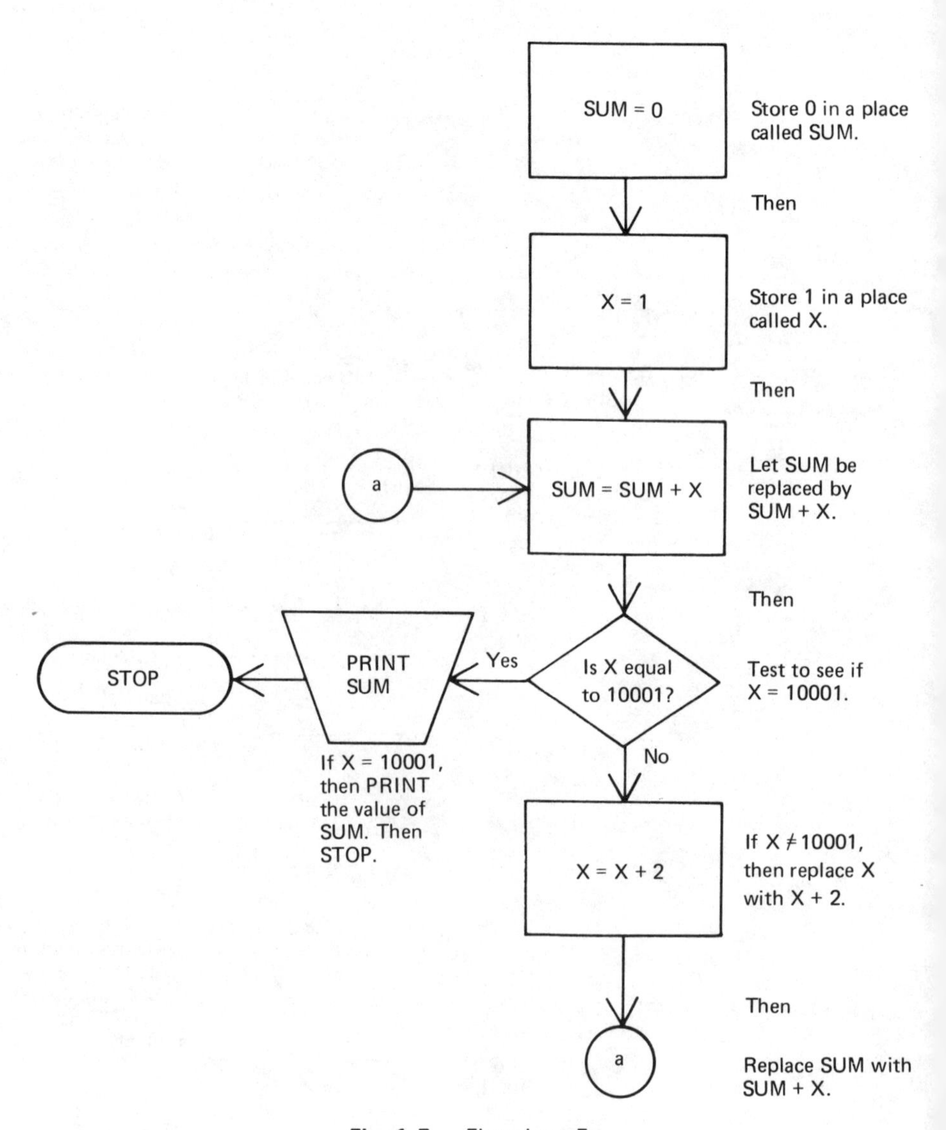

Fig. 1.5. Flowchart 5

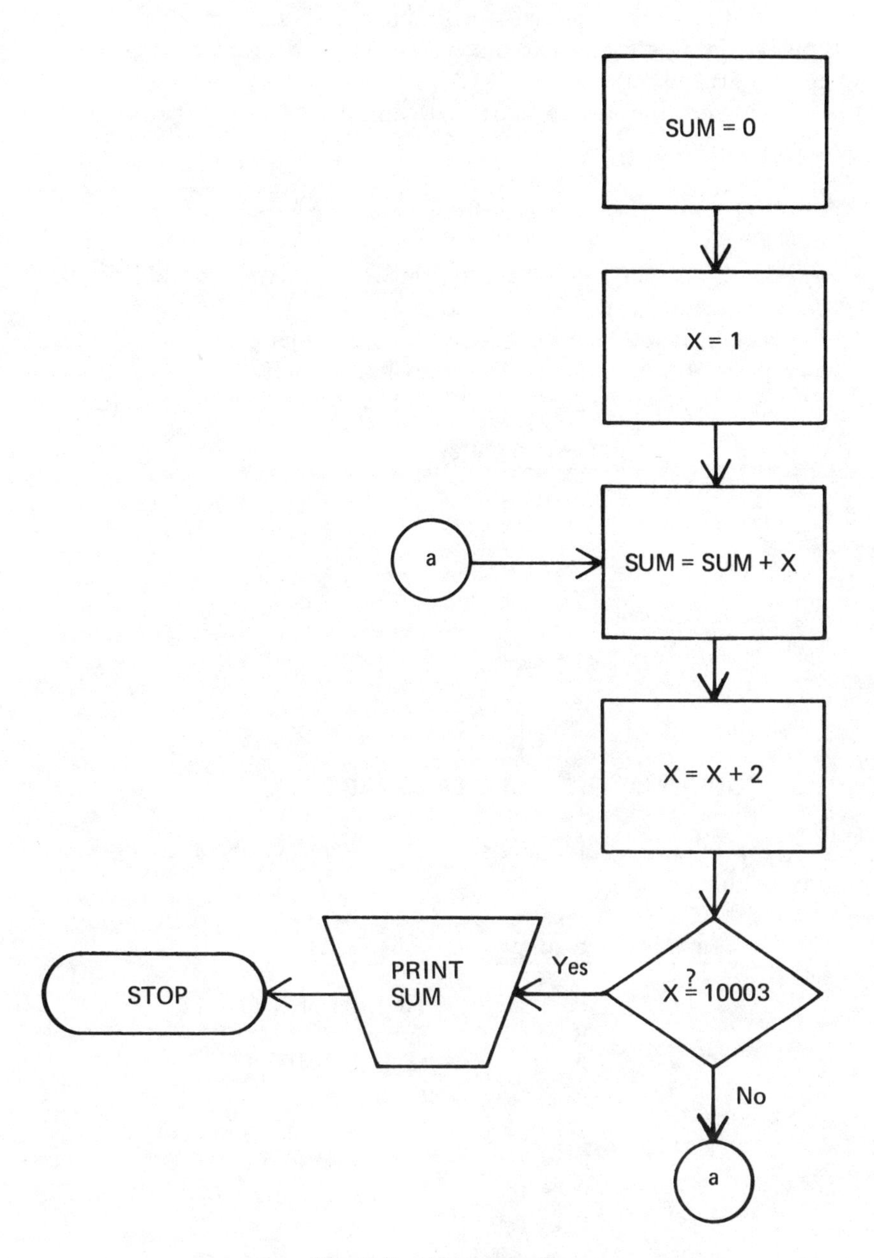

Fig. 1.6. Flowchart 6

3. Revise the flowchart in Problem 2 so as to print out a list of names and ages for people older than 50, and for people between the ages of 30 and 50.

4. Construct a flowchart that can be used to write a program to compute and print out

$$\text{SUM} = \frac{2 + 4 + \ldots + 782}{97}$$

5. Construct a flowchart to compare two numbers and print them out in numerical order. That is, given: A, B, A, B, . . . , A, B. Wanted: A flowchart that will take each A and B in the order given, compare them, then print out the numbers from smallest to largest.

For example, given 3, 6, 4, 1, 2, 1, 4, 4, −9, −8, −62, −100.

Desired Output:

3	6
1	4
1	2
4	4
−9	−8
−100	−62
. . . .	. . . .

1.5 CODING IN THE BASIC LANGUAGE

The following table contains some of the symbols for operations and relations in Basic:

Symbol	Example	Meaning
+	R + T	The sum of R and T
−	R − T	Subtract T from R
*	R * T	Multiply R by T
/	R / T	Divide R by T
↑	T ↑ 3	Raise T to the third power
=	R = T	R is equal to T
>	R > T	R is greater than T
<	R < T	R is less than T
>=	R >= T	R is greater than or equal to T
<=	R <= T	R is less than or equal to T
<>	R <> T	R is not equal to T

Internal Functions

An internal function is one whose meaning has been incorporated in the Basic compiler. The following expressions are some internal functions that can be evaluated, given the proper argument R. The argument R is the value you place in the parentheses and which the computer is to use when applying the function's rule.

Function	Meaning	Example
SIN (R)	Find the sine of R radians	SIN (.5321)
COS (R)	Find the cosine of R radians	COS (.8104)
TAN (R)	Find the tangent of R radians	TAN (.7110)
ABS (R)	Find the absolute value of R	ABS (−3)
SQR (R)	Find the square root of R	SQR (57.23)
INT (R)	Find the largest integer less than or equal to R	INT (27.2)
LOG (R)	Find the natural log of R	LOG (23.3)
EXP (R)	Find the natural exponent of R (e^R)	EXP (3)
SGN(R)	Assumes the value	$\begin{cases} +1 \text{ if } R > 0 \\ 0 \text{ if } R = 0 \\ -1 \text{ if } R < 0 \end{cases}$

Simple Variables

In Basic a simple variable is named by a single letter or a single letter followed by a single digit. For example,

Acceptable variables: T, A, X3, B6
Unacceptable variables: SUM, X97, TAVERAGE

Numerical Constants

In most Basic compilers, numerical constants may contain from one to seven significant digits.

Example	Written in Basic
−55.34	−55.34
32.333333	32.33333
0.000000237	2.37E−7 = $(2.37)(10^{-7})$
295,000,000	2.95E + 8 = $(2.95)(10^{8})$

Suppose we had a number that consisted of more than seven significant digits, such as

123.45674 or 123.45678

Some computers would truncate (chop off) the extra digits, while other computers would round off to the nearest significant digit. For example,

Number	Truncated to 7 Sig. Digits	Rounded Off to 7 Sig. Digits
123.45674	123.4567	123.4567
123.45678	123.4567	123.4568

Natural Order of Steps

When a program is run on a computer, the computer will execute the first executable statement first; the second executable statement second, and so on, until the last executable statement is executed, unless one of the statements is a command to branch out of this sequence. In a Basic program, each statement must be numbered. The statements must be numbered in ascending order, from smallest to largest. When the programmer numbers the statements, he usually does not number the steps of the program consecutively because in many instances he may have inadvertently omitted required steps in the program. The non-consecutiveness of the numbering of the steps will allow him, if necessary, to type in the omitted steps at the end of the program.

The END Statement

The last statement in any Basic programming must be the END statement.

General Form. Line number END; for example, 100 END. If you want the computer to stop executing a program, have it execute the END statement.

Output

In Basic we use the PRINT statement to obtain numerical values stored in the computer's memory or to direct the computer to print out symbols contained on its keyboard.

General Form	Example
Line number PRINT expression, expression, . . .	5 PRINT A, B2, X
Line number PRINT "symbols," "symbols," . . .	27 PRINT "HELLO IT'S ME"
Line number PRINT "symbols," "expressions," . . .	52 PRINT "T=", T

Let us consider the following Basic programs.

	Basic Program	Output	Comments
(a)	2 Print "HELLO" 4 END	HELLO	The computer will print out the symbols within the quotation marks.
(b)	5 PRINT 2 + 3, 3 − 5 10 END	5 −2	The computer will perform calculations in the PRINT statement. If we use a comma between different desired outputs, the computer will skip a definite number of spaces, dependent on the Basic compiler used. Many Basic compilers will skip 15 spaces if a comma is used.
(c)	5 PRINT 2 + 3; 3 − 5 8 END	5 −2	The computer will skip a definite number of spaces between different output if a semi-colon is used. Many Basic

	Basic Program	Output	Comments
			compilers will skip 7 spaces if a semi-colon is used.
(d)	5 PRINT 2 + 3 10 PRINT 3 − 5 15 END	5 −2	The computer will return the carriage before printing when it executes another PRINT statement.
(e)	11 PRINT "SUM1 = "; 2+3 13 PRINT "SUM2 = "; 3−5 15 END	SUM 1 = 5 SUM2 = −2	Symbols and numerical values can be outputted using the same PRINT statement.

(f)

```
15 PRINT "SUM1 =" 2 + 3; "SUM2 =" 3−5
20 END
OUTPUT: SUM1 = 5   SUM2 = −2
```

It is important to note that 5 PRINT "T" commands the computer to print out the *symbol* T, while 5 PRINT T commands the computer to print out the *numerical value* stored in a location called T.

Problem Set 1.2

1. Write a Basic program to print out your name and today's date on the same line.
2. Write a Basic program to print out your name and today's date on separate lines.
3. Write a Basic program to print out your name, date of birth, and age on separate lines.

The LET Statement

In Basic we use the LET statement to represent an arithmetic statement, which allows us to compute and store the numerical value of some variable.

General Form. Line number LET variable = arithmetic expression. Before we consider some examples, remember that the algebraic expressions

$x + y$, $x - y$, xy, x/y, x^3 are written

X + Y, X − Y, X*Y, X/Y, X↑3

in the Basic language.

Arithmetic Statement	Written in Basic
$a = 3 + 2$	5 LET A=3 + 2
$c = 2 - \frac{5}{X}$	8 LET C=2−5/X
$z = 5 + 7 - 9^4$	29 LET Z=5 + 7 − 9↑4
$x = 3a + 2c$	91 LET X=3*A + 2*C
$y = 3x^2 - 5x + 7$	21 LET Y=3*X↑2 − 5 * X + 7
$\text{Average} = \frac{T + R + S}{3}$	88 LET A=(T+R+S)/3
$t_2 = \sqrt{957.22}$	11 LET T2=SQR (957.22)
$t_3 = \sin x$	13 LET T3=SIN (X)

Note: There is only one variable to the left of the equals sign. Do not use a LET statement in a program unless all the variables on the right side of the equation have been previously defined. Also, the following hierarchy of five algebraic operations must be adhered to when writing your LET statement:

1. All work inside a parenthesis.
2. Raising to a power.
3. One-term operation (take the negative of an expression $-(+2) = -2$, $-(-2) = 2$).
4. Multiplication or division.
5. Addition or subtraction.

Now let us consider the following Basic programs.

	Basic Program	Output	Comments
(a)	2 LET T = 5 4 LET R = 7 6 LET S = 9 8 LET A = T + R + S/3 10 PRINT A 12 END	15	The computer first divided the value of S by 3, and then added the values of T and R.
(b)	2 LET T = 5 4 LET R = 7 6 LET S = 9 8 LET A = (T + R + S)/3 10 PRINT A 12 END	7	The computer first added the values of T, R, and S, and then divided by 3.

Problem Set 1.3

1. Write LET statements for each of the following:

(a) $x = 5 - 2^3$

(b) $y = 2 - 17^3 + \frac{27}{101}$

(c) $T = \frac{8 + 7^2}{21}$

(d) $\text{SUM} = \text{A} + \text{B} - 3\text{C}^2$

(e) $\text{AVERAGE} = \frac{\text{T1} + \text{T2} + \text{T3} + \text{T4}}{4}$

(f) $\text{Y} = \sqrt{57.321}$

(g) $\text{CURVE} = \sin x + 3 \cos y$

(h) $\text{FUNCTION} = 2\sqrt{x^2 - 1} + 2x - 7$

(i) $\text{SNOOPY} = 2 + 7 - 3^3 + \sqrt{94}$

(j) $\text{HOBO} = 1^2 + 3^2 + 5^2 + 7^2 + 9^2$

2. Write a Basic program to compute and print out the variables T, Y, SNOOPY, and HOBO defined in Problem 1.

INPUT

In Basic we have two different statements which allow us to send data into the computer's memory. One statement is the READ statement, which requires an additional DATA statement. The other statement is the INPUT statement, which does not require another statement.

General Forms. Line number READ variable, variable, . . . ;
Line number INPUT variable, variable, ...

Examples

```
5  READ A, B, C2
29  INPUT X, Y3
```

Note: Every READ statement requires a DATA statement.

General Form. Line number DATA number, number, ...

Example

```
77  DATA 5, 2, 3, -7, 14.51, 27.1
```

Note: The DATA statement may appear anywhere in a Basic program except that it cannot be the last statement in any program (END statement must be the last statement) and it cannot be the first statements in some programs (programs requiring a DIM statement). The DATA statement is a nonexecutable statement; that is, it does not command the computer to perform any internal operation. The DATA statement is used primarily as a place to store numbers.

Example

```
5  READ A, B, C2
10  LET Y = A + B + C2          Output
15  PRINT "SUM =" Y             SUM= 8
20  DATA 1, 3, 4
25  END
```

The INPUT statement does not require a DATA statement. When the computer executes the INPUT statement in a Basic program, it will ask the programmer to type in a numerical value for each of the variables in the INPUT statement. Sample programs using the INPUT statement will be illustrated later.

The GO TO Statement

In Basic we use the GO TO statement for an *unconditional branch* to a certain line in the program.

General Form. Line number GO TO line number.

Example

```
15 GO TO 97
```

When the computer executes line 15 it will branch to and execute line 97 next.

The IF-THEN Statement

In Basic we use the IF-THEN statement for a *conditional branch* to a certain line number in the program.

General Form.

Line number IF expression relation expression THEN line number.

Examples

(1)
```
5 IF X = 999 THEN 27
```

that is, if X = 999, the computer will execute line 27 next. If X $\neq$ 999, the computer will execute the line immediately following line 5 next.

(2)
```
12 IF T + R< = A + 27 THEN 100
```

that is, if T + R is less than or equal to A + 27, the computer will execute line 100 next. If T + R is greater than A + 27, the computer will execute the line immediately following line 12 next.

Sample Problem

Wanted: A Basic program (Fig. 1.7) to print out the even integers from 2 to 12 and their squares.

Solution

```
10 LET X = 2
12 PRINT X; X↑2
14 LET X = X + 2
16 IF X < = 12 THEN 12
18 END
```

Output	
2	4
4	16
6	36
8	64
10	100
12	144

THE FOR AND NEXT STATEMENTS

Any Basic program that involves initializing a variable at some numerical value and incrementing the variable by some set number until it reaches some maximum value can make use of the FOR and NEXT statements.

General Form. Line number FOR $l = m$ TO n STEP r
(l is a simple variable; m, n, r are constants, variables, or expressions)

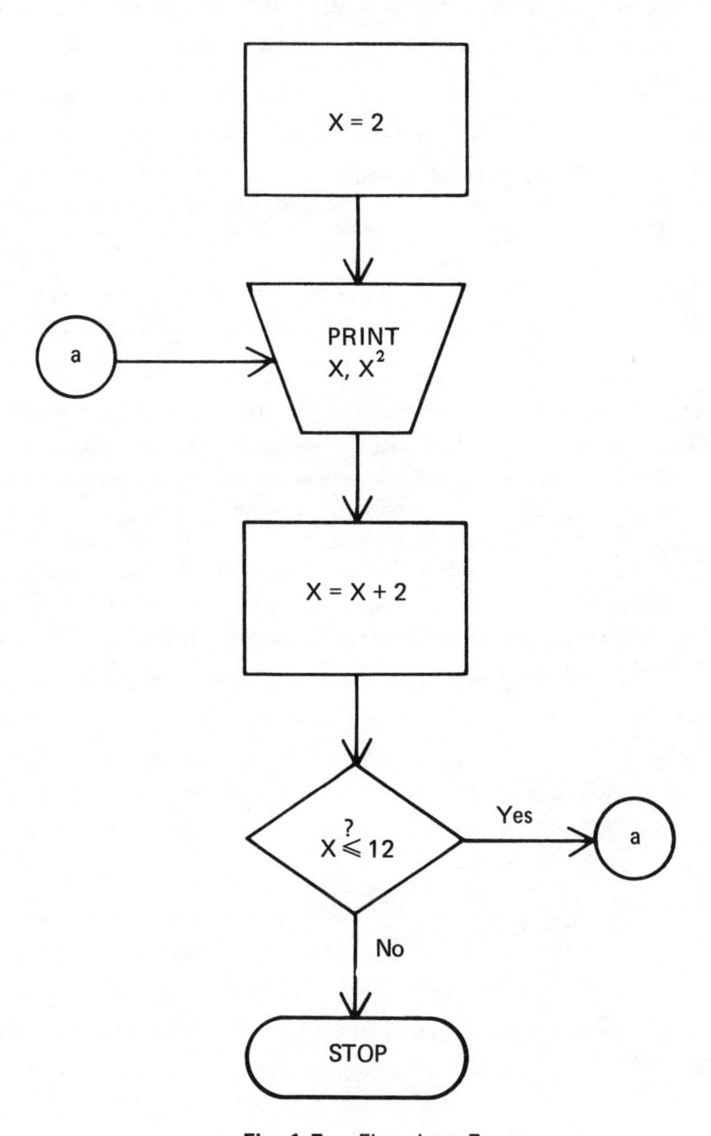

Fig. 1.7. Flowchart 7

where m is the initial value for l, n is the maximal value for l, and r is the increment.

Example

5 FOR X = 2 TO 12 STEP 2

Note: Every FOR statement requires a NEXT statement.
General Form. Line number NEXT variable; l is a simple variable.

Example

20 NEXT X

Consider the following general program:

```
5   FOR l = m TO n STEP r
10  .....)
15  .....}        10, 15, 20 are any executable statements.
20  .....)
25  NEXT 1
```

The computer will execute statements 10, 15, 20 with $l = m$; then execute statements 10, 15, 20 with $l = m + r$; then execute statements 10, 15, 20 with $l = m + 2r$, etc., . . . until it executes statements 10, 15, 20 with $l = n$; then and only then will the computer execute the next statement after line 25. This procedure, by which the computer executes the same sequence of steps more than once in the same program, is called *looping*.

For example, let us consider the program in the sample problem for the IF-THEN statement.

```
                                 Output
                              ____________
10  LET X = 2                  2       4
12  PRINT X; X ↑ 2             4      16
14  LET X = X + 2              6      36
16  IF X < = 12 THEN 12        8      64
18  END                       10     100
                              12     144
```

This could be rewritten using FOR and NEXT statements as follows:

```
                                     Output
                                  ____________
5  FOR X = 2 TO 12 STEP 2         2        4
10  PRINT X; X↑2                  4       16
15  NEXT X                        6       36
20  END                           8       64
                                  10     100
                                  12     144
```

We have already seen that every READ statements requires a DATA statement. If the same computation is to be computed more than once, a GO TO statement is required. Consider the following Basic program:

```
                                 Output

10  READ A,B
15  LET T = A + B                  8
20  PRINT T
25  DATA 3, 5, -2, 7, 4, 9
30  END
```

Only one value of T is computed and printed out because after the computer read 3 and 5 for A and B and then computed and printed out T, it executed the next executable statement, the END statement, which tells the machine to stop executing the program.

Now, let us add a GO TO statement after the PRINT statement.

```
Basic Program                  Output

10  READ A,B                     8
15  LET T = A + B                5
20  PRINT T                      13
22  GO TO 10
25  DATA 3, 5, -2, 7, 4, 9     OUT OF DATA LINE 10
30  END
```

After the computer reads the initial values for A and B and computes the initial value of T, and then prints the initial value of T, statement 22 (22 GO TO 10) commands the computer to execute statement 10 again so as to read two other values for A and B. Then the computer will compute the new value of T and print the new value of T, and so on, until there are no more data in the DATA statement. This procedure is another example of *looping* in computer programming.

The computer will read the first two numbers in the DATA statement, then the next two numbers in the DATA statement, and so on until there are no more data. Then the computer will tell you it is out of data and stop executing your program, even though it never executes the END statement.

The preceding program could be rewritten so as to have the computer stop executing the program by reaching the END statement. Consider the following revised program:

Basic Program	Output
10 READ A,B,C	8
12 IF C = 1 THEN 27	5
15 LET T = A + B	13
20 PRINT T	END OF PROGRAM
22 GO TO 10	
25 DATA 3,5,0, −2,7,0,4,9,0,0,0,1	
27 PRINT "END OF PROGRAM"	
30 END	

In this revised program the computer will execute the statement 27 if C = 1 and then reach the END statement to stop executing the program. This alteration is unnecessary unless a comment is required upon completion of the program.

Now, if we change line 20 in the program above, we could have our answers printed out in various different forms.

	Line 20	Output
(a)	20 PRINT T,	8 5 13 END OF PROGRAM

By placing a comma after T, we can have the machine print on the same line and skip 15 spaces between the different values of T.

	Line 20	Output
(b)	20 PRINT T;	8 5 13 END OF PROGRAM

By placing a semicolon after T, we can have the machine print on the same line and skip 7 spaces between the different values of T.

	Line 20	Output
(c)	20 PRINT "T="T	T= 8 T= 5 T= 13 END OF PROGRAM

	Line 20	Output
(d)	20 PRINT "T="T;	T= 8 T= 5 T= 13 END OF PROGRAM

THE TAB FUNCTION

We have seen that the use of a comma in a PRINT statement allows us to skip about 15 spaces, and the use of a semicolon in a PRINT statement allows us to skip about 7 spaces. The TAB function allows us to move to a desired position in a line.

General Form. TAB (X), where X is any positive real number, commands the machine to move to position X if X=0,1,2,3, . . ., or to move to the greatest integer space less than X if X≠0,1,2,3 . . .. On most computer terminal printers the spaces are numbers 0,1,2, . . ., 71. If you exceed 71 spaces on a line, the computer will start at the 0 position on the next line.

Example

```
5  FOR X = 0 to 5
10  PRINT "F."; TAB (X+2);"S."
15  NEXT X
20  END
```

Output

```
F.S.
F. S.
F.  S.
F.   S.
F.    S.
F.     S.
```

Note: If line 10 were printed as 10 PRINT "F.", "S.", then the computer would skip about 15 spaces between F. and S. But if line 10 were printed as shown here, the TAB (X+2); function commands the machine to skip X spaces between F. and S.

Problem Set 1.4

1. Code the sample flowchart in Fig 1.4. Use the following:

   DATA 1,3,5,0,97,88,21,0,999,275,87,0,0,0,0,1

2. Code the sample flowchart in Fig. 1.5:
   (a) without using FOR and NEXT statements.
   (b) using FOR and NEXT statements.
3. Give the computer output for the following programs:

(a)
```
5 LET A = 1
10 LET B = A + 4
15 LET C = 2*B + A + 5
20 LET D = 2*(B+A)+5
25 PRINT A,B,C,2*D
30 END
```

(b)
```
10 LET E = 1
20 LET E = 3↑2
25 LET F = (E + 2)↑3 + 1
35 PRINT E,F, SQR (E), F/9
40 END
```

(c)
```
10 READ A,B,C,D
20 LET X = A + B - D
30 LET Y = A - B + C
32 PRINT X, Y
36 DATA 1,2,3,4,5,6,8,9
40 END
```

(d) Program (c) with the following additional line:

34 GO TO 10

4. Code the Fig. 1.8 flowchart in the Basic language.
5. Code the Fig. 1.9 flowchart in the Basic language.
6. Give the computer output for the following program:

```
7 PRINT "USING THE TAB FUNCTION"
9 LET X = 5
11 PRINT X; TAB (6); X↑2
13 PRINT 2*X; TAB (8); (2*X)↑2
15 END
```

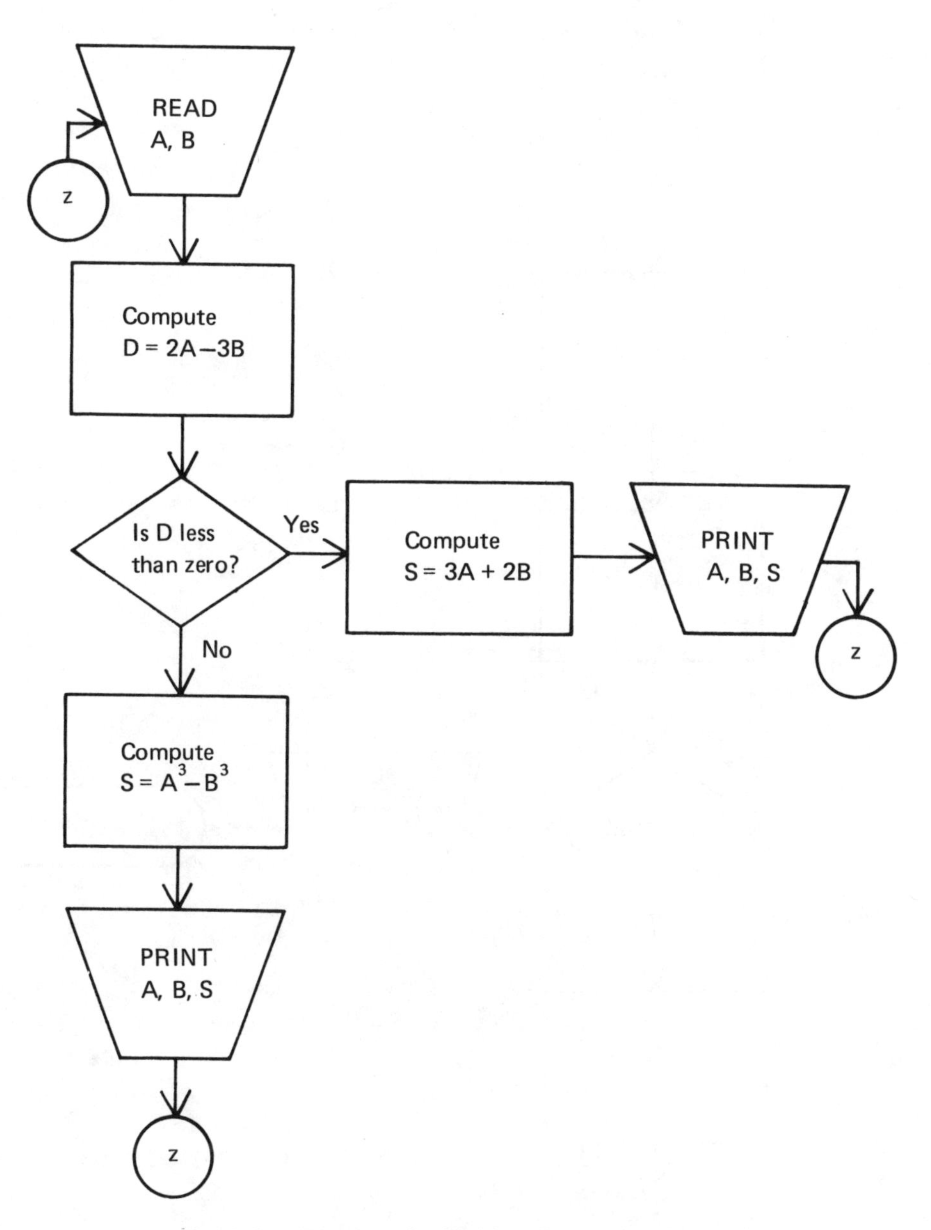

Fig. 1.8. Flowchart for Problem Set 1.4, Problem 4

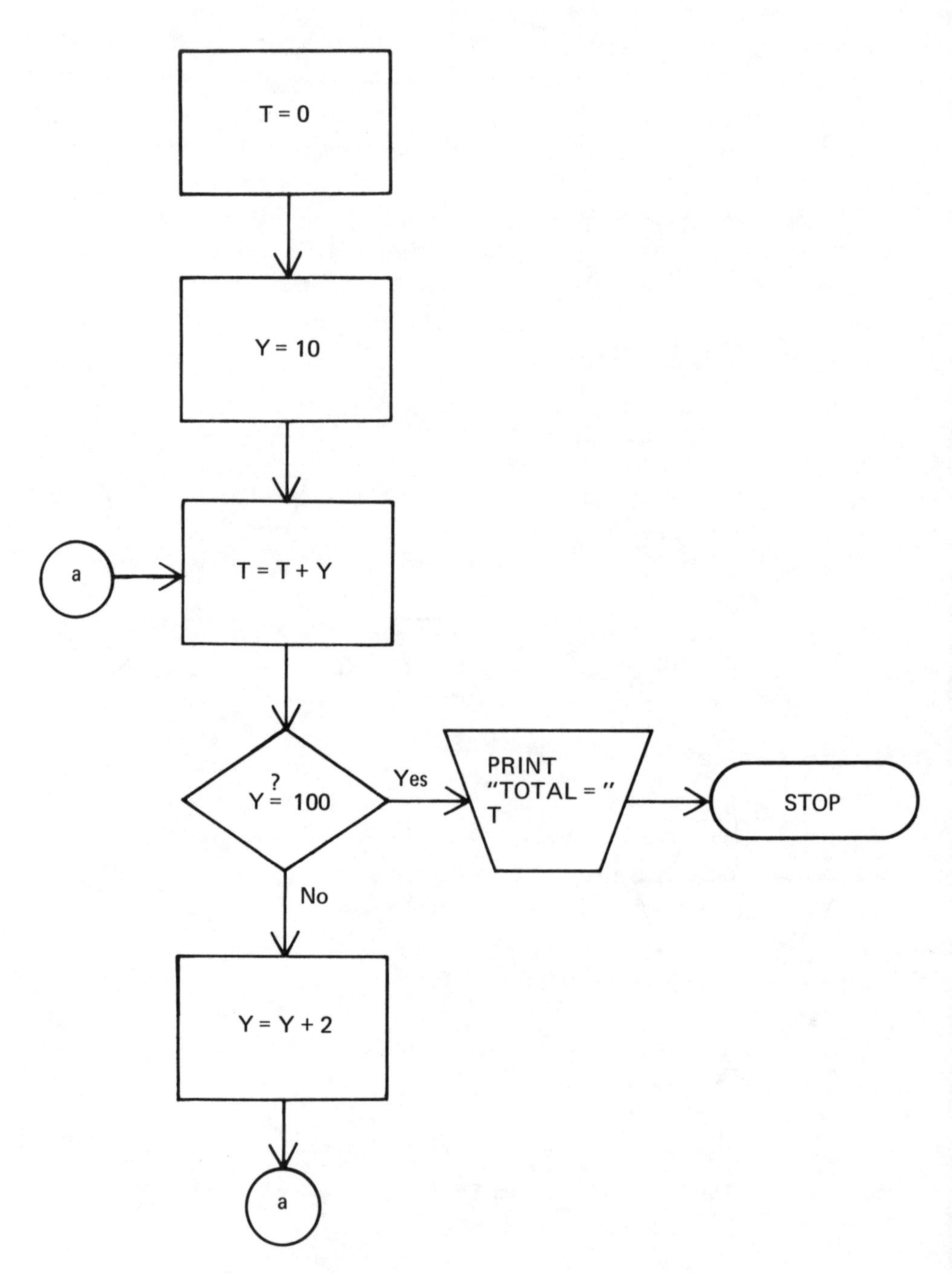

Fig. 1.9. Flowchart for Problem Set 1.4, Problem 5

1.6 OPERATING A COMPUTER TERMINAL

A computer terminal (Fig. 1.10) is an alternative to operating a computer directly. Basically, a computer terminal is a Teletype machine that communicates with a computer by means of a telephone line. Therefore, the computer may be located miles from a computer terminal. This allows many people to share the cost of one large computer. There are three integral parts to a computer terminal: the keyboard, the paper tape unit, and the acoustic coupler.

The Keyboard

The keyboard can be used to type a program directly to the computer once a telephone connection has been made, or to prepunch a program on paper tape and then contact the computer and have it read the paper tape. The latter procedure is preferred because it will reduce the costly computer time used.

The Paper Tape Unit

The paper tape unit can be used in the following ways:

(a) *Off Line* to prepunch a program on paper tape.
(b) *On Line* to have the computer read the program on paper tape.
(c) *On Line* to have the computer place a program from its memory onto paper tape.

The Acoustic Coupler

The acoustic coupler is the device that allows us to transmit information to a computer and receive information from a computer by means of a telephone line.

Procedure for Running a Program on a Terminal

1. Prepunch the coded program on paper tape.
2. Contact the computer by telephone.
3. Have the computer read the paper tape.
4. Have the computer run the program.
5. Debugging or receiving results.
6. Sign off procedure.

Fig. 1.10. Computer terminal

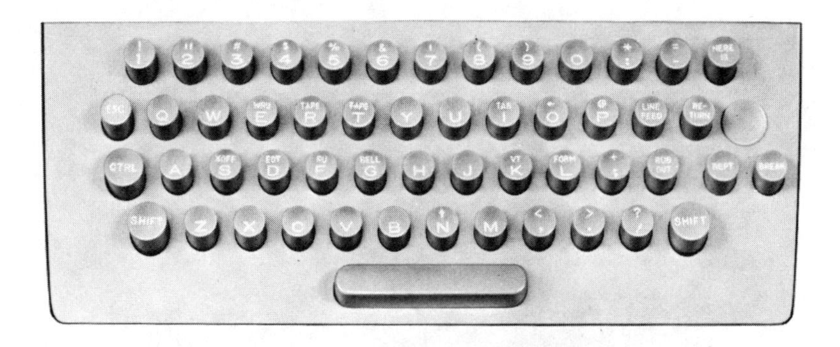

Fig. 1.11. Computer keyboard

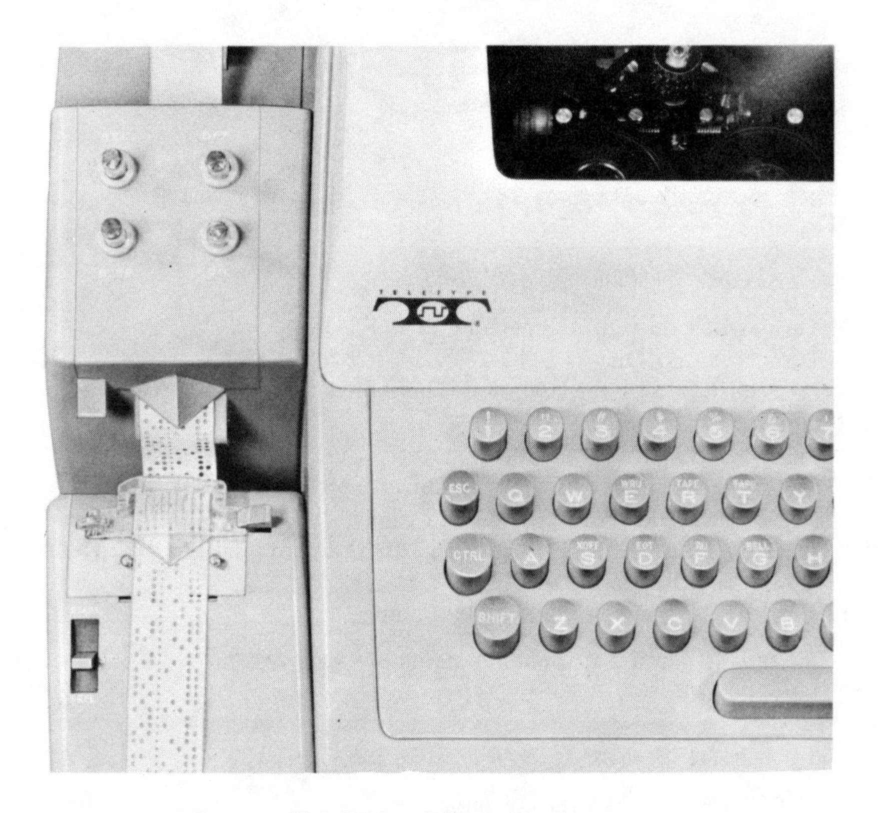

Fig. 1.12. Paper tape unit

Fig. 1.13. Acoustic coupler

Procedure for Punching a Paper Tape

1. Turn switch on right of machine to LOCAL.
2. Turn tape unit On.
3. Press HERE IS on the keyboard. This will allow for some space on the tape for threading.
4. Punch each line in your program exactly the way it appears on your coded sheet. Punch one line at a time and press RETURN then LINE FEED on the keyboard at the end of each line.
5. After punching the last statement in the program punch HERE IS on the keyboard again.
6. Shut off the machine and the tape unit.

Note: If you are to punch a statement such as

5 READ A,B,C,

and you make an error such as

5 READ AB (forget to punch the comma between A and B)

there are two ways to correct your error:

(a) Press RETURN and LINE FEED on the keyboard and retype

the complete statement again. (When the computer reads the tape it will erase the first line 5 and replace it with the corrected line 5.)

(b) Press the symbol $\overleftarrow{0}$ in upper shift on the keyboard. The symbol← will command the computer to erase one character to the left of it. Therefore, in the sample error given above we would punch

5 READ AB←, B, C

If we were to punch

10 LET X = 3 * A + 2 * B ↑ 2

and we punched

10 LET X = 3 * B + 2

and then noticed our error, we could correct by erasing more than one character:

10 LET X = 3 * B + 2←←←A + 2 * B ↑ 2

Procedure for Contacting Machine

1. Turn switch on right of machine to LINE.
2. Turn acoustic coupler ON and set the duplex switch on FULL.
3. Dial a system phone number. The system will answer with a high-pitched sound.
4. Place the phone in the acoustic coupler.
5. The computer will ask for your identification number. Type in your ID number and press RETURN.
6. The computer will print out some information such as the system's name, date, time, and any news. Then the computer will type READY. At this time you can punch a program directly to the computer or have the computer read a program from a paper tape.

Procedure for Running a Program That Has Been Punched on Paper Tape

1. Follow the procedure for contacting the computer.
2. Place the tape reader switch in the FREE position.
3. Place your paper tape in the tape reader.
4. Type in TAPE on the keyboard.
5. Move the tape reader switch to the START Position. (The computer will read the paper tape and store the program in its memory.)
6. Type in LIST on the keyboard. This commands the computer to list the program with any corrections that might have been made while punching on tape.

7. Type in RUN on the keyboard. This commands the computer to execute the steps in your program. It will either print out your results or tell you that you have an error.

Example

If the computer read the following from a paper tape,

```
5 READ AB←,B,C,←
10 LET X = 3*B+2←←←A+2*B↑2-C
15 PRINT A;B;C,"X="X
20 DATA 1,1,1,1,2,3
```

type in LIST and press RETURN on the keyboard. The computer will print:

```
5 READ A,B,C
10 LET X = 3*A+2*B↑2-C
15 PRINT A;B;C,"X="X
20 DATA 1,1,1,1,2,3
```

Type in RUN and press RETURN on the keyboard. Some computers will print:

ERROR

Press RETURN and the computer will print END STATEMENT NOT THE LAST STATEMENT IN PROGRAM (or some equivalent message). You can correct this error by typing:

25 END (return)

Then line 25 will be added to your program in the computer's memory. Summarizing this procedure:

Type	Press	Computer's Response
LIST	RETURN	5 READ A,B,C 10 LET X = 3*A+2*B↑2-C 15 PRINT A;B;C,"X="X 20 DATA 1,1,1,1,2,3 25 END
RUN	RETURN	1 1 1 X= 4

Note: This program does not have a language error in logic. The DATA statement contains another set of values for A,B, and C. If the computer is to read them, we need to add statement 17 GO TO 5.

Type	Press	Computer's Response
17 GO TO 5	RETURN	
LIST	RETURN	5 READ A,B,C 10 LET X = 3*A+2*B↑2-C 15 PRINT A;B;C,"X="X 17 GO TO 5 20 DATA 1,1,1,1,2,3 25 END
RUN	RETURN	1 1 1 X= 4 1 2 3 X= 8 OUT OF DATA IN LINE 5

At this time you can either sign off by typing in some command such as BYE and pressing the RETURN button or you can allow another person to use the computer by typing in some command, such as SCRATCH or NEW, and pressing the RETURN button (a command such as SCRATCH or NEW erases the previous program from the computer's memory).

Type	Press	Computer Response
SCR	RETURN	
TAPE	RETURN	The computer will read a new tape if it has been placed in the tape reader. Assume it reads the following tape:

```
5 INPUT A
10 IF A = 9999 THEN 35
15 FOR X = 1 TO 3←A
20 PRINT X;X↑2
25 NEXT A
30 GO TO 5
35 PRINT "JOB DONE BY SNOOPY"
40 END
```

RUN	RETURN RETURN	ERROR UNDECIPHERABLE OPERAND LINE 25 (should be NEXT X)
25 NEXT X	RETURN	
RUN 3	RETURN RETURN	? (You must now enter a numerical value for A. Let's assume you type 3.) 1 1 2 4 3 9 ? (You must now enter the next value for A. Let's assume you type 7.)
7	RETURN	1 1 2 4 3 9 4 16 5 25 6 36 7 49 ? (Let's assume you type 9999.)
9999	RETURN	JOB DONE BY SNOOPY
BYE	RETURN	OFF AT (time) _______ (minutes of computer time)

At this time you will lose your telephone connection to the computer.

Note: A partial listing of possible Basic compiler error messages is included in Appendix B.

Problem Set 1.5

1. Punch the following program on paper tape:

```
10 READ X,Y,Z
15 LET W = 3 * X + Y ↑ 2
20 LET T = Z + Y + X/2
23 PRINT X,Y,Z
30 PRINT "W=" W ; "T ="T
```

```
40  DATA 2,4,6,12,0,1,18,-7,-3
45  END
```

2. Contact the computer by telephone and have it read the tape for Problem 1 and ask it to RUN the program.

3. Add statement 35 GO TO 10 to the program and ask the computer to LIST the program and RUN the program.

4. Have the computer SCRATCH (erase) the program for Problem 1 from its memory.

5. Type in the following program ON LINE to the computer:

```
2  INPUT A,B,C,D
4  IF D = 0 THEN 10
6  PRINT "AVERAGE =" (A+B+C)/3
8  GO TO 2
10  END
```

Use the following data:

A	B	C	D
5	3	9	1
22	99	88	1
67	27	92	1
0	0	0	0

6. Each of the following programs contains statements that have errors in the Basic language. If you can find them, correct them. In either case, type each of them (separately) into the computer. If you have made the proper correction, the computer will execute the program; if not, the computer will tell you which statements contain errors. Then try to correct the errors.

(a)

```
10  PRINT FIND THE ERRORS
20  READ A,B
30  LET S= (A+B ↑ 2)
40  PRINT, "ERRORS", A,B,S
55  IF A = B THEN 80
60  PRINT SUM
65  GO TO 20
75  DATA 2,5,7,8,9,3
80  END
```

(b)

```
5  READ A,B
7  LET Y = 3*A-7*B+2
10  GO TO 5
15  DATA 1,2,7,9,8,27
```

(c)
```
10 READ A,B
20 PRINT A ↑ 2; TAB(10); B ↑ 2
30 GO TO 5
35 DATA 3,4,5,6,7,8
40 END
```

1.7 SUBSCRIPTED VARIABLES

In Basic a simple variable can be named by a single letter or a single letter followed by a single digit. Therefore, given a group of scores, we could define them by X0, X1, X2, . . ., X9. What if we had more than ten scores? We could use subscripted variables by placing the subscripts in parentheses. That is, X(0), X(1), X(2), . . . , X(10), X(11), ... When using subscripted variables in Basic, you as the programmer must reserve space for them in the computer's memory. This is done by means of a *dimension* statement.

General Form: Line number DIM variable (*r*), variable (*s*), . . . , where *r* is the number of spaces reserved for a single row of scores for the first subscripted variable, and *s* is the number of spaces reserved for a single row of scores for the second subscripted variable. For example: 2 DIM X(5), Y(12).

- The DIM statement *must be* the first statement in a Basic program.
- It is *acceptable to overdimension* (reserving more places than necessary) for a subscripted variable.
- It is *not acceptable to underdimension* (not reserving enough places) for a subscripted variable. For example, suppose we wanted to write one program that could be used to compute the average score on a statistics exam for each class section in your school. If the number of students in each section varied from 15 students in the smallest section to 36 students in the largest section, we would have to DIM variable (36) if the program is to handle all sections.

Note: It is possible to have double-subscripted variables, such as A(1,2), X(3,5). These are used in matrix algebra, a topic that will not be discussed in this textbook.

There are other advantages in using subscripted variables.

Sample Problem

Given: N, X(1), X(2), . . ., X(N), where $N \leq 30$. Sample data:

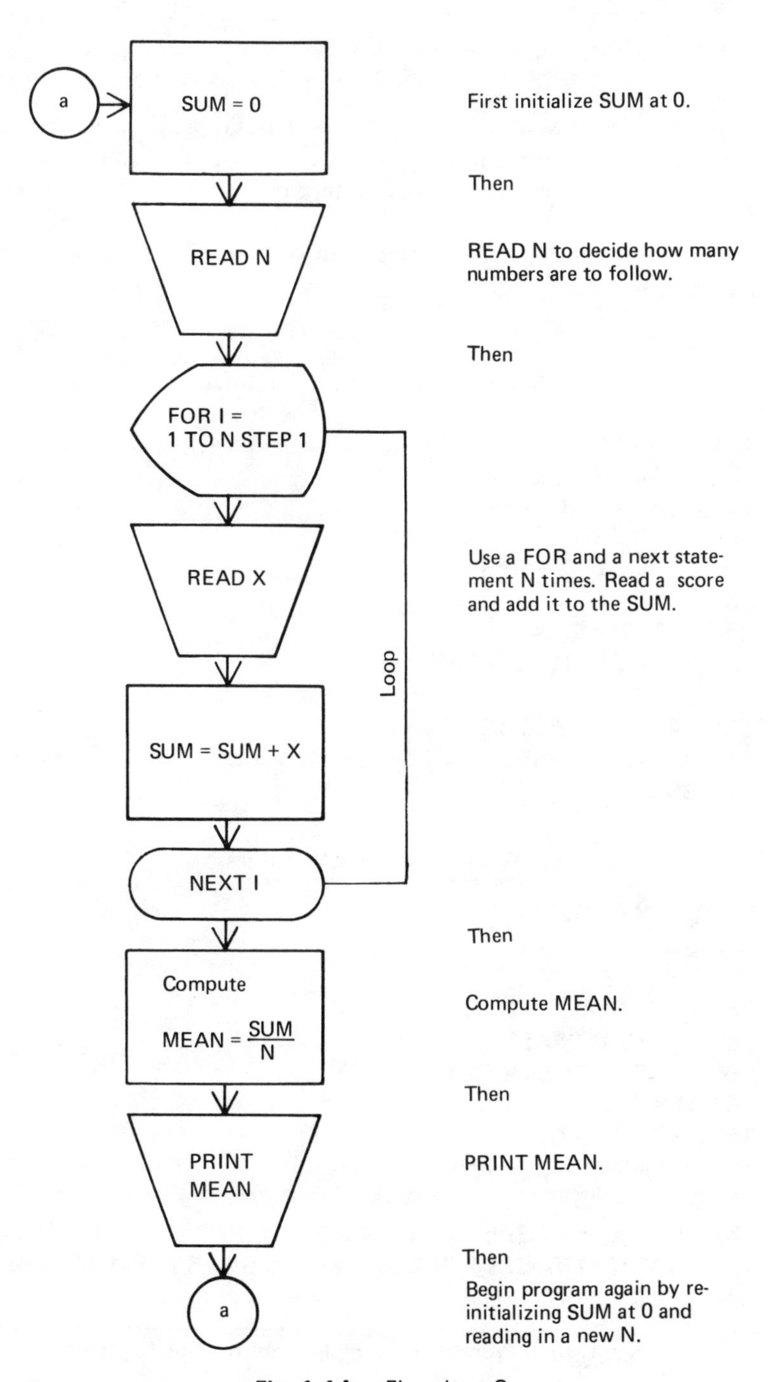

Fig. 1.14. Flowchart 8

DATA 5, 75, 82, 71, 52, 91
DATA 10, 52, 75, 76, 74, 81, 82, 87, 94, 51, 81

Wanted: A flowchart (Fig. 1.14) and a Basic program to compute and print out the arithmetic mean (average) of N given numbers. (Note: The first number in each DATA statement tells how many scores are to follow.)

This program could be written with or without subscripted variables, as shown in Fig. 1.15.

Without Subscripted Variables

```
5 LET S=Ø
1Ø READ N
15 FOR I= 1 TO N
2Ø     READ X
25     LET S=S+X
3Ø NEXT I
35 LET M=S/N
4Ø PRINT "MEAN="M
45 GO TO 5
5Ø DATA 5,75,82,71,52,91
55 DATA 1Ø,52,75,76,74,81,82,87,94,51,8Ø
6Ø END
```

With SubscriptedVariables

```
2 DIM X(3Ø)
5 LET S=Ø
1Ø READ N
15 FOR I= 1 TO N
2Ø    READ X(I)
25    LET S=S+X(I)
3Ø NEXT I
35 LET M=S/N
4Ø PRINT "MEAN="M
45 GO TO 5
1ØØ DATA 5,75,82,71,52,91
1Ø5 DATA 1Ø,52,75,76,74,81,82,87,94,51,8Ø
2ØØ END
```

Fig. 1.15. Program for sample problem (Fig. 1.14)

Both programs in Fig. 1.15 would produce the following output:

```
MEAN= 74.2
MEAN= 75.2

OUT OF DATA IN LINE 10
```

However, in the program without subscripted variables, the only score saved is the last one read. In the program with subscripted variables, each of the N scores has been saved in the computer's memory.

Sample Problem

Now, consider the following change in the preceding program. Given: N, X(1), X(2), . . . , X(N), assume $N \leq 30$. Wanted: A Basic program to compute the arithmetic mean of N numbers and then subtract the mean from each of the scores (that is, find out how much and in what direction, positive or negative, each score deviates from the mean). If we were given

DATA 5, 75, 82, 71, 52, 91

the desired output would be as follows:

N	X(N)	X(N) – MEAN
MEAN =	74.2	
1	75	0.8
2	82	7.8
3	71	–3.2
4	52	–22.2
5	91	16.8

The program with the subscripted variables in Fig. 1.15 could be modified as follows to obtain the desired results.

```
1 DIM X(30),D(30)
3 PRINT "N","X(N)","X(N)-MEAN"
5 LET S = 0
10 READ N
15 FOR I = 1 TO N
```

```
20 READ X(I)
25 LET S = S+X(I)
30 NEXT I
35 LET M=S/N
40 PRINT "MEAN="M
45 FOR I = 1 TO N
50 LET D(I) = X(I)-M
55 PRINT I,X(I),D(I)
60 NEXT I
65 GO TO 3
100 DATA 5,75,82,71,52,91
101 DATA 10,52,75,76,74,81,82,87,94,51,80
200 END
```

1.8 PREPACKAGED PROGRAMS

Prepackaged programs are programs that have been stored in the memory of a computer. Each program is stored under a specific name. The user of a prepackaged program calls the program out of memory by name, then puts in data in the proper place, and commands the machine to run the program with the given data. In the succeeding chapters this textbook will use many prepackaged statistical programs to solve problems. The user of a prepackaged program is told the name of the program, what it does, where to place the data, and any other pertinent information.

For example, suppose the program just completed in Article 1.7, which computed the arithmetic mean of N numbers and the deviations of each score from the mean or average, was stored under the program name DEMO. You might receive an instruction sheet as follows:

Name: DEMO

Description: This program computes the arithmetic mean of N numbers and the deviation of each number from the arithmetic mean.

Instructions: Enter the data, beginning in line 100 as follows:

100 DATA N, X(1), X(2), . . .,X(N)

where

N = number of raw scores

X(1), X(2), . . ., X(N) = raw scores

Note: Line 200 is the END statement. If your data statements exceed line 199, the user must type in a new END statement.

Example

```
GET-DEMO
100 DATA 2, 50, 70
RUN
N  X(N)  X(N)-MEAN
MEAN= 60
1  50  -10
2  70  +10
N  X(N)  X(N)-MEAN
OUT OF DATA IN LINE 10
```

In some Basic systems, the command to call a prepackaged program out of memory is as follows:

GET-name   (of the prepackaged program)

Note: This command may change from one system to another. Other systems use commands such as ENTER or LOAD. If an instructor wants to store an original program to be used later on, the command is one such as:

SAVE-name of program

Therefore, if we want to execute the program that computes the deviations of a number of scores from their mean, using the prepackaged program DEMO, the procedure will be as follows:

1. Contact the computer and give your identification number. Then, when the computer responds with a statement like READY,

2. Type in:

```
GET-DEMO
100  DATA5, 75, 82, 71, 52, 91
101  DATA10, 52, 75, 76, 74, 81, 82, 87, 94, 51, 80
RUN
```

3. The output will be as follows:

```
DEMO
N      X(N)      X(N)-MEAN
MEAN=  74.2
1      75               .8
2      82              7.8
3      71             -3.2
4      52            -22.2
5      91             16.8
N      X(N)      X(N)-MEAN
MEAN=  75.2
1      52            -23.2
```

```
2    75       -.2
3    76        .8
4    74      -1.2
5    81       5.8
6    82       6.8
7    87      11.8
8    94      18.8
9    51     -24.2
10   80       4.8
N    X(N)    X(N)-MEAN
OUT OF DATA IN LINE 10
```

Note: If you would like to see how a prepackaged program looks, command the machine to LIST it as shown in Article 1.6.

The use of prepackaged programs and additional topics in Basic will be discussed in detail as they are used to solve problems in succeeding chapters.

Problem Set 1.6

1. Give the computer output of the following Basic program:

```
2  DIM A(10), B(10)
4  READ N
6  FOR I = 1 TO N
8  READ A(I)
10  LET B(I) = A(I)↑2 - 1
12  PRINT I; A(I); B(I)
14  NEXT I
16  GO TO 4
18  DATA 2, 10, -10
20  DATA 5, 1, 2, 4, 8, -9
22  END
```

2. (a) State the three errors in the following Basic program:

```
5  DIM X(5)
10  FOR I = 1 TO 5
15  READ X(I)
20  LET Y(I) = X(I)↑3
25  PRINT, I, X(I), Y(I)
30  NEXT X
```

```
32  GO TO 10
35  DATA 1, 2, 3, 4, 5
40  DATA 6, 7, 8, 9, 10
45  END
```

(b) Now write corrections of these three errors so that this program will run.

3. After your instructor has saved the DEMO program previously mentioned, call it out of memory and run it with the following data:

```
DATA 4, 55, 65, 72, 81
DATA 9, 61, 52, 89, 94, 71, 75, 88, 97, 50
```

4. What statement is missing in the following Basic program?

```
10  FOR N = 1 TO 7
20  READ A(N), B(N)
30  LET C(N) = A(N) - B(N)
40  NEXT N
50  FOR N = 1 TO 7
60  PRINT C(N);
70  NEXT N
80  DATA 2, 9, -7, 3, 14, 22, -81
90  DATA 19, -8, 3, .5, 27, -52, 8
100  END
```

5. Add the missing statement in Problem 4 and run the program on a computer terminal.

2

Descriptive Statistics and Related Prepackaged Computer Programs

2.1 INTRODUCTION

Descriptive statistics is that area of mathematics which deals with the collection and analysis of numerical data. The data are sometimes called measurements or observations. For example, if we were given the daily prices of one share of Southern Oil Corporation stock during a certain week as $31.85, $31.00, $31.00, $30.15, $30.50, we might want to describe a central price and the deviation from this central price during the week (method 1). Or we might be told the number of days during an entire year for which one share of Southern Oil Corporation stock sold for a price ranging from $20 to $24, $25 to $29, $30 to $34, $35 to $39, and $40 to $44 (method 2). In the first method the data are not grouped because daily prices are listed "as is" and not put into certain class intervals. You can clearly see that in the second method the data are grouped and placed into progressive numerical class intervals.

Inferential statistics allows us to make predictions or decisions about a population, based on information contained in a sample of that population. For example, if we wanted to make a prediction about the outcome of a national election, it would be impractical to question every person of voting age in the United States (the population). However, we could randomly select (as our sample) a representative population, using some arbitrary method and not some set pattern. We could select 1000 voters from the eastern states, 1000 voters from the western states, 1000 voters from the northern states, and 1000 voters from the southern states, and could question each one of them on their choice of candidate. Then we could use the information obtained from our sample to make our prediction.

If we wanted to decide whether one method of teaching statistics yielded better results than another, it might be impossible to use the total population of all students enrolled in statistics courses. As a sample, therefore, we might randomly select 200 students enrolled in a statistics course, then randomly place 100 of them in sections teaching statistics by method 1, and place the remaining 100 students in sections teaching statistics by method 2. Then, by administering a test at the beginning and end of the course to each student, we might be able to decide whether one method yields better results than the other method.

It is important to note that the basic tools of inferential statistics are descriptive statistics and elementary probability.

2.2 DESCRIPTIVE STATISTICS—UNGROUPED DATA

If the data under investigation are not arranged into different classes or groups, but left "as is," we call this information *ungrouped data*. The ungrouped data are sometimes called measurements, observations, or raw scores. Variables, such as X, are often used to represent raw scores. For example, suppose the price of one share of Southern Oil Corporation stock on the American Stock Exchange for the week of March 18, 1974, were as follows:

Mon.:	$X_1 = \$31.85$	Thurs.:	$X_4 = \$30.15$
Tues.:	$X_2 = \$31.00$	Fri.:	$X_5 = \$30.50$
Wed.:	$X_3 = \$31.00$		

There are several ways to describe the central price for the week of March 18, 1974. Let us consider the definitions and methods in Article 2.3.

2.3 THREE MEASURES OF CENTRAL TENDENCY—UNGROUPED DATA

THE MEAN

Given the measurements $X_1, X_2, \ldots X_n$, their *arithmetic mean* ($\overline{X}$) is denoted by

$$\overline{X} = \frac{X_1 + X_2 + X_3 + \cdots + X_n}{n}$$

where n is the total number of measurements in our sample.

Sigma Notation

Another way to represent the sum of n raw scores ($X_1 + X_2 + X_3 \cdots + X_n$) is by using the *sigma notation*

$$\sum_{i=1}^{n} X_i$$

It is read as the sum of X_i as i goes from 1 to n by increments of 1. Therefore, we could write the formula for the arithmetic mean as follows:

$$\overline{X} = \frac{\sum_{i=1}^{n} X_i}{n}$$

In the sample problem in Article 2.1 concerning the mean price of one share of stock for the week of March 18

$$\overline{X} = \frac{\sum_{i=1}^{5} X_i}{5} = \frac{X_1 + X_2 + X_3 + X_4 + X_5}{5}$$

$$= \frac{\$31.85 + \$31.00 + \$31.00 + \$30.15 + \$30.50}{5}$$

$$= \frac{\$154.50}{5}$$

$$= \$30.90$$

This mean price of one share of stock for the week of March 18 does not tell us much about the mean price of one share of stock for the entire year of 1974, because our sample of one week or five daily prices is too small. You can see that the mean or average is a measure of central tendency, which depends directly on the size of the measurements in the sample. If the sum of our five daily stock prices were larger, then the mean would be greater; and if the sum of our five daily stock prices were smaller, then the mean would be less.

The Median

Given the set of measurements $X_1, X_2, X_3, \ldots, X_n$ arranged in numerical order, from smallest to largest or largest to smallest, their *median*, which is denoted by the abbreviation "Md," is the middle score if n is odd or is the average of the two middle scores if n is even. In our sampling of the five daily stock prices \$31.85, \$31.00, \$31.00, \$30.15, \$30.50, n is odd. If we arrange them from smallest to largest as \$30.15,

\$30.50, \$31.00, \$31.00, \$31.85, the third score is the median score, or Md = \$31.00.

If we were given the measurements 97, 55, 32, 21, 95, 34, 87, 52, 65, 91, we could arrange them from smallest to largest: 21, 32, 34, 52, 55, 65, 87, 91, 95, 97. There is an even number of scores ($n = 10$); therefore, the fifth and the sixth scores would be the middle scores, and their average would be the median score. Thus,

$$\text{Md} = \frac{55 + 65}{2} = \frac{120}{2} = 60$$

In writing a computer program we can make use of the following formula, which utilizes *subscripted variables*, to direct the computer to evaluate the median of n measurements. Given the set of n measurements $X_1, X_2, X_3, \ldots, X_n$ arranged in numerical order, their *median* is denoted as

$$\text{Md} = \begin{cases} X_{(n+1)/2} & \text{if } n \text{ is odd} \\ \dfrac{X_{n/2} + X_{(n+2)/2}}{2} & \text{if } n \text{ is even} \end{cases}$$

For example, we could use subscripted variables when arranging the five daily stock prices in numerical order from smallest to largest:

$$\begin{array}{ll} X_1 = \$30.15 & X_4 = \$31.00 \\ X_2 = \$30.50 & X_5 = \$31.85 \\ X_3 = \$31.00 & \end{array}$$

Since n is odd ($n = 5$),

$$\text{Md} = X_{(n+1)/2} = X_{(5+1)/2} = X_3 = \$31.00$$

If we use subscripted variables while arranging the measurements of 97, 55, 32, 21, 95, 34, 87, 52, 65, 91 in numerical order from smallest to largest, we write

$$\begin{array}{ll} X_1 = 21 & X_6 = 65 \\ X_2 = 32 & X_7 = 87 \\ X_3 = 34 & X_8 = 91 \\ X_4 = 52 & X_9 = 95 \\ X_5 = 55 & X_{10} = 97 \end{array}$$

Since n is even ($n = 10$),

$$\text{Md} = \frac{X_{n/2} + X_{(n+2)/2}}{2}$$

$$= \frac{X_{10/2} + X_{(10+2)/2}}{2} = \frac{X_5 + X_6}{2} = \frac{55 + 65}{2}$$

$$= \frac{120}{2} = 60$$

THE MODE

Given the measurements $X_1, X_2, X_3, \ldots, X_n$. Their *mode*, denoted by the abbreviation Mo, is the measurement or measurements that occur most frequently in the sample. In our example of the five daily stock prices of \$31.85, \$31.00, \$31.00, \$30.15, \$31.50, Mo = \$31.00.

Now consider the following three sets of data:

I. 70, 80, 70, 80, 95, 80; then Mo = 80.
II. 70, 80, 70, 80, 95, 96; then Mo = 70 and 80.
III. 70, 51, 32, 80, 95, 97; then Mo does not exist.

Suppose five students in each of two sections of statistics obtained the following scores on their final exam:

Section I. 70, 72, 68, 73, 67; then $\overline{X} = 70$.
Section II. 70, 50, 90, 100, 40; then $\overline{X} = 70$.

Deviation from the Mean

It is obvious that each section scored a mean grade of 70 on the final exam, but it would be incorrect to say that the performance of all sections was similar. Why? As you can see, all scores in the first section were close to the mean score, where $\overline{X} = 70$, whereas the scores in the second section were not close to the mean score, where $\overline{X} = 70$. What we want to be able to do is to describe somehow the *average deviation* from the mean. If we subtract the mean from each score and add the deviations from the mean for each section, we obtain the following results.

For Section I

X	$X - \overline{X}$
70	0
72	2
68	−2
73	3
67	−3
	SUM = 0

For Section II

X	$X - \overline{X}$
70	0
50	−20
90	20
100	30
40	−30
	SUM = 0

Now, if we divide each zero sum by the number of measurements in each section (that is, $n = 5$) we find the average deviation from the mean to be zero. But certainly this is not useful because measurements in each section *do* deviate an actual amount from the mean. Therefore we square each deviation from the mean before summing and before dividing by n. Then how can we obtain the average deviation from the mean in each section? Statisticians have developed formulas called *measures of deviation*. These are known as variance, standard deviation, and range.

2.4 MEASURES OF DEVIATION—UNGROUPED DATA

VARIANCE

Given the measurements $X_1, X_2, X_3, \ldots, X_n$, their *variance* ($s^2$) is denoted by

$$s^2 = \frac{(X_1 - \bar{X})^2 + (X_2 - \bar{X})^2 + \cdots + (X_n - \bar{X})^2}{n}$$

or

$$s^2 = \frac{\sum_{i=1}^{n} (X_i - \bar{X})^2}{n}$$

where n is the total number of measurements in the sample and $\bar{X}$ is their mean.

Consider the scores for the two sections of statistics given previously in Article 2.3 for student grades.

For Section I

X	$X - \bar{X}$	$(X - \bar{X})^2$
70	0	0
72	2	4
68	−2	4
73	3	9
67	−3	9
$\Sigma X = 350$		$\Sigma(X - \bar{X})^2 = 26$

For Section II

X	$X - \bar{X}$	$(X - \bar{X})^2$
70	0	0
50	−20	400
90	20	400
100	30	900
40	−30	900
$\Sigma X = 350$		$\Sigma(X - \bar{X})^2 = 2600$

$$\bar{X} = \frac{350}{5} = 70 \qquad \bar{X} = \frac{350}{5} = 70$$

$$\text{Variance} = s^2 = \frac{26}{5} = 5.2 \qquad \text{Variance} = s^2 = \frac{2600}{5} = 520$$

The variance in each section gives us the average of the squares of the deviations from the mean. Squaring the deviations from the mean eliminates the problem of summing the deviations from the mean and obtaining zero. In order to obtain the average deviation from the mean we now define standard deviation.

STANDARD DEVIATION

Given the measurements $X_1, X_2, X_3, \ldots, X_n$, the *standard deviation* is denoted by

$$s = \sqrt{\frac{(X_1 - \bar{X})^2 + (X_2 - \bar{X})^2 + \cdots + (X_n - \bar{X})^2}{n}} = \sqrt{\text{variance}}$$

$$s = \sqrt{\frac{\sum_{i=1}^{n} (X_i - \bar{X})^2}{n}}$$

where n is the total number of measurements in the sample and $\bar{X}$ is their mean. (See Table F.1 in Appendix.)

For Sections I and II, we can compute

I. Variance $= s^2 = 5.2$
   Standard deviation $= s = \sqrt{5.2} = 2.28$

II. Variance $= s^2 = 520$
   Standard deviation $= s = \sqrt{520} = 22.8$

Now we can see that although both sections had a mean score of 70 on the final examination, the standard deviation of 2.28 for Section I was much smaller than the standard deviation of 22.8 for Section II. Therefore, the two sections did not perform similarly on the final exam.

Again let us consider the prices for one share of Southern Oil Corporation stock for the week of March 18, 1974.

X	$X - \bar{X}$	$(X - \bar{X})^2$
\$31.85	+0.95	0.9025
\$31.00	+0.10	0.01
\$31.00	+0.10	0.01
\$30.15	−0.75	0.5625
\$30.50	−0.40	0.1600
$\Sigma X = \$154.50$		$\Sigma(X - \bar{X})^2 = \$1.645$

$$\overline{X} = \frac{154.50}{5} = \$30.90$$

$$\text{Variance} = s^2 = \frac{\$1.645}{5} = \$0.329 \approx \$0.33$$

Note: The symbol $\approx$ is read "approximately equal to." Therefore,

$$\begin{aligned}\text{Standard deviation} = s &= \sqrt{\text{variance}} \\ &= \sqrt{0.329} \\ &\approx \$0.5733 \approx \$0.57 \\ &\text{or } 57¢\end{aligned}$$

Alternate Formula for Computing Variance

The preceding formula for computing the variance,

$$s^2 = \frac{\sum_{i=1}^{n} (X_i - \overline{X})^2}{n}$$

can be shown to be equivalent to

$$s^2 = \frac{\sum_{i=1}^{n} (X_i^2) - \left[\left(\sum_{i=1}^{n} X_i\right)^2 / n\right]}{n}$$

For example, consider the final examination grades of students for each of Sections I and II mentioned previously.

Section I		Section II	
X	X²	X	X²
70	4900	70	4900
72	5184	50	2500
68	4624	90	8100
73	5329	100	10000
67	4489	40	1600
$\Sigma X = 350$	$\Sigma X^2 = 24526$	$\Sigma X = 350$	$\Sigma X^2 = 27100$

Using the alternate formula,

Section I	Section II
$s^2 = \dfrac{24{,}526 - (350)^2/5}{5}$	$s^2 = \dfrac{27{,}100 - (350)^2/5}{5}$
$= \dfrac{24{,}526 - 24{,}500}{5}$	$= \dfrac{27{,}100 - 24{,}500}{5}$
$s^2 = \dfrac{26}{5} = 5.2$ and	$s^2 = \dfrac{2600}{5} = 520$
$s = \sqrt{5.2} = 2.28$	$s = \sqrt{520} = 22.8$

Notice that the results are equal to those obtained by using the first formula given for the variance.

Using the first formula, we also determined the variance for the Southern Oil Corporation stock price as $s^2 = \$0.329 \approx \$0.33$. Let us use the alternate formula to compute the variance for this stock price.

X	X^2
\$31.85	\$1014.4225
31.00	961.0000
31.00	961.0000
30.15	909.0225
30.50	930.2500
$\Sigma X = \$154.50$	$\Sigma X^2 = \$4775.6950$

$$s^2 = \frac{(4775.6950) - (154.50)^2/5}{5}$$

$$s^2 = \frac{(4775.6950) - (23{,}870.2500)/5}{5}$$

$$= \frac{4775.6950 - 4774.05}{5} = \frac{1.6450}{5} = \$0.3290 \approx \$0.33$$

Range

Given the measurements $X_1, X_2, X_3, \ldots, X_n$, their *range* = (largest measure) − (smallest measure).

Consider the three previously discussed sample problems:

1. Five daily prices of one share of Southern Oil stock: \$31.85, \$31.00, \$31.00, \$30.15, \$30.50.

$$\text{Range} = \$31.85 - \$30.15 = \$1.70$$

2. Section I: 70, 72, 68, 73, 67.

$$\text{Range} = 73 - 67 = 6$$

3. Section II: 70, 50, 90, 100, 40.

$$\text{Range} = 100 - 40 = 60$$

The range tells us only how much the measurements vary from highest to lowest, and is therefore unstable because it depends on only two measurements in the entire sample.

Problem Set 2.1

1. Ten students scored the following number of correct responses on a 12-item true-false test: 4, 7, 6, 9, 6, 7, 7, 10, 9, 11. Compute the following information for the given distribution: (a) mean ($\overline{X}$), (b) median (Md), (c) mode (Mo), (d) range, (e) variance (s^2), and (f) standard deviation (s).

2. From a lot of 1000 radio tubes, 100 tubes were selected and tested for defects. This procedure was repeated for five consecutive days. The number of defectives found in the 100 tubes that were tested for the five consecutive days were as follows: 5, 3, 4, 2, 7. Compute the following information for the number of defective tubes: (a) mean ($\overline{X}$), (b) median (Md), (c) mode (Mo), (d) standard deviation (s).

3. The following table lists the final exam grades for two sections of an elementary statistics course.

Section 1	Section 2
75	75
70	55
82	95
81	80
68	81
70	82
70	70
75	91
81	67
82	52

(a) Compute the measures of central tendency ($\overline{X}$, Md, Mo) for each section.
(b) Compute the measures of deviation (s^2, s, range) for each section.

4. (a) Without performing any calculations, determine which one of these two samples has the larger standard deviation:

Sample 1	Sample 2
5	3
5	5
5	8
6	14
6	29
6	53
7	

(b) Explain the reason for your choice.

5. Find the mean ($\overline{X}$), standard deviation (s), and range for *each* of the following distributions:

Distribution A	Distribution B
26	36
25	36
24	35
21	32
20	30
18	29
17	28
14	25
14	25
12	22
11	18
10	15
8	
4	
1	

6. Compute the mean, median, and mode for *each* of the following three sets ot data:
(a) 26, 25, 24, 21, 20, 18, 17, 17, 14, 12, 11, 10
(b) 18, 25, 29, 30, 33, 35, 37, 37, 39, 42, 46
(c) 35, 30, 28, 28, 28, 25, 23, 23, 23, 18

7. Find the mean, median, mode, range, and standard deviation for *each* of the following score distributions:

(a)	(b)	(c)	
26	23	18	40
24	21	22	43
23	18	26	46
19	17	30	46
15	15	32	47
13	14	33	49
12	14	35	51
9	12	35	53
9	9	35	
8	7	37	
6	5	38	

2.5 A PREPACKAGED PROGRAM FOR UNGROUPED DATA

Name: STAT1

Description: This program computes the arithmetic mean, variance, and standard deviation of N numbers.

Instructions: Enter the data beginning in line 150 as follows:

150 DATA N, X_1, X_2, X_3, . . . ,X_n

where

N = number of raw scores

$X_1, X_2, X_3, \ldots, X_n$ = raw scores

Note: Line 200 is the END statement.

Example

User's Input

```
GET-STAT1

150 DATA 5,70,90,100,40,50
151 DATA 5,70,68,72,67,73
RUN
```

Computer Output

```
MEAN= 70    VARIANCE= 520  ST. DEV.= 22.8035
MEAN= 70    VARIANCE= 5.2  ST. DEV.= 2.28035

LINE      5:    END OF DATA

TIME 0.0 SECS.
```

Let us look at a development of this prepackaged program.

Given: N, X_1, X_2, . . . , X_n.

Wanted: A Basic program to compute and print out the arithmetic mean, variance, and standard deviation of the N measurements.

Desired Output: MEAN = (?) VARIANCE = (?) ST. DEV. = (?)

The program could be constructed using either of the two formulas for variance.

Using

$$s^2 = \frac{\sum_{i=1}^{n} (X_i - \overline{X})^2}{n}$$

we can construct the flowchart of Fig. 2.1.

Using

$$s^2 = \frac{\sum_{i=1}^{n} (X_i^2) - \left[\left(\sum_{i=1}^{n} X_i\right)^2 / n\right]}{n}$$

we can formulate the flowchart of Fig. 2.2, for which the coding is as follows:

```
5 READ N
10 LET S1=0
15 LETS2=0
20 FOR I=1 TO N
25     READ X
30     LET S1=S1+X
35     LET S2=S2+X↑2
40 NEXT I
45 LET M=S1/N
50 LET V=(S2-(S1)↑2/N)/N
55 LET D=SQR(V)
60 PRINT"MEAN="M;"VARIANCE="V;"ST. DEV.="D
65 GO TO 5
150 DATA
200 END
```

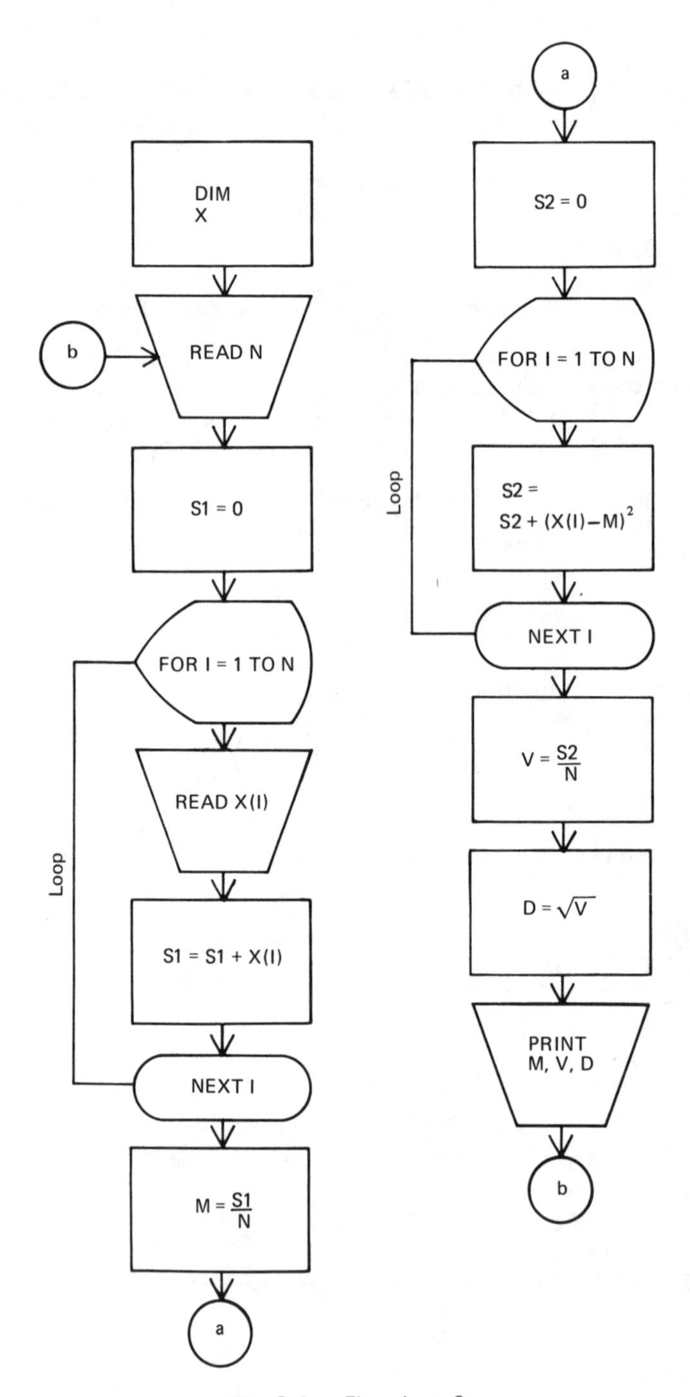

Fig. 2.1. Flowchart 9

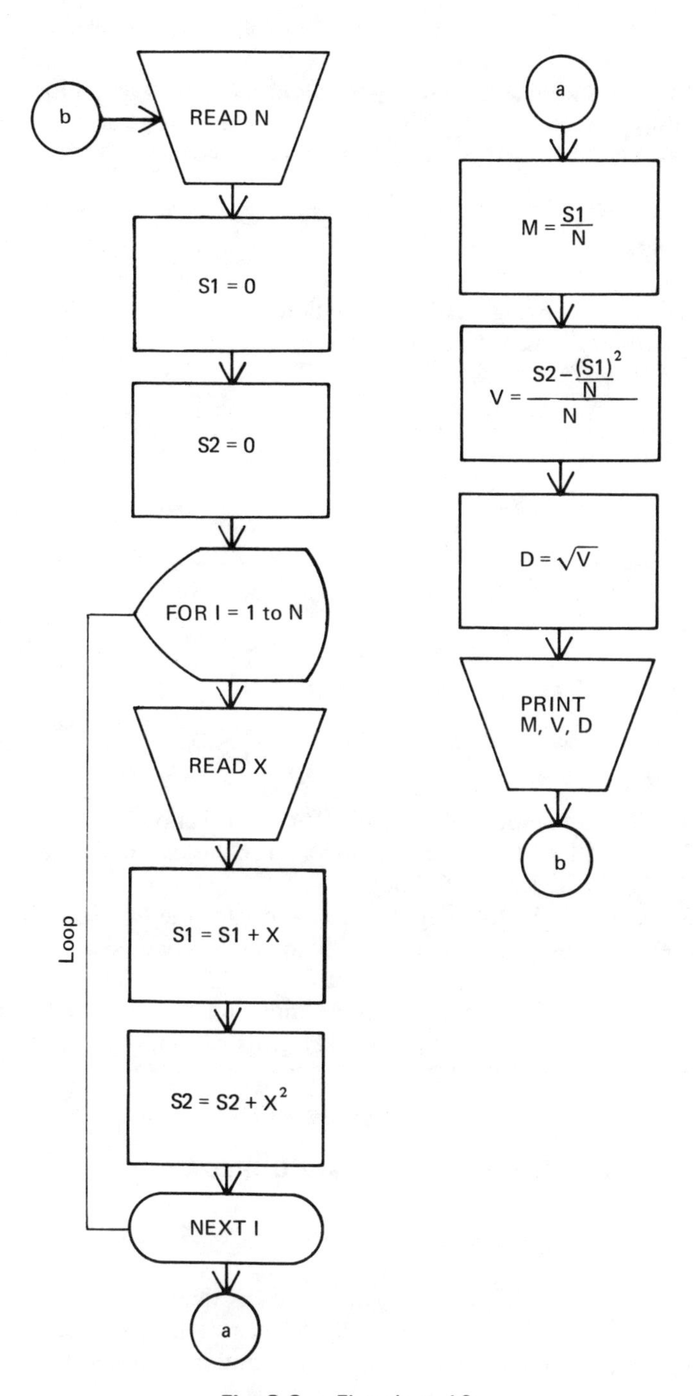

Fig. 2.2. Flowchart 10

Notice that the actual coding was done using the second formula for computing the variance. This formula is easier to use when coding because it eliminates the use of subscripted variables.

Problem Set 2.2

1. The following final examination scores for two different sections of statistics were listed in Problem Set 2.1, Problem 3.

Section 1	Section 2
75	75
70	55
82	95
81	80
68	81
70	82
70	70
75	91
81	67
82	52

Use STAT1 to compute $\overline{X}$, s^2, and s for each section.

2. Use STAT1 to compute $\overline{X}$, s^2, s for the five daily stock prices ($31.85, $31, $31, $30.15, $31.50).

3. Use STAT1 to compute $\overline{X}$, s^2, and s for the following sampling of 20 senior grade-point averages at Orange University: 3.2, 2.5, 3.1, 2.1, 3.3, 3.7, 1.9, 2.2, 2.2, 2.6, 2.1, 2.9, 2.8, 3.6, 3.0, 2.0, 3.5, 2.2, 2.6, 3.1.

4. Use STAT1 to compute $\overline{X}$, s^2, and s for the following amounts, which represent a truck driver's weekly gross earnings for the past nine weeks: $257, $315, $291, $301, $315, $288, $313, $352, $299.

2.6 DESCRIPTIVE STATISTICS—GROUPED DATA

Frequency Distributions

Suppose the Office of Testing and Evaluation at your college wanted to study the academic performance of its freshman students during the past school year. They might randomly select 40 freshman grade-point averages from their files. Let us assume the following grade-point averages were chosen:

1.9	2.3	2.3	3.3
3.2	1.6	1.7	2.0
2.0	2.1	1.6	3.0
2.4	1.7	2.9	2.1
1.9	2.5	2.2	2.5
3.1	1.6	2.3	2.6
2.0	1.5	1.8	1.8
2.4	2.0	1.9	2.2
3.0	2.4	2.3	1.7
1.9	2.1	2.8	2.9

The data may be analyzed by grouping the averages into specific class intervals. A *class interval* is a subdivision of measurement limits, which classifies measurements into an arbitrarily ordered group. For example, we are going to classify the grade-point averages given above according to the class intervals in Table 2.1

Table 2.1

Class	Class Interval	Class Boundaries
1	1.3–1.5	1.25–1.55
2	1.6–1.8	1.55–1.85
3	1.9–2.1	1.85–2.15
4	2.2–2.4	2.15–2.45
5	2.5–2.7	2.45–2.75
6	2.8–3.0	2.75–3.05
7	3.1–3.3	3.05–3.35

Although the development of class intervals is flexible, we should adhere to the following guidelines:

1. *The class intervals must accommodate all data* (consider the highest score 3.3 and the lowest score 1.5). Notice that each class interval consists of a minimal score called the *lower class limit* and a maximal score called the *upper class limit*. For example, in class 7 the lower class limit is 3.1 and the upper class limit is 3.3. The *class boundaries* allow us to place scores like 1.67 or 2.54 in the proper class in the proper class interval. Each class has a *lower class boundary* and an *upper class*

boundary. In the seventh class, the lower class boundary is 3.05, and the upper class boundary is 3.35. Consider the class intervals in Table 2.2.

Table 2.2

Class	Class Intervals	Class Boundaries
1	40–49	39.5–49.5
2	50–59	49.5–59.5
3	60–69	59.5–69.5
4	70–79	?
5	80–89	?
6	90–99	?
7	100–109	?
8	110–119	?

2. *Decide how many class intervals to use and the width of each class interval.* The width of a class interval = upper class boundary − lower class boundary. In the example above, all class widths are the same. Using class 1, we calculate the width = 49.5 − 39.5 = 10. In Table 2.1 each of the class widths are the same; using class 7, we can compute width = 3.35 − 3.05 = 3.0. If possible, it is a good practice to have equal class widths. It is practical to use at least 6 class intervals, but not more than 16 class intervals. Try to make the width of each interval an odd number so that there will be no need to approximate the *midpoint* of a class interval.

3. *Each piece of data must correspond to and be counted in one and only one class interval.*

Consider the following extension of Table 2.1.

Table 2.3

Class	Class Intervals	Class Boundaries	(f) Frequencies	(cf) Cumulative Frequencies	(x) Midpoint
1	1.3–1.5	1.25–1.55	1	1	1.4
2	1.6–1.8	1.55–1.85	8	9	1.7
3	1.9–2.1	1.85–2.15	11	20	2.0
4	2.2–2.4	2.15–2.45	9	29	2.3
5	2.5–2.7	2.45–2.75	3	32	2.6
6	2.8–3.0	2.75–3.05	5	37	2.9
7	3.1–3.3	3.05–3.35	3	40	3.2

Notes:

1. The *frequency column (f)* in Table 2.3 tells us how many scores are contained in each class interval. (For example, there are three scores [2.5, 2.5, 2.6] that are within the class boundaries of the fifth class.)

2. The *cumulative frequency (cf) column* tells us how many of the scores are less than or equal to the upper class boundary in the given interval. In Table 2.3, we are told that in class 2 there are a total of nine scores less than or equal to 1.85. What are these nine scores? 1.5, 1.6, 1.6, 1.6, 1.7, 1.7, 1.7, 1.8, and 1.8.

3. *The midpoint* (x) of a class interval is a score that we assign to each of the frequencies contained in a certain class. For example, in the grade-point average in Table 2.3 consider:

Class	Class Interval	Class Boundaries
4	2.2–2.4	2.15–2.45

Class width = 2.45 − 2.15 = 0.3 (an odd number)
Midpoint = 2.3 (the middle score in the class interval 2.2, 2.3, 2.4)

The midpoint of a class interval is the average of the two middle scores if the class width is *even*. From Table 2.2 consider:

Class	Class Interval	Class Boundaries
5	80–89	79.5–89.5

Class width = 89.5 − 79.5 = 10 (an even number)
Midpoint = (84 + 85)/2 = 169/2 = 84.5
(the average of the two middle scores in the class interval: 80, 81, 82, 83, 84, 85, 86, 87, 88, 89)

In many instances we will be given a frequency distribution without being given the actual measurements. We will assign the midpoint score to each of the frequencies in a given class interval. This arbitrary designation will allow us to approximate measures of central tendency and measures of variation from the information contained in the given frequency distribution.

2.7 GRAPHING FREQUENCY DISTRIBUTIONS

A two-dimensional coordinate system can be used to graph the measurements of a frequency distribution. A graph of a frequency distribution illustrates how the measurements are distributed within their range. Consider the following table of grouped data, where the class intervals represent weekly salaries for a random sampling of 25 employees at the Hoboken Auto Body Shop.

Class	Class Interval	Class Boundaries	(x) Midpoint	(f) Frequencies	(cf) Cumulative Frequencies
1	150–154	149.5–154.5	152	2	2
2	155–159	154.5–159.5	157	3	5
3	160–164	159.5–164.5	162	4	9
4	165–169	164.5–169.5	167	8	17
5	170–174	169.5–174.5	172	1	18
6	175–179	174.5–179.5	177	2	20
7	180–184	179.5–184.5	182	3	23
8	185–189	184.5–189.5	187	2	25

2.8 FREQUENCY HISTOGRAM

The frequency *histogram* (Fig. 2.3) is a geometric figure that consists of adjoining rectangles to represent the frequencies contained in each of the class intervals of a given distribution. The frequency histogram in Fig. 2.3 is based on the data of the 25 auto shop employees' salaries. To plot a particular point on a frequency histogram, we move horizontally midway between the class boundaries of a particular class and then move vertically up to the number of frequencies contained in that class. You will notice the space between 0 and 149.5 on the horizontal axis, which is used to indicate that the lowest class boundary is not 0 and that the class boundaries start with a minimal value of 149.5. To plot the first point, we moved horizontally midway between 149.5 and 154.5, and vertically to frequency 2. To plot the second point, we moved horizontally midway between 154.5 and 159.5, and vertically to frequency 3.

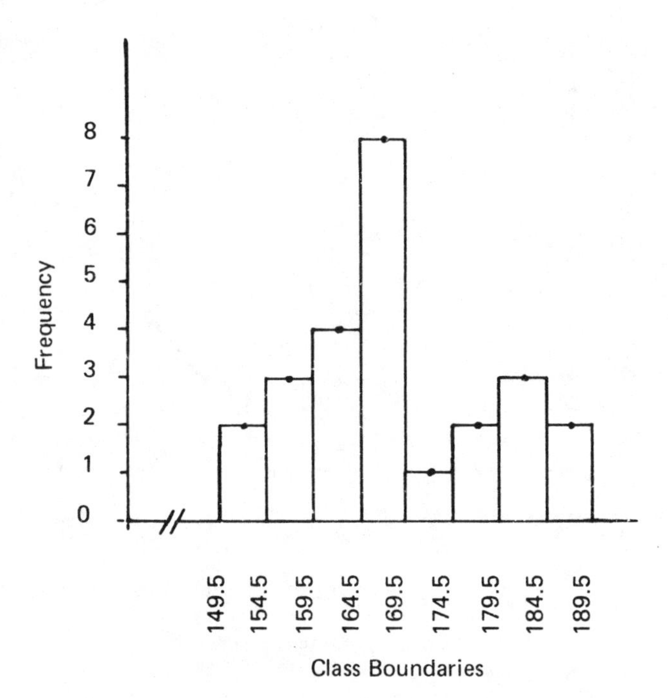

Fig. 2.3. The frequency histogram

2.9 FREQUENCY POLYGON

The frequency *polygon* is a geometric figure constructed by joining pairs of points that represent the frequencies to which a certain midpoint score is assigned. The frequency polygon in Fig. 2.4 is based on the data of the 25 auto shop employees' salaries. To plot a particular point on a frequency polygon, we move horizontally to the midpoint of a particular class interval and then move vertically to the number of frequencies (to which we are assigning that midpoint score) contained in that interval. Then we connect consecutive points with straight lines.

You will notice that at the beginning and end of our horizontal axis there are midpoints (147 and 192) which do not actually exist for the eight class intervals of our auto shop employees' salaries. These designations allow us to plot the points (147,0) and 192,0) so that we will have a *closed curve*. The point (152,2) indicates that we are assigning a score of 152 to two of the frequencies in the distribution. The point

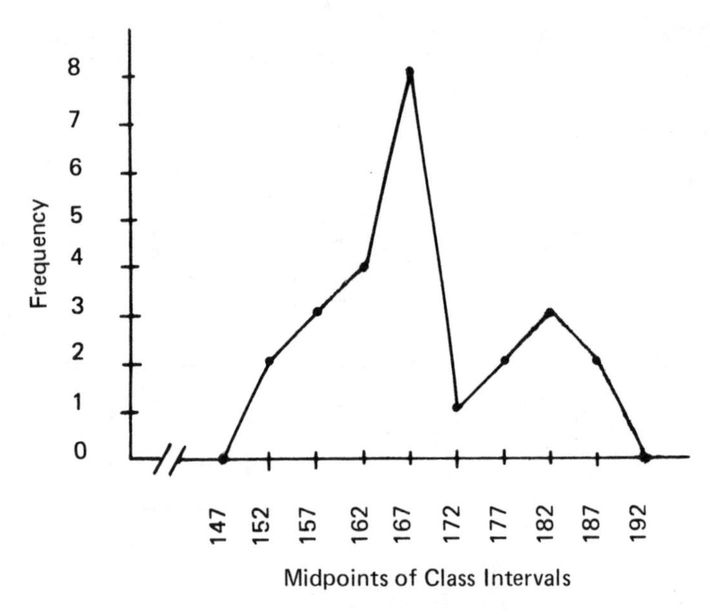

Fig. 2.4. The frequency polygon

(167,8) indicates that we are assigning a score of 167 to eight of the frequencies in the distribution. A frequency polygon illustrates whether the frequencies are increasing or decreasing from one interval to another.

2.10 CUMULATIVE FREQUENCY GRAPH

The *cumulative frequency graph* is constructed by joining pairs of points that represent the total frequencies that are less than or equal to each class boundary in a distribution. The cumulative frequency graph in Fig. 2.5 is constructed using the distribution of the 25 auto shop employees' salaries. To plot a particular point, we move horizontally to a class boundary and then vertically to the total frequency of scores that are less than or equal to that class boundary. We can see the amount of increase in the total frequencies as we move from one class boundary to another class boundary. For example, the point (159.5,5) indicates that there are a total of five frequencies less than or equal to 159.5 in the given distribution. What does the point (184.5,23) indicate?

2.11 PERCENT GRAPHS

To construct *percent graphs*, whether they be histograms or polygons, you must change the vertical axis from frequencies or cumulative frequencies to percentage of frequencies or percentage of cumulative frequencies.

For example, let us consider the addition of a percent of frequency column to our distribution of 25 auto shop employees' salaries.

Class	Class Interval	(x) Midpoint	(f) Frequency	(cf) Cumulative Frequencies	($\%f$) Percent of Frequency
1	150–154	152	2	2	$(2/25)(100) = 8$
2	155–159	157	3	5	$(3/25)(100) = 12$
3	160–164	162	4	9	$(4/25)(100) = 16$
4	165–169	167	8	17	$(8/25)(100) = 32$
5	170–174	172	1	18	$(1/25)(100) = 4$
6	175–179	177	2	20	$(2/25)(100) = 8$
7	180–184	182	3	23	$(3/25)(100) = 12$
8	185–189	187	2	25	$(2/25)(100) = 8$

2.12 PERCENT OF FREQUENCY POLYGON GRAPH

The *percent of frequency polygon graph* is a polygon graph that indicates each of the midpoints of class intervals in a distribution and the corresponding percentage of measurements assigned to each midpoint. Figure 2.6 is the percent of frequency graph for the 25 salaries. The point (157,12) indicates that 12% of the measurements are assigned a salary of $157 for evaluating measures of central tendency and measures of deviation. The point (172,4) indicates that 4% of the measurements are assigned a salary of $172.

Problem Set 2.3

1. The Western Tire Corporation is in the process of testing the longevity of an experimental tire. Thirty test cars reported the following number of months of wear before the experimental tire was replaced.

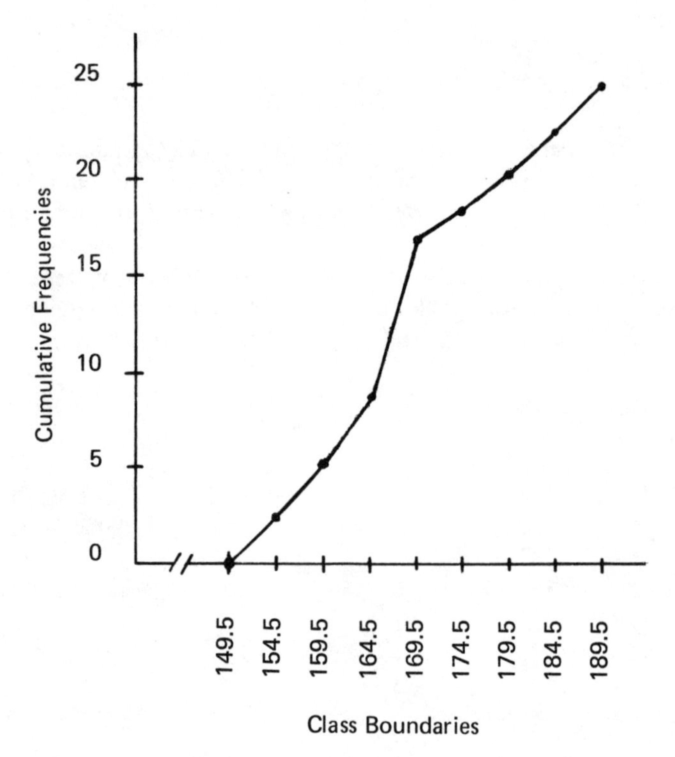

Fig. 2.5. Cumulative frequency graph

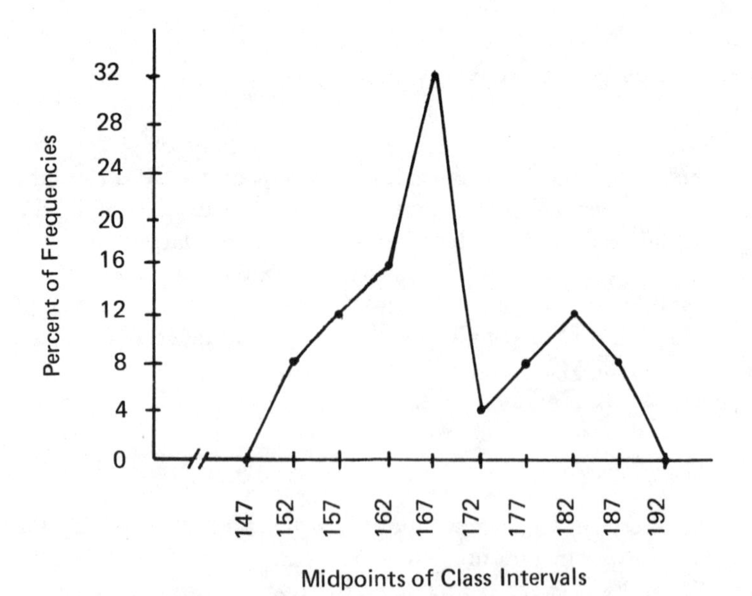

Fig. 2.6. Percent of frequency graph

27	18	26	20	25
23	26	19	20	31
23	32	15	24	21
25	25	27	17	20
24	24	24	28	24
30	25	30	29	25

Using the reported data,

(a) Construct a frequency distribution using class intervals 15–17, 18–20, etc.
(b) Construct a frequency histogram graph.
(c) Construct a cumulative frequency graph.
(d) Construct a frequency polygon.
(e) Construct a percent of frequency polygon.

2. Given the following distribution of freshman grade-point averages at Las Vegas University:

Class	Class Interval	Class Boundaries	(x) Midpoint	f	cf	$\%f$	$\%cf$
1	1.4–1.6			4			
2	1.7–1.9			20			
3	2.0–2.2			35			
4	2.3–2.5			40			
5	2.6–2.8			11			
6	2.9–3.1			20			
7	3.2–3.4			20			

(a) Fill in the missing information.
(b) Construct a cumulative frequency graph.
(c) Construct a percent of frequency graph.
(d) Construct a frequency histogram.

3. Given the following distribution of ages of patients who were treated for tuberculosis at the Hope Clinic during 1974:

Class	Class Interval	f
1	10–20	10
2	21–31	15
3	32–42	25
4	43–53	10
5	54–64	10
6	65–75	15

(a) Construct a frequency polygon.
(b) Construct a cumulative frequency graph.
(c) Construct a percent of frequency graph.

4. (a) Complete the following tabulation of scores on a statistics test.

Class	Class Interval	Class Boundaries	(x) Mipoint	f	cf	$\%f$	$\%cf$
1	65–69			2			
2	70–74			2			
3	75–79			9			
4	80–84			19			
5	85–89			14			
6	90–94			3			
7	95–99			1			

(b) Construct a cumulative frequency graph.
(c) Construct a percent of frequency graph.
(d) Construct a frequency histogram.

5. For the distribution of weights of 100 elementary school pupils,
(a) Construct a frequency histogram graph.
(b) Construct a cumulative frequency graph.
(c) Construct a frequency polygon.
(d) Construct a percent of frequency polygon, using the following data:

Class	Class Interval	f
1	50–59	3
2	60–69	6
3	70–79	15
4	80–89	22
5	90–99	24
6	100–109	15
7	110–119	10
8	120–129	4
9	130–139	1

6. Given the following 60 time lapses (in minutes) between eruptions of Liberty Geyser in Monmouth National Park, measured from the beginning of one eruption to the beginning of the next:

75	77	75	70	72
73	86	83	78	74
97	84	96	99	85
86	83	80	87	80
80	88	85	80	83
89	86	79	82	84
87	82	80	81	79
82	76	89	65	68
69	78	92	77	88
84	87	84	83	84
79	80	84	74	80
86			74	
83			85	
			84	

(a) Construct a frequency distribution using class intervals 65–69, 70–74, 75–79, etc.
(b) Construct a frequency histogram graph.
(c) Construct a cumulative frequency graph.
(d) Construct a frequency polygon.
(e) Construct a percent of frequency polygon.

2.13 MEASURES OF CENTRAL TENDENCY FOR GROUPED DATA

Reconsider the sampling of 25 employees' weekly salaries at the Hoboken Auto Body Shop.

Class	Class Interval	Class Boundaries	(x) Midpoint	(f) Frequency	(cf) Cumulative Frequency
1	150–154	149.5–154.5	152	2	2
2	155–159	154.5–159.5	157	3	5
3	160–164	159.5–164.5	162	4	9
4	165–169	164.5–169.5	167	8	17
5	170–174	169.5–174.5	172	1	18
6	175–179	174.5–179.5	177	2	20
7	180–184	179.5–184.5	182	3	23
8	185–189	184.5–189.5	187	2	25

Notice that we do not know the *actual* salaries. Therefore, in order to approximate measures of central tendency and measures of deviation, we will assign to each frequency in a class the midpoint of that class. For example, in the distribution given above we will *assume* there are three weekly salaries of \$157, . . . , two weekly salaries of \$187.

1. The *arithmetic mean in a distribution of grouped data* is denoted by

$$\overline{X} = \frac{\sum_{i=1}^{n} x_i f_i}{n}$$

where $x_1, x_2, \ldots, x_n$ are the midpoints of each class interval, $f_1, f_2, \ldots, f_n$ are the corresponding frequencies in each class interval, and n represents the total frequencies.

For the salary distribution in the sampling tabulation, we obtain

Class	x	f	cf	xf
1	152	2	2	304
2	157	3	5	471
3	162	4	9	648
4	167	8	17	1336
5	172	1	18	172
6	177	2	20	354
7	182	3	23	546
8	187	2	25	374
				$\Sigma xf = 4205$

Thus,

$$\overline{X} = \frac{\sum_{i=1}^{n} X_i f_i}{n} = \frac{4205}{25} = \$168.20$$

(2). The *mode* in a distribution of grouped data, denoted by Mo, is the midpoint of the class containing the *largest* frequency. In the salary distribution listed above, the fourth class contains the largest frequency; therefore, Mo = \$167.

(3). The *median* in a distribution of grouped data is often called the *50th percentile* and denoted by the symbol P_{50}. The median or 50th percentile is a score P_{50} such that approximately 50% of the measurements in a distribution are less than P_{50}, and 50% of the measurements in the distribution are greater than P_{50}.

Given the measurements $x_1, x_2, \ldots, x_n$, the x(th) percentile is a score, denoted by P_x, such that approximately $x/100$ of the measurements are less than P_x, and approximately $1 - (x/100)$ of the measurements are greater than P_x. For example, the 50th percentile (or P_{50}) is a score such that 50/100 or one half of the measurements are less than P_{50} and 1–1/2 or one half of the measurements are greater than P_{50}. The 20th percentile (P_{20}) is a score such that 20/100 or one fifth of the measurements are less than P_{20}, and 1–1/5 or four fifths of the measurements are greater than P_{20}.

Now let us discuss a method for *approximating* the x(th) percentile. In a distribution containing n measurements arranged in ascending order, *the position of the* x*th percentile* is denoted by $p = n\ (x/100)$.

For example, in the preceding salary distribution tabulation, there are a total of $n = 25$ frequencies; therefore, the position of the 50th percentile $p = 25\ (50/100) = 12.5$; thus, the 50th percentile, or P_{50}, lies in the fourth class (it is more than nine cumulative frequencies, so it is not in the third class) between the cumulative frequencies of 9 and 17. The position of the 20th percentile $p = 25\ (20/100) = 5$; thus, the 20th percentile, or P_{20}, lies at the end of the second class, which has a total of five cumulative frequencies. Now we can determine the value of the x(th) percentile as

$$P_x = l + \frac{w(p - cf)}{f}$$

where

l = lower class boundary of the class containing the position of the desired percentile
w = class interval width
p = position of desired percentile, where $p = n(x/100)$
cf = number of cumulative frequencies in the class one step below the class containing the position of the desired percentile
f = frequency in the class containing the position of the desired percentile

Let us again consider the distribution of 25 employees' weekly salaries.

Class	Class Interval	Class Boundaries	f	cf
1	150–154	149.5–154.5	2	2
2	155–159	154.5–159.5	3	5
3	160–164	159.5–164.5	4	9
4	165–169	164.5–169.5	8	17
5	170–174	169.5–174.5	1	18
6	175–179	174.5–179.5	2	20
7	180–184	179.5–184.5	3	23
8	185–189	184.5–189.5	2	25

The position of the *median* (Md), or $p = 25(50/100) = 12.5$, lies in the fourth class between the total of 9 cumulative frequencies at the end of class 3, and the total of 17 cumulative frequencies at the end of class 4. Therefore, the value of the median or

$$\begin{aligned} \text{Md} = P_{50} &= 164.5 + \frac{5(12.5 - 9)}{8} \\ &= 164.5 + \frac{5(3.5)}{8} \\ &= 164.5 + 2.19 \\ &= 166.69 \approx 166.7 \end{aligned}$$

The position of the 25th percentile, or $p = 25\ (25/100) = 6.25$, lies in the third class between the total of five cumulative frequencies at the end of class 2 and the total of nine cumulative frequencies at the end of class 3. Therefore, the value of the 25th percentile, or

$$\begin{aligned} P_{25} &= 159.5 + \frac{5(6.25 - 5)}{4} \\ &= 159.5 + \frac{5(1.25)}{4} \\ &= 159.5 + 1.56 \\ &= 161.06 \approx 161.1 \end{aligned}$$

2.14 MEASURES OF DEVIATION FOR GROUPED DATA

Given the *ungrouped* measurements $x_1, x_2, \ldots, x_n$, we saw that their variance (s^2) could be computed using the following formula:

$$s^2 = \frac{\sum_{i=1}^{n} (X_i)^2 - \left[\left(\sum_{i=1}^{n} X_i\right)^2 / n\right]}{n}$$

1. In many distributions of *grouped data*, we do not know what the actual measurements are. Therefore, we define the *variance of a distribution of grouped data* as follows:

$$s^2 = \frac{\sum_{i=1}^{n} (X_i^2 f_i) - \left[\left(\sum_{i=1}^{n} X_i f_i\right)^2 / n\right]}{n}$$

where

$x_1, x_2, \ldots, x_n$ = *midpoints* of each class interval
$f_1, f_2, \ldots, f_n$ = frequency in each corresponding class interval
n = *total* of frequencies in the distribution

2. The *standard deviation of a distribution of grouped data* is denoted by $s = \sqrt{\text{variance}}$.

Again let us examine the distribution of grouped data for 25 employees' weekly salaries at the Hoboken Auto Body Shop.

Class	Class Interval	(x) Midpoint	f	cf	xf	x^2	x^2f
1	150–154	152	2	2	304	23104	46208
2	155–159	157	3	5	471	24649	73947
3	160–164	162	4	9	648	26244	104976
4	165–169	167	8	17	1336	27889	223112
5	170–174	172	1	18	172	29584	29584
6	175–179	177	2	20	354	31329	62658
7	180–184	182	3	23	546	33124	99372
8	185–189	187	2	25	374	34969	69938
					$\Sigma(xf) = 4205$		$\Sigma(x^2f) = 709795$

$$\text{Variance} = s^2 = \frac{709795 - [(4205)^2/25]}{25}$$

$$= \frac{709795 - (17682025/25)}{25}$$

$$= \frac{2514}{25}$$

$$= 100.57$$

$$\text{Standard deviation} = s = \sqrt{100.57} = 10.0279 \approx 10.03$$

Problem Set 2.4

1. Compute the mean, median, mode, variance, and standard deviation for the frequency distribution longevity of a car tire, which you constructed in Problem Set 2.3, Problem 1.

2. Compute the mean, median, mode, variance, and standard deviation for the frequency distribution of grade-point averages given in Problem Set 2.3, Problem 2.

3. Every day for 30 consecutive days the Shifty Testing Corporation tested 1000 speakers for the Loud Sound Radio Corporation. They recorded the following frequency distribution, where each class interval represents the number of speakers found to be defective out of the lot of 1000 speakers tested daily.

Class	Class Interval	f
1	0–2	10
2	3–5	10
3	6–8	5
4	9–11	3
5	12–14	1
6	15–17	1

Compute the mean, median, mode, variance, and standard deviation for the given distribution.

4. Compute the mean, median, mode, variance, and standard deviation for the frequency distribution of scores on a statistics test given in Problem 4 of Problem Set 2.3.

5. Compute the mean, median, variance, and standard deviation for the frequency distribution of weights of 100 elementary school pupils given in Problem 5 of Problem Set 2.3.

6. Compute the mean, median, and standard deviation for the frequency distribution of time elapses (in minutes) between eruptions of Liberty Geyser given in Problem 6 of Problem Set 2.3.

Note: Retain your calculations for this problem set for future comparison with results obtained in Problem Set 2.5.

2.15 A PREPACKAGED PROGRAM FOR GROUPED DATA

Name: STAT2

Description: This program computes the mean, median, variance, and standard deviation for a set of *grouped data*.

Instructions: Enter the data beginning in line 150 as follows:

150 DATA N, F(1), X(1), F(2), X(2), . . . , F(N), X(N)

where

N = number of class intervals, $N \leq 16$

$F(1), F(2), \ldots, F(N)$ = number of frequencies in each class

$X(1), X(2), \ldots, X(N)$ = midpoints in each class

Note: Line 200 is the END statement.

Sample Problem

Class Int	Class Bound	f	(x) Midpoint
1.4–1.6	1.35–1.65	4	1.5
1.7–1.9	1.65–1.95	20	1.8
2.0–2.2	1.95–2.25	35	2.1
2.3–2.5	2.25–2.55	40	2.4
2.6–2.8	2.55–2.85	11	2.7
2.9–3.1	2.85–3.15	20	3.0
3.2–3.4	3.15–3.45	20	3.3

User's Input

```
GET-STAT2

150 DATA 7,4,1.5,20,1.8,35,2.1,40,2.4,11,2.7,20,3,20,3.3
RUN
```

Computer Output

```
MEAN                MEDIAN              VARIANCE            ST.DEV.
 2.448               2.37               .250898             .500898


LINE   10:   END OF DATA

TIME 0.0 SECS.
```

Let us take a look at the flowcharting of STAT2 (Fig. 2.7).
The coding of STAT2 is as follows:

```
5 DIM F(16),X(16),C(16)
10 READ N
15 LET S1 = 0
20 LET S2 = 0
25 LET C(0) = 0
30 FOR I= 1 TO N
35     READ F(I), X(I)
40     LET S1 = S1 + F(I)*X(I)
45     LET S2 = S2 + F(I)*X(I)↑2
50     LET C(I) = C(I-1) + F(I)
55 NEXT I
60 LET M = S1/C(N)
65 LET V = (S2-(S1)↑2/C(N))/C(N)
70 LET D = SQR(V)
75 FOR I = 1 TO N
80     IF C(I) > C(N)/2 THEN 90
85 NEXT I
90 LET W = X(2)-X(1)
95 LET L = X(I)-W/2
100 LET M1 = L-W*(C(I-1)-C(N)/2)/F(I)
105 PRINT "MEAN","MEDIAN","VARIANCE","ST.DEV."
110 PRINT M,M1,V,D
115 PRINT
120 GO TO 10
150 DATA
200 END
```

Problem Set 2.5

Use STAT2 to compute the mean, median, variance, and standard deviation of the following frequency distributions:

1.

Class	Class Interval	Class Boundaries	Midpoint	f
1	15–17	14.5–17.5	16	2
2	18–20	17.5–20.5	19	5
3	21–23	20.5–23.5	22	3
4	24–26	23.5–26.5	25	12
5	27–29	26.5–29.5	28	4
6	30–32	29.5–32.5	31	4

2.

Class	Class Interval	Class Boundaries	Midpoint	f
1	1.4–1.6	1.35–1.65	1.5	4
2	1.7–1.9	1.65–1.95	1.8	20
3	2.0–2.2	1.95–2.25	2.1	35
4	2.3–2.5	2.25–2.55	2.4	40
5	2.6–2.8	2.55–2.85	2.7	11
6	2.9–3.1	2.85–3.15	3.0	20
7	3.2–3.4	3.15–3.45	3.3	20

3.

Class	Class Interval	f
1	0–2	10
2	3–5	10
3	6–8	5
4	9–11	3
5	12–14	1
6	15–17	1

4.

Class	Class Interval	Class Boundaries	Midpoint	f	cf
1	65–69	64.5–69.5	67	2	2
2	70–74	69.5–74.5	72	2	4
3	75–79	74.5–79.5	77	9	13
4	80–84	79.5–84.5	82	19	32
5	85–89	84.5–89.5	87	14	46
6	90–94	89.5–94.5	92	3	49
7	95–99	94.5–99.5	97	1	50

5.

Class	Class Interval	*f*	*cf*
1	50–59	3	3
2	60–69	6	9
3	70–79	15	24
4	80–89	22	46
5	90–99	24	70
6	100–109	15	85
7	110–119	10	95
8	120–129	4	99
9	130–139	1	100

6.

Class	Class Interval	*f*	*cf*
1	65–69	3	3
2	70–74	6	9
3	75–79	10	19
4	80–84	23	42
5	85–89	14	56
6	90–94	1	57
7	95–99	3	60

Note: The computer output should be the same as your manual calculations for these statistics in each of the six problems in Problem Set 2.4.

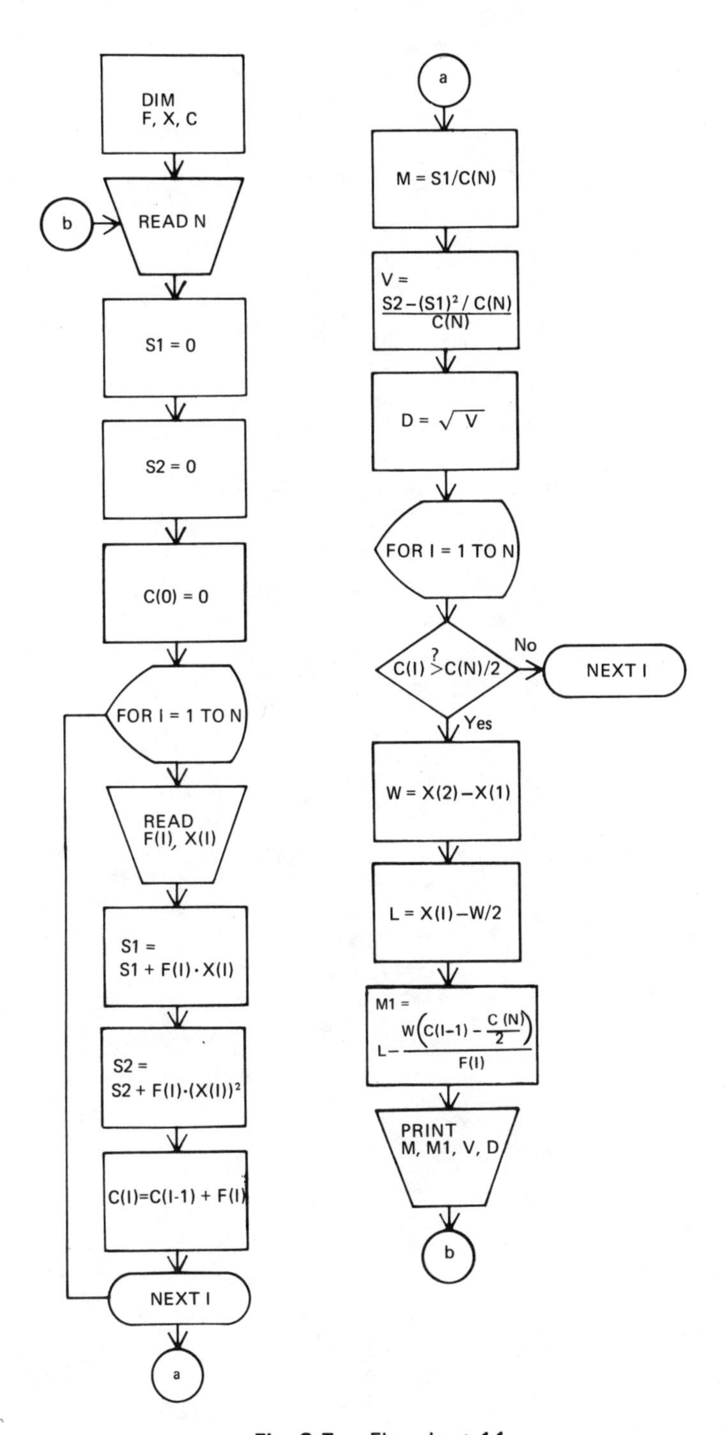

Fig. 2.7. Flowchart 11

3

Sets, Permutations, Combinations, and the Binomial Theorem, with a Prepackaged Computer Program

3.1 INTRODUCTION TO SETS

As a fundamental notion of mathematics the concept of set is essential to the study of probability theory.

A *set* is a well-defined collection of objects; that is, it is clearly indicated whether a given object does or does not belong to a given set. We customarily use capital letters to denote sets and list the objects contained in a set within a pair of braces. For example:

Listing Method	Verbal Description
$A = \{1,2,3,4\}$	Set A contains the numbers 1,2,3, and 4.
$C = \{1,2,3 \cdots\}$	Set C contains all the counting numbers.
$F = \{1,8,27 \cdots\}$	Set F contains the cubes of all the counting numbers.
$G = \{P_1, P_2, P_3\}$	Set G contains the objects P_1, P_2, P_3, which may also be sets themselves.

The objects that belong to a set are called *elements* or *members* of the set and are denoted by lower-case letters with the following notation:

$a \in A$ is read: "a is an element of the set A," or "a belongs to the set A."

$b \notin B$ is read: "b is not an element of the set B," or "b does not belong to the set B."

The symbol "/" generally means "not" when it is put through or slashes another mathematical symbol. For example $d \notin D$ (d is not an element of set D), $3 \neq 2$ (3 is not equal to 2).

Besides the listing and verbal methods of describing sets, we may indicate membership in a particular set by designating the elements in terms of one or more properties, or by giving a *rule for set-builder notation.*

Examples

1. $F = \{1,8,27, \cdots\}$ may be denoted
$F = \{X^3 | X \text{ is a counting number}\}$

We read: "set F is the set of all X^3 such that (the meaning of the vertical line) X is a counting number."

2. $C = \{1,2,3 \cdots\}$ may be denoted by the rule
$C = \{c | c \text{ is a counting number}\}$

3. $E = \{2,4,6,8 \cdots\}$ may be written in set builder notation as
$E = \{e | e \text{ is an even positive integer}\}$

4. $H = \{(1,2), (2,1) (2,2)\}$.

Notice that each element of this set consists not of numbers but of *ordered pairs* of numbers, where the order of the components of each element in set H is most significant. You know that the ordered pair (1,2), with the first component 1 and the second component 2, is distinct from the ordered pair (2,1). Set-builder notation is extremely helpful in defining set operations and set relationships.

A set may have a *finite* or an *infinite* number of elements. A set is called *finite* if the elements of the set can be counted, and the counting process comes to an end. In the *infinite* set there are an unlimited number of elements. For example, the set of days in a week is certainly countable or finite, but the set of all points on a number line constitutes an infinite set. Set $A = \{1,2,3,4\}$, which has exactly four elements, is finite; whereas set $B = \{1,8,27, \ldots\}$, which has no last element, is an infinite set.

When we speak of the *universal set*, or universe, we mean the set of all elements that we are considering for a particular discussion, and denote it by U. When you are given a problem, it is important that you grasp clearly the particular universe with which you are dealing. Sometimes a universe U is a very small set; for example, $U = \{2\}$ may be our universe of discourse. Or it may be an infinite set, such as $U = \{1,2,3,4, \cdots\}$.

3.2 SUBSETS

Definition 1

A set X is said to be a *subset* of a set Y if and only if every element of X is also an element of Y. We write this relationship as $X \subset Y$ and read it as "X is a subset of Y," or "X is included in Y." For example, let E = the set of positive *even* integers and I = the set of positive integers. Then the set of positive even integers is a subset of the positive integers, and we write $E \subset I$. Can we say that the set of positive *odd* integers is a subset of the positive integers?

If we let

U = {married couples in the U.S.A.}
C = {married couples with one child in the U.S.A.}
T = {married couples with two children in the U.S.A.}

then C and T are subsets of U, and we write $C \subset U$ and $T \subset U$.

Definition 2

If two sets C and D have the same or *identical* elements, they are said to be equal. That is, $C = D$ if and only if $C \subset D$ and $D \subset C$. For example, let $C = \{1,2,3,4,5\}$ and $D = \{5,4,3,2,1\}$. Then, by our definition, $C \subset D$ and $D \subset C$ (every element of C is an element of D and every element of D is an element of C), and we write $C = D$.

Definition 3

If two finite sets A and B have the *same number* of elements, they are said to be equivalent. Thus, A = {Monday, Wednesday, Friday} and $B = \{a, b, c\}$ are *equivalent* sets because each set has the same number (three) of elements.

Definition 4

The *empty* or *null* set is a set that contains no elements and is denoted by the symbol ϕ or { }. If we try to list the elements of the null set, we find that there are none. For example, can we list the members of the set consisting of all women who have served as president of the United States? Or can we list the members of the set of human beings with four eyes? The sets in these two examples may be called the null or empty sets.

The null set is considered to be a subset of *every* set. For example, if we are given the set $E = \{2,4,6,8,10\}$, we might ask, "What is $\{x \varepsilon E \mid x$ is odd$\}$?" In other words, what is the subset that contains all odd integers in E? Since E has no odd elements, this set must be the empty set ϕ. We write $F = \{x \varepsilon E \mid x \text{ is odd}\} = \phi$.

Definition 5

Any subset of given set B which does not contain all the elements of B is called a *proper* subset of B. That is, A is a proper subset of B if and only if A is a subset of B and at least one element of B is not an element of A. For example, if $A = \{2,3\}$ and $B = \{1,2,3,4\}$, then A is a proper subset of B. You can see that there is at least one element of B (element 4) such that $4 \varepsilon B$ and $4 \not\varepsilon A$.

The universal set and the null set are referred to as *improper* subsets; the other subsets are referred to as proper. For example, if we list all the subsets of the set $U = \{a,b,c,\}$, we would find there are eight subsets: $\{a\}$, $\{b\}$, $\{c\}$, $\{a,b\}$, $\{a,c\}$, $\{b,c\}$, $\{a,b,c\}$, and ϕ. The set $\{a,b,c\}$, which is the entire set U, and the set ϕ are improper subsets of U; the other six subsets are proper subsets of U.

Definition 6

Two sets, A and B, are *disjoint* if they have no common elements. For example, the set of positive integers and the set of negative integers are disjoint because the elements in A are *mutually exclusive* of the elements in B. Would you say that the set of students studying French and the set of students studying mathematics in your school are disjoint?

3.3 OPERATIONS WITH SETS

Definition

The *complement* of a set A, denoted by A', is defined as follows: If A is a subset of a given universe U, then A' is another subset containing all the elements of the universe that are not in A. Thus,

$$A' = \{x \mid x \varepsilon U \text{ and } x \not\varepsilon A\}$$

To determine A', we list all the elements contained in the universal set that are *not* contained in set A. For example, if $U = \{1,2,3, \cdots\}$, $A = \{1,2,3,4,5\}$, and $B = \{6,8,9\}$, then $A' = \{6,7,8,9, \cdots\}$ and $B' = \{1,2,3,4,5,7,10,11,12,13, \cdots\}$.

VENN DIAGRAMS

A good way to show or illustrate the relationships between sets and operations with sets is to use sketches known as *Venn diagrams*. We usually represent the universe by the region within some plane geometric figure such as a rectangle, and identify subsets of the universe by smaller regions within the rectangle. The Venn diagram in Fig. 3.1 illustrates that $A \subset U$. Do you see that Fig. 3.1 shows us that every element contained in A is also contained in U?

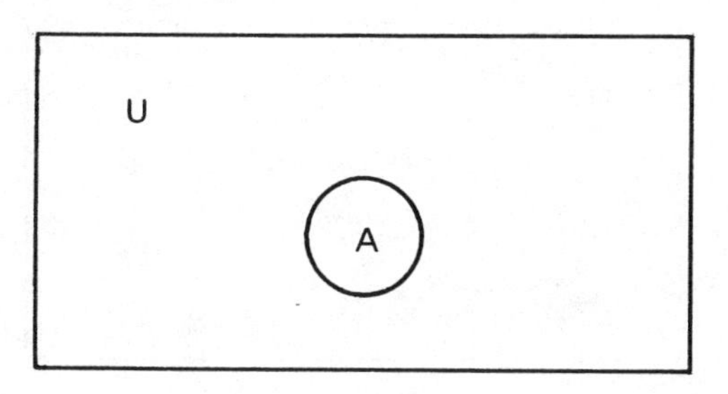

Fig. 3.1. Venn diagram: $A \subset U$

If we want to show that $A \subset B$ and $B \subset U$, we use the Venn diagram in Fig. 3.2.

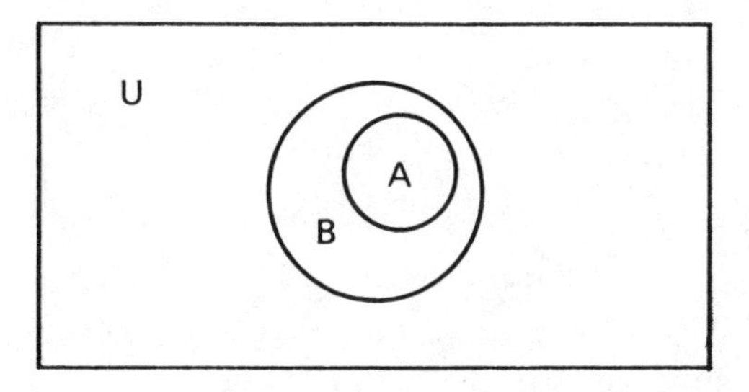

Fig. 3.2. Venn diagram: $A \subset B$

We could illustrate by the Venn diagram in Fig. 3.3 that A and B are *disjoint* subsets (have no elements in common) of U.

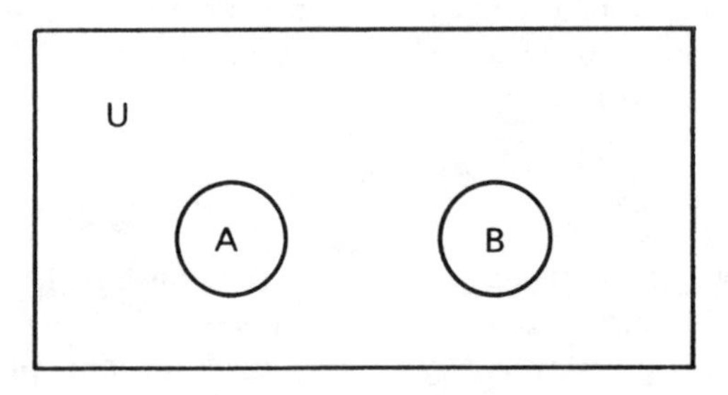

Fig. 3.3. Venn diagram: disjoint subsets

We illustrate in Fig. 3.4 that A' is the complement of A, where A and A' are subsets of U.

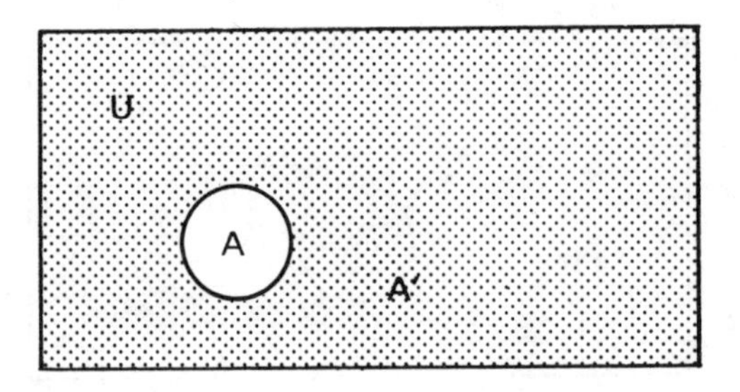

Fig. 3.4. Venn diagram: A' as complement of A

The Union of Two Sets

Definition

The *union* of two sets A and B, written $A \cup B$, is the *set of all elements* that belong to A or to B or to both A and B. That is, $A \cup B = \{x \mid x \varepsilon A \text{ or } x \varepsilon B\}$. For example, if we are given that $U = \{1,2,3,4, \cdots, 10\}$, $A = \{1,2,3\}$, and $B = \{4,5\}$, then $A \cup B = \{1,2,3,4,5\}$.

We make the Venn diagram in Fig. 3.5 illustrate this example of the union of A and B by shading the part that indicates the set whose elements are either in A or in B, or in both.

Note: Given sets $A = \{1,2,3\}$ and $C = \{3,7,8\}$, then $A \cup C = \{1,2,3,7,8\}$. Notice that the common element 3 is listed *only once* in the union of sets A and B.

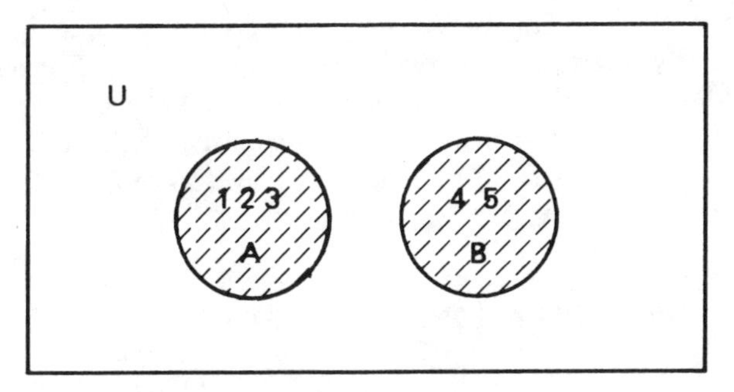

Fig. 3.5. $A \cup B = \{1, 2, 3, 4, 5\}$

THE INTERSECTION OF TWO SETS

Definition

The *intersection* of two sets A and B, written $A \cap B$, is the set of elements that belong to *both A and B*. That is, $A \cap B = \{x \mid x \varepsilon A$ and $x \varepsilon B\}$.

For example, if we are given that

$$U = \{1,2,3, \cdots, 10\}, \quad A = \{1,2,3,4,5\}, \quad B = \{2,5,7,8,9\}$$

then $A \cap B = \{2,5\}$.

We could illustrate this example of the intersection of A and B by shading part of the Venn diagram, as in Fig. 3.6, which indicates the set whose elements are in A and in B.

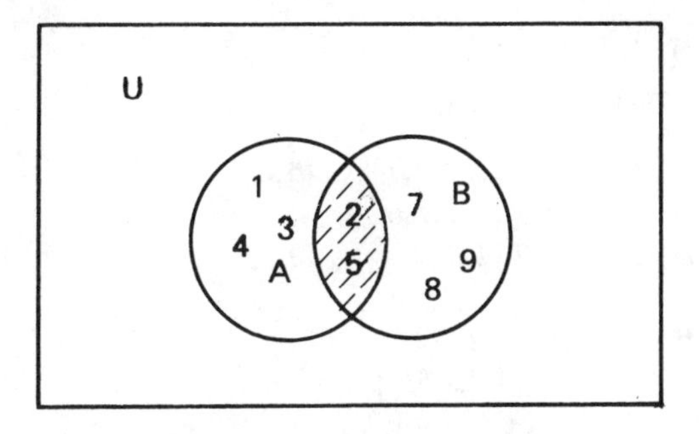

Fig. 3.6. $A \cap B = \{2,5\}$

MUTUALLY EXCLUSIVE (OR DISJOINT) SETS

Definition

Two sets X and Y are said to be *mutually exclusive* if and only if

$X \cap Y = \phi$ or { }. That is, the two sets have no elements in common, or equivalently, their intersection contains no elements. For example, if A, B, and C are subsets of U, with $A = \{3,4,6\}$, $B = \{5,7,8\}$, and $C = \{7,8,9\}$, then A and B are *mutually* exclusive because $A \cap B = \phi$. Notice that B and C are not mutually exclusive, since $B \cap C = \{7,8\}$, a set that is not the empty set.

3.4 MULTIPLE OPERATIONS WITH SETS

If we are given $U = \{1,2,3, \cdots, 10\}$, $A = \{1,2,3,7,5\}$, $B = \{1,5,3\}$, and $C = \{1,6\}$, we could perform the following operations.

1. If we wish to determine the complement of $(B \cup C)$, denoted by $(B \cup C)'$, we first find $B \cup C = \{1,3,5,6\}$. Then we determine the complement of $(B \cup C)$, or $(B \cup C)' = \{2,4,7,8,9,10\}$.

Note: In multiple set operations, the operation(s) inside parentheses is (are) performed first.

2. To find $(C \cap B)'$, we first determine $C \cap B = \{1\}$ and then $(C \cap B)' = \{2,3,4,5,6,7,8,9,10\}$.

3. $C' \cap B' = \{2,3,4,5,7,8,9,10\} \cap \{2,4,6,7,8,9,10\}$
$= \{2,4,7,8,9,10\}$

4. $C' \cup B' = \{2,3,4,5,7,8,9,10\} \cup \{2,4,6,7,8,9,10\}$
$= \{2,3,4,5,6,7,8,9,10\}$

Note: In operations 2, 3, and 4, we can see that $(C \cap B)' \neq C' \cap B'$, but $(C \cap B)' = C' \cup B'$. We could also show that $(C \cup B)' = C' \cap B'$. In mathematics the general relationships

$$(X \cup Y)' = X' \cap Y' \qquad \text{and} \qquad (X \cap Y)' = X' \cup Y'$$

are known as *DeMorgan's laws.*

5. $(A \cup B) \cup C = \{1,2,3,5,7\} \cup \{1,6\}$
$= \{1,2,3,5,6,7\}$

6. $A \cup (B \cup C) = \{1,2,3,7,5\} \cup \{1,3,5,6\}$
$= \{1,2,3,5,6,7\}$

7. $A \cup (B \cap C) = \{1,2,3,7,5\} \cup \{1\}$
$= \{1,2,3,5,7\}$

8. $(A \cup B) \cap (A \cup C) = \{1,2,3,5,7\} \cap \{1,2,3,5,6,7\}$
$= \{1,2,3,5,7\}$

Note: In operations 5, 6, 7, and 8,

$$(A \cup B) \cup C = \mathrm{A} \cup (\mathrm{B} \cup \mathrm{C}) \text{ and } A \cup (B \cap C) = (A \cup B) \cap (A \cup C)$$

In the algebra of sets, we can show that, given any universal set U

and any subsets A, B, and C of U, the following operational laws are true:

1. Commutative laws:

$$A \cup B = B \cup A$$
$$A \cap B = B \cap A$$

2. Associative laws:

$$(A \cup B) \cup C = A \cup (B \cup C)$$
$$(A \cap B) \cap C = A \cap (B \cap C)$$

3. Distributive laws:

$$A \cup (B \cap C) = (A \cup B) \cap (A \cup C)$$
$$A \cap (B \cup C) = (A \cap B) \cup (A \cap C)$$

4. Identity laws:

$$A \cup U = U \qquad A \cup \phi = A$$
$$A \cap U = A \qquad A \cap \phi = \phi$$

5. Complement laws:

$$A \cup A' = U \qquad (A')' = A$$
$$A \cap A' = \phi$$

6. De Morgan's laws:

$$(A \cup B)' = A' \cap B'$$
$$(A \cap B)' = A' \cup B'$$

7. Idempotent laws:

$$A \cup A = A$$
$$A \cap A = A$$

Problem Set 3.1

1. List the elements of the following sets:
   (a) A = set of all people in your immediate family.
   (b) B = set of all odd counting numbers less than 10.
   (c) C = set of all presidents of the United States since President Kennedy.
2. If $U = \{1,2,3, \cdots\}$, then list the elements of the following sets:
   (a) $E = \{x^2 \mid x \,\varepsilon\, U\}$
   (b) $F = \{x \mid x \text{ is divisible by 2 and } x \,\varepsilon\, U\}$

(c) $G = \{x \mid x + 7 = 15 \text{ and } x \varepsilon U\}$
(d) $H = \{x \mid x + 7 = 8 \text{ and } x \varepsilon U\}$
(e) $I = \{x \mid x^2 = 4 \text{ and } x \varepsilon U\}$
(f) $J = \{x \mid x^2 = 2 \text{ and } x \varepsilon U\}$

3. If $U = \{1,2,3,4,5\}$, $B = \{2,4\}$, $C = \{3,5\}$, $D = \{1,2,4\}$, $E = \{5,3\}$, $F = \{4,2\}$, $G = \{1\}$, $H = \{5,3,2,1\}$, find

(a) $B \cup C$
(b) $B \cap C$
(c) E'
(d) G'
(e) U'
(f) $H' \cup B$
(g) $(F \cup E)'$
(h) $F' \cup E'$
(i) $F' \cap E'$
(j) $(E \cup H) \cup C$
(k) $E \cup (H \cup C)$
(l) $(F \cap H) \cap E$
(m) $F \cap (H \cap E)$
(n) $F \cup (H \cap E)$
(o) $(F \cup H) \cap E$

4. For the sets given in problem 3, is

(a) $B \supset D$?
(b) $D \supset B$?
(c) $B \supset F$?
(d) $F \supset B$?
(e) $B \subset D$?
(f) $D \subset B$?
(g) $B \subset F$?
(h) $F \subset B$?
(i) $B = D$?
(j) $D = B$?
(k) $B = F$?
(l) $F = B$?

5. Describe in words or symbols the shaded areas in the Venn diagrams shown in Fig. 3.7.

(a)

(b)

(c)

(d)

(e)

(f)

(g)

6. (a) Write all the subsets of $U = \{a,b,c,d\}$.
(b) Are there any disjoint subsets of U with one element each? If so, list these subsets.
(c) Are there any disjoint subsets of U with two elements each? If so, list them.

7. Given $U = \{2,4,6,8,10,12,14\}$

$$A = \{4,8,10\}$$
$$B = \{6,10,14\}$$
$$C = \{2,12\}$$

Show that

(a) $A \cup (B \cap C) = (A \cup B) \cap (A \cup C)$
(b) $A \cap (B \cup C) = (A \cap B) \cup (A \cap C)$
(c) $(A \cap B)' = A' \cup B'$
(d) $A \cap (B \cap C) = (A \cap B) \cap C$
(e) $(A')' = A$
(f) $A \cup \phi = A$

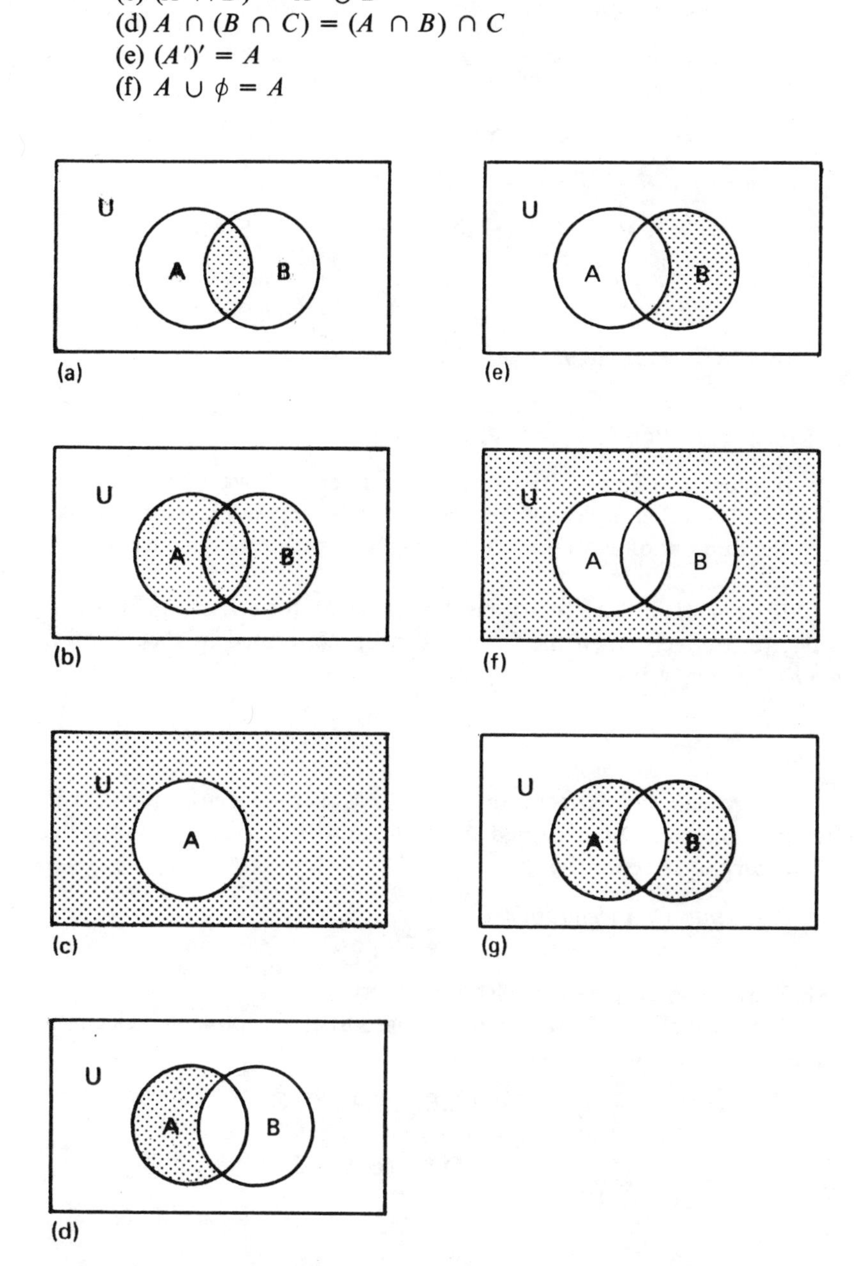

Fig. 3.7. Problem Set 3.1, Problem 5

3.5 SOPHISTICATED COUNTING

The Cartesian Product

The Cartesian product of two sets A and B, denoted by $A \times B$, is defined as follows:

$$A \times B = \{(x,y) \mid x \varepsilon A \text{ and } y \varepsilon B\}$$

For example, if $A = \{1,2,3\}$ $B = \{5,6\}$, then

$$A \times B = \{(1,5), (1,6), (2,5), (2,6), (3,5), (3,6)\}$$

and

$$B \times A = \{(5,1), (5,2), (5,3), (6,1), (6,2), (6,3)\}$$

Is $A \times B$ equal to $B \times A$? Why not?

Fundamental Principle of Counting

If a task can be performed in any number of n_1 ways and after it is completed another task can be performed in any number of n_2 ways, then the number of ways the two tasks can be performed in succession is $n_1 \cdot n_2$ ways.

In general, if $1,2,3,\cdots,k$ tasks can be performed in $n_1, n_2,\cdots,n_k$ ways, respectively, then the k tasks can be performed in succession in $n_1 \cdot n_2 \cdot \cdots \cdot n_k$ ways.

Examples

1. A class is to select a president first, then a vice-president, from five possible candidates named A,B,C,D, and E. In how many possible ways can the selection be made?

President		*Vice-president*	
5	$\cdot$	4	= 20 possible selections

What are these 20 possible selections?

Let us look at a *tree diagram* (Fig. 3.8) of the possible selections. Thus

$$U = \{(A,B), (A,C), (A,D), (A,E), (B,A), (B,C), (B,D), (B,E), (C,A), (C,B), (C,D), (C,E), (D,A), (D,B), (D,C), (D,E), (E,A), (E,B), (E,C), (E,D)\}$$

or

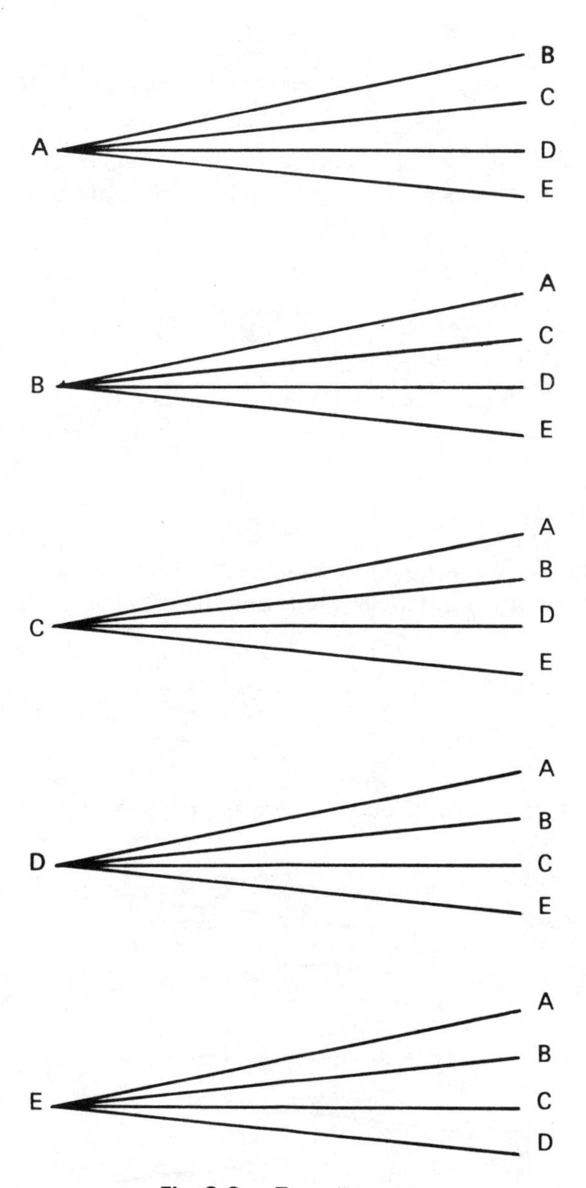

Fig. 3.8. Tree diagram

$U = \{(x,y) \mid x \in T, y \in T;$ x is president, y is vice-president$\}$

where:

$$T = \{A,B,C,D,E\}$$

Note: The order of selection (selecting a president before selecting a vice-president) makes a difference.

2. If two tires are tested and labeled G for good or D for defective, how many possibilities of labeling are there? What are they?

For tire 1		*For tire 2*	
2	·	2	= 4 possibilities

Examine the tree diagram in Fig. 3.9, which shows us these four possibilities. Thus,

$$U = \{(G,G), (G,D), (D,G), (D,D)\}$$

or

$$U = \{(x,y) \mid x \in C \text{ and } y \in C\}$$

where $C = \{G,D\}$.

3. How many different three-digit numbers can be formed using the digits 7,8,9 if repetitions of digits are allowed?

For hundred's digit		*For ten's digit*		*For one's digit*	
3	·	3	·	3	= 27 three-digit numbers

We illustrate these 27 possibilities of numbers with the tree diagram in Fig. 3.10.

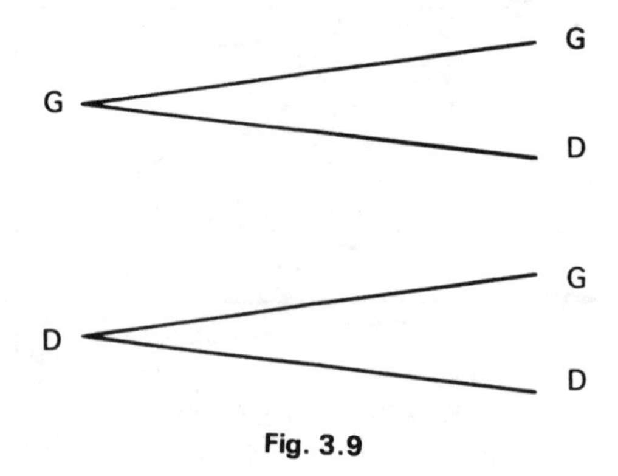

Fig. 3.9

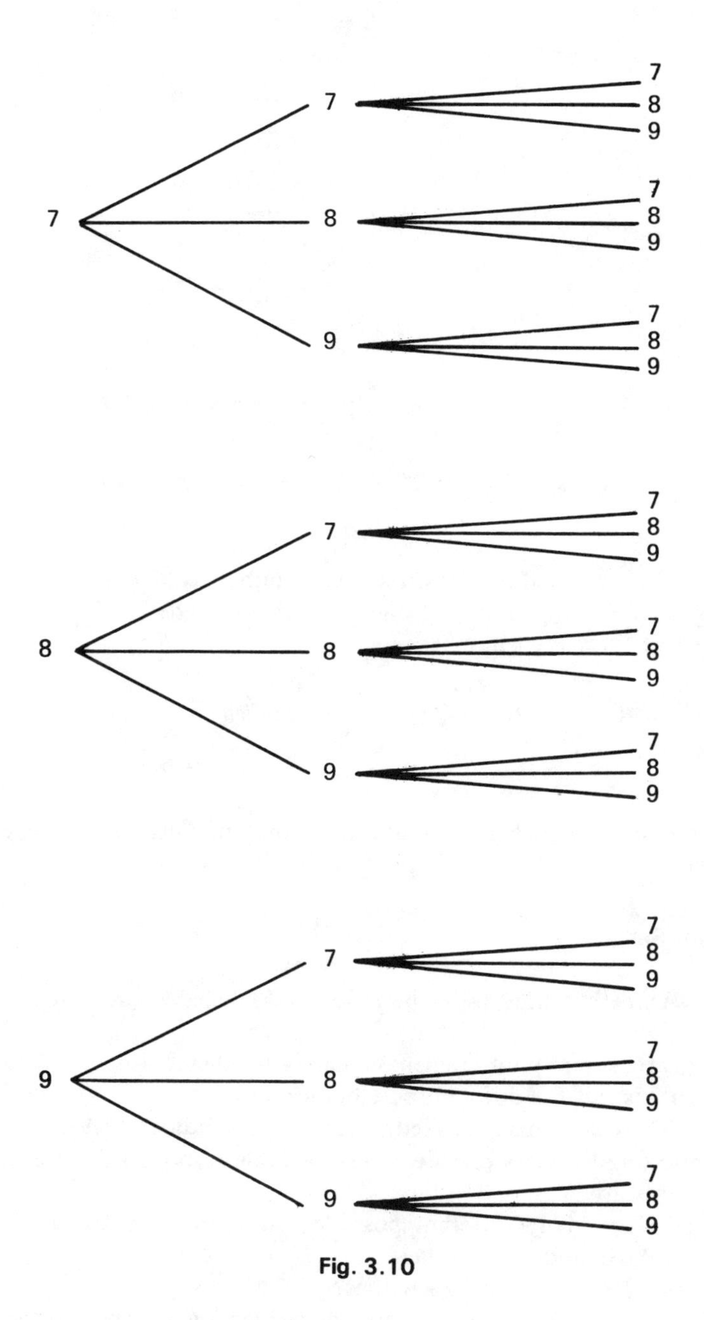

Fig. 3.10

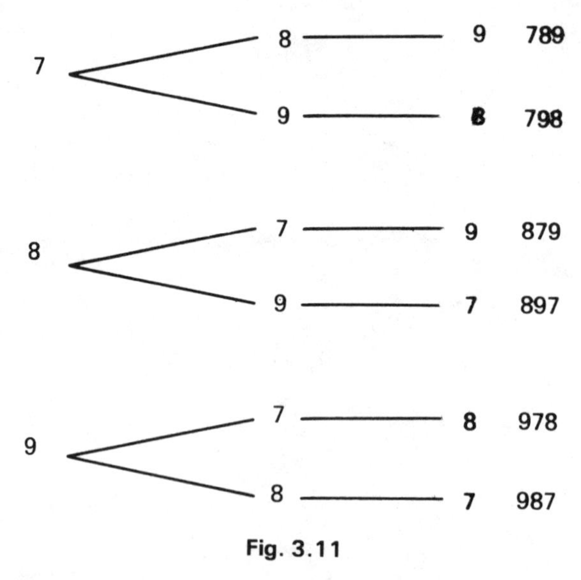

Fig. 3.11

4. How many different three-digit numbers can be formed using the digits 7,8,9 if repetitions of digits are *not* allowed? Again we list all the possible three-digit numbers.

For hundred's digit	*For ten's didgit*	*For one's digit*	
3	2	1	= 6

Our tree diagram in Fig. 3.11 illustrates the composition of these six numbers.

Problem Set 3.2

1. Three batteries are to be tested and labeled A (accepted) or R (rejected).
   (a) How many different possibilities are there?
   (b) List all the different possibilities.
2. Three students are asked which of two candidates (A or B) they are supporting for class president. The possible responses from each of the students are A,B, or U (undecided).
   (a) How many different possible responses from the group of three students are there?
   (b) What are these possibilities?
3. The names of 20 students are written on different pieces of

paper and placed in a box. Two names are to be selected from the box and awarded first and second prizes.

(a) How many different possibilities of selection are there?

(b) If we know that John was awarded first place, in how many different ways can the second prize winner be selected?

4. A patient has contracted pneumonia. His physician decides to prescribe 3 of 15 possible antibiotics. How many possible prescriptions can the doctor write?

5. A subscriber to a theater club has the choice of selecting in some order four of nine possible play productions. In how many different ways can she make her selection?

3.6 ORDER VERSUS NO ORDER

Let us discuss briefly the selection of r objects from n objects. In one type of selection the *order* of selection will make a difference. For example, if we are to select a president and then a vice-president from five candidates (A,B,C,D, or E), the selection (A,B) will be different from the selection (B,A). Why? This type of selection is called a *permutation* of two objects, selected from five objects.

However, if we are to select a committee of any two people from the five candidates (A,B,C,D, or E), the selection (A,B) will not be different from the selection (B,A). Why not? This type of selection is called a *combination* of two objects selected from five objects.

3.7 FACTORIAL NOTATION

Factorial notation allows us to express the product of consecutive integers in a simple form. The symbol $x!$ is read as "x factorial." By definition $x! = x(x - 1)(x - 2) \cdots (1)$. For example,

$$\begin{aligned}
1! &= 1 \\
2! &= 2 \cdot 1 = 2 \\
3! &= 3 \cdot 2 \cdot 1 = 6 \\
4! &= 4 \cdot 3 \cdot 2 \cdot 1 = 24 \\
5! &= 5 \cdot 4 \cdot 3 \cdot 2 \cdot 1 = 120
\end{aligned}$$

Note that $n! = n(n - 1)!$

$$\begin{aligned}
6! &= 6(5!) = 6(120) = 720 \\
7! &= 7(6!) = 7(720) = 5040, \text{ etc.}
\end{aligned}$$

Also note that $0! = 1$ by definition.

We will use factorial notation in our formulas for counting the number of permutations and the number of combinations of r objects selected from n objects. Therefore, we must be able to evaluate different expressions involving factorials, such as:

$$4! - 2! = 4 \cdot 3 \cdot 2 \cdot 1 - 2 \cdot 1 = 24 - 2 = 22$$

$$(4 - 2)! = 2! = 2$$

$$\frac{10!}{8!} = \frac{10 \cdot 9 \cdot \overset{1}{\cancel{8!}}}{\underset{1}{\cancel{8!}}} = 90$$

$$\frac{10!}{3!7!} = \frac{\overset{1}{10} \cdot \overset{3}{\cancel{9}} \cdot \overset{4}{\cancel{8}} \cdot \overset{1}{\cancel{7!}}}{\underset{1}{\cancel{3}} \cdot \underset{1}{\cancel{2}} \cdot 1 \cdot \underset{1}{\cancel{7!}}} = 120$$

(See Table F.2 in Appendix.)

3.8 PERMUTATIONS

When asked to select r objects from n distinct objects, where the order of selection makes a difference, we are computing the number of permutations of n distinct objects taken r at a time. For example, if we are asked how many different ways we can select a president and then vice-president and then a treasurer from five candidates, we are selecting three people from five people where order makes a difference, and we write:

$${}_5P_3 = 5 \cdot 4 \cdot 3 = 60 \text{ possible selections}$$

This computation is a direct application of the fundamental principle of counting.

In general, the number of permutations of r objects selected from n distinct objects is denoted by ${}_nP_r = n(n - 1)(n - 2) \cdots (n - (r - 1))$, or ${}_nP_r = n(n - 1)(n - 2) \cdots (n - r + 1)$.

We may also write

$${}_8P_8 = 8 \cdot 7 \cdot 6 \cdot 5 \cdot 4 \cdot 3 \cdot 2 \cdot 1 = \frac{8!}{(8-8)!} = \frac{8!}{0!} = 8!$$

$${}_8P_7 = 8 \cdot 7 \cdot 6 \cdot 5 \cdot 4 \cdot 3 \cdot 2 = \frac{8!}{(8-7)!} = \frac{8!}{1!} = 8!$$

$${}_8P_6 = 8 \cdot 7 \cdot 6 \cdot 5 \cdot 4 \cdot 3 = \frac{8!}{(8-6)!} = \frac{8!}{2!}$$

$${}_8P_5 = 8 \cdot 7 \cdot 6 \cdot 5 \cdot 4 = \frac{8!}{(8-5)!} = \frac{8!}{3!}$$

$${}_8P_4 = 8 \cdot 7 \cdot 6 \cdot 5 = \frac{8!}{(8-4)!} = \frac{8!}{4!}$$

$${}_8P_3 = 8 \cdot 7 \cdot 6 = \frac{8!}{(8-3)!} = \frac{8!}{5!}$$

$${}_8P_2 = 8 \cdot 7 = \frac{8!}{(8-2)!} = \frac{8!}{6!}$$

$${}_8P_1 = 8 = \frac{8!}{(8-1)!} = \frac{8!}{7!}$$

$${}_8P_0 = 1 = \frac{8!}{(8-0)!} = \frac{8!}{8!}$$

(There is only one way we can select 0 objects from 8 objects.)

Thus, *in general*,

$${}_nP_r = n(n-1)(n-2) \cdots (n-r+1) \text{ or equivalently}$$

$${}_nP_r = \frac{n!}{(n-r)!}$$

In the sample problem above where we selected a president and then a vice-president and then a treasurer from five people, we compute

$${}_5P_3 = \frac{5!}{(5-3)!} = \frac{5!}{2!} = \frac{5 \cdot 4 \cdot 3 \cdot \overset{1}{\cancel{2}} \cdot \overset{1}{\cancel{1}}}{\underset{1}{\cancel{2}} \cdot \underset{1}{\cancel{1}}} = 60$$

Now suppose we were asked to find how many different arrangements could be made of all letters in the word "SIT." We could write

$${}_3P_3 = \frac{3!}{(3-3)!} = \frac{3!}{0!} = 3! = 6$$

The tree diagram in Fig. 3.12 illustrates the six different arrangements.

However, if we were asked to find the number of different arrangements of all letters in the word "TOO," there would *not* be six different arrangements. Examine the tree diagram in Fig. 3.13. In the word "SIT," which contains three different letters, any two of the different letters can be arranged in $2 \cdot 1$ or $2!$ ways, which are included in the arrangement of all three different letters as $3 \cdot 2 \cdot 1$, or ${}_3P_3$. However, in the word "TOO," the two letters O and O can be arranged in *only one* different way. Therefore, the *different* arrangements of all the letters in the word 'TOO' are

$$\frac{{}_3P_3}{2!} = \frac{3!}{2!} = 3$$

Note: We must cancel out the 2! arrangements of two different objects in ${}_3P_3$ ways because the two O's are *not* different and can be arranged in *only* one way.

Thus, *in general*, the number of permutations of n objects where a_1 are alike, a_2 are alike, ... , a_m are alike is

$$\frac{{}_nP_n}{a_1!a_2! \cdots a_m!} = \frac{n!}{a_1!a_2! \cdots a_m!}$$

For example, if a comedian is to make ten appearances during a

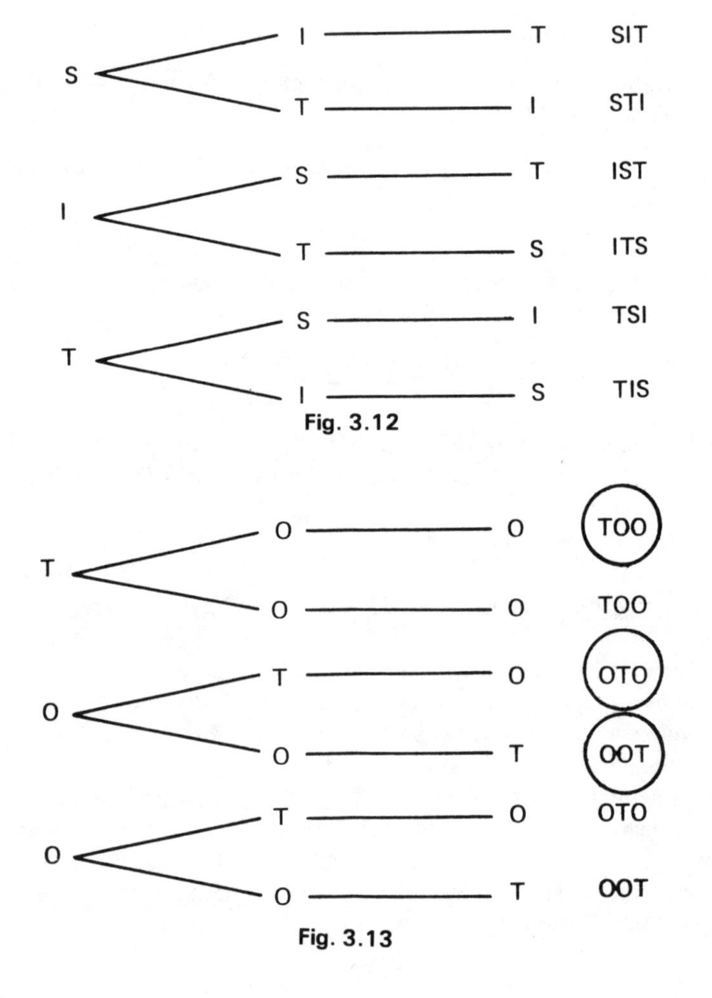

Fig. 3.12

Fig. 3.13

particular month and has three different routines of which he wants to use the first routine five times, the second routine three times, and the third routine two times, these routines can be done in

$$\frac{10!}{5!\,3!\,2!} = \frac{10 \cdot 9 \cdot \overset{4}{\cancel{8}} \cdot 7 \cdot \overset{1}{\cancel{6}} \cdot \overset{1}{\cancel{5}}!}{\underset{1}{\cancel{5}}!\,\underset{1}{\cancel{3}}!\,\underset{1}{\cancel{2}}!} = 2520 \text{ different ways}$$

Problem Set 3.3

1. Compute:
   (a) $5!$ (c) $6! + 3!$ (e) ${}_nP_n$
   (b) $(6!)(3!)$ (d) $6!/3!$ (f) ${}_nP_0$

2. A disc jockey has 5 time slots left in his broadcast to play 5 new songs, which he can select from 12 new songs on file. How many different ways can this be done?

3. If the disc jockey in Problem 2 has 12 time slots to play the 12 new songs, how many ways could this be done?

4. A student must include one language course, one mathematics course, one science course, one general education course, and one physical education course in his/her program for next semester. The student may select these required courses from a list of three language, two mathematics, two science, five general education, and six physical education courses. In how many different ways can the student select his/her program for next semester?

5. An oil company has 13 salesmen's positions to fill. Twenty-five men apply for the 13 positions. Twelve of the 25 applicants have previous sales experience.
   (a) In how many different ways can the 13 positions be filled by using any of the 25 applicants?
   (b) In how many different ways can the 13 positions be filled by using only applicants with previous sales experience?
   (c) In how many different ways can the 13 positions be filled with the 25 applicants if 5 of the positions require previous sales experience and the other 8 positions do not require previous sales experience?

6. How many different arrangements can be made of all the letters in each word
   (a) Home
   (b) Nine
   (c) Letters
   (d) Togetherness

7. A TV producer wishes to place five different taped TV

programs into five different time slots. In how many different ways can this arrangement be done?

3.9 COMBINATIONS

When asked to select r objects from n distinct objects, where the order of selection does *not* make a difference, we are computing the number of combinations of n distinct objects taken r at a time. For example, if we are asked to select a committee of any three people of five people (A,B,C,D, or E), the order in which we select the committee does not make a difference. That is, the committee consisting of persons A,B, and C would be the same as the committee consisting of persons A,C, and B.

Recall that when we were to select a president, and then a vice-president and then a treasurer from five people (A,B,C,D, or E), the order *did* make a difference. We could select these three people in ${}_5P_3$ ways = 5!/(5 − 3)!, or 60 ways.

Let us consider the different committees of any three people selected from among persons A,B,C,D, and E. We can list these groups as (A,B,C,), (A,B,D), (A,B,E), (A,C,D), (A,C,E), (A,D,E), (B,C,D), (B,C,E), (B,D,E), (C,D,E). As you can see, there are ten possible committees of three selected from among five different people, and we write ${}_5C_3 = 10$.

Note: $${}_5C_3 = \frac{{}_5P_3}{3!} = \frac{5!}{(5-3!)}\Big/3! = \frac{5!}{3!\,(5-3)!} = \frac{5!}{3!2!} = 10$$

(See Table F.3 in Appendix)

Thus, *in general*, the number of combinations of r objects selected from n distinct objects is denoted by

$${}_nC_r = \frac{{}_nP_r}{r!} \qquad \text{or} \qquad {}_nC_r = \frac{n!}{r!(n-r)!}$$

Sample Problem

In how many different ways can a committee of five be selected from among five men and four women if this committee is to contain at least three women? Notice that the committee selected may include three women or four women.

$$\begin{array}{lcl} \text{(3 women, 2 men)} & \text{or} & \text{(4 women, 1 man)} \\ ({}_4C_3 \cdot {}_5C_2) & + & ({}_4C_4 \cdot {}_5C_1) \\ = (4 \cdot 10) & + & (1 \cdot 5) \\ = 40 + 5 = 45 \text{ different ways} & & \end{array}$$

3.10 THE BINOMIAL THEOREM

In later chapters we will have use for the binomial theorem, which is usually presented in an intermediate algebra course. By using the binomial theorem, we can raise the sum of two terms, or the difference between two terms, to an indicated power. Thus

$$(g + d)^1 = g + d$$
$$(g + d)^2 = (g + d)(g + d) = g^2 + 2gd + d^2$$
$$(g + d)^3 = (g + d)(g + d)(g + d)$$
$$= g^3 + 3g^2d + 3gd^2 + d^3$$

In general, the binomial theorem states that

$$(g + d)^n = g^n + \frac{n}{1}g^{n-1}d + \frac{n(n-1)}{1(2)}g^{n-2}d^2 + \cdots + d^n$$

where $n\varepsilon\{1,2,3,\ldots\}$, or

$$(g + d)^n = {}_nC_ng^n + {}_nC_{n-1}\, g^{n-1}\, d + {}_nC_{n-2}g^{n-2}d^2 + \cdots + {}_nC_0d^n$$

where $n\varepsilon\{1,2,3,\ldots\}$.

Examples:

1. $(a + b)^4 = {}_4C_4a^4 + {}_4C_3a^3b + {}_4C_2a^2b^2 + {}_4C_1ab^3 + {}_4C_0b^4$
   $= a^4 + 4a^3b + 6a^2b^2 + 4ab^3 + b^4$
2. $(x - y)^3 = {}_3C_3x^3 + {}_3C_2x^2(-y)^1 + {}_3C_1x(-y)^2 + {}_3C_0(-y)^3$
   $= x^3 - 3x^2y + 3xy^2 - y^3$
3. $(2x + y)^5 = {}_5C_5(2x)^5 + {}_5C_4(2x)^4y + {}_5C_3(2x)^3y^2 + {}_5C_2(2x)^2y^3 + {}_5C_1(2x)^1y^4 + {}_5C_0y^5$
   $= (1)(32x^5) + 5(16x^4)y + 10(8x^3)y^2 + 10(4x^2)y^3 + 5\,(2x)y^4 + (1)y^5$
   $= 32x^5 + 80x^4y + 80x^3y^2 + 40x^2y^3 + 10xy^4 + y^5$

Again consider the general statement of the binomial theorem:

$$(g + d)^n = {}_nC_ng^n + {}_nC_{n-1}g^{n-1}d + {}_nC_{n-2}g^{n-2}d^2 + \cdots + {}_nC_0d^n$$

where $n\varepsilon\{1,2,3,\ldots\}$.

Note: The term containing g^r in the expansion of $(g + d)^n$ is equal to ${}_nC_rg^rd^{n-r}$. Thus,

(a) The term containing g^7 in $(g + d)^{11} = {}_{11}C_7g^7d^{11-7} = 330g^7d^4$.
(b) The term containing a^9 in $(a + r)^{20} = {}_{20}C_9a^9r^{20-9} = 167{,}960a^9r^{11}$.
(c) The term containing s^8 in $(s + f)^{10} = {}_{10}C_8s^8f^{10-8} = 45s^8f^2$.
(d) The term containing $(0.8)^4$ in $(0.8 + 0.2)^5 = {}_5C_4(0.8)^4\,(0.2)^{5-4} = 5(0.4096)\,(0.2) = 0.4096$.

See Table F.3 in the Appendix.

Problem Set 3.4

1. Compute
   (a) ${}_6C_3$
   (b) ${}_6C_3 + {}_6C_4$
   (c) ${}_nC_n$
   (d) ${}_nC_0$
2. Show that ${}_6C_4 = {}_5C_3 + {}_5C_4$
3. How many different five-card hands can be formed using a standard deck of 52 playing cards?
4. How many different five-card hands dealt from a standard deck of 52 playing cards contain
   (a) Exactly three aces?
   (b) At least three aces?
   (c) At most three aces?
5. (a) How many different committees of six people can be chosen from seven women and five men?
   (b) How many of these committees of six people will contain exactly four men?
   (c) At least four men?
   (d) Exactly four women?
6. A true-false test consists of ten questions. In how many different ways can a student get
   (a) Three questions wrong?
   (b) At least eight questions right?
   (c) Two questions right?
7. Use the binomial theorem to
   (a) Expand $(a + b)^5$.
   (b) Expand $(2a - b)^4$.
   (c) Expand $(x - y)^3$.
   (d) Find the term containing a^3 in $(a + r)^7$.
   (e) Find the term containing g^9 in $(g + d)^{13}$.
   (f) Find the term containing $(0.9)^3$ in $(0.9 + 0.1)^6$.
   (g) Find the term containing $(0.99)^2$ in $(0.99 + 0.01)^3$.
8. Lillian's teacher stated to her class that the term "combination lock" is actually a misnomer, or wrong name. She added that such a lock should more accurately be termed "a lock of permutations." Do you agree or disagree with her teacher? Why?

3.11 A PREPACKAGED COUNTING PROGRAM

Name: STAT3

Description: This program computes the number of permutations of r objects selected from n distinct objects or the number of combinations of r objects selected from n distinct objects.

Instructions: Enter the data beginning in line 150 as follows:

150 DATA N,R,D

where

N = total number of distinct objects

R = number of objects to be selected from the N distinct objects

D = 0 if we want ${}_nP_r$; 1 if we want ${}_nC_r$

Note: (1) Line 200 is the END statement. (2) STAT3 will not compute ${}_nC_0 = 1$.

Sample Problem

Find ${}_8P_3$ and ${}_8C_3$.

User's Input

```
GET-STAT3

150 DATA 8, 3, 0
151 DATA 8, 3, 1
RUN
```

Computer Output

```
P( 8      , 3      )= 336
C( 8      , 3      )= 56

LINE      5:    END OF DATA

TIME 0.0 SECS.
```

The flowchart for STAT3 is shown in Fig. 3.14.

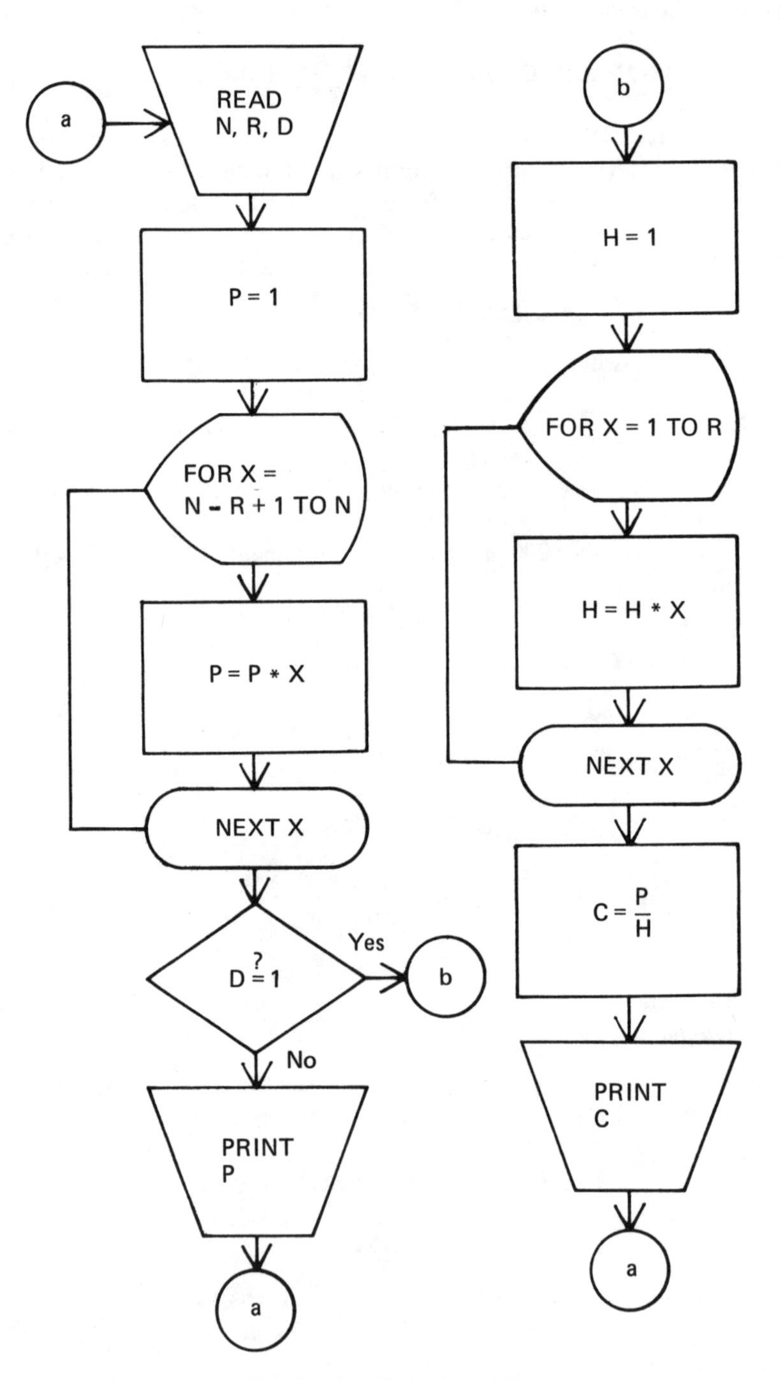

Fig. 3.14. Flowchart 12

The coding for STAT3 is as follows:

```
5 READ N,R,D
10 LET P = 1
15 FOR X = N-R+1 TO N
20     LET P = P*X
25 NEXT X
30 IF D = 1 THEN 45
35 PRINT "P("N;",";R;")="P
40 GO TO 5
45 LET H = 1
50 FOR X = 1 TO R
55     LET H = H*X
60 NEXT X
65 LET C = P/H
70 PRINT "C("N;",";R;")="C
75 GO TO 5
150 DATA
200 END
```

Problem Set 3.5

1. Use STAT3 to compute the following:
   (a) ${}_{11}P_3$ (d) ${}_{10}C_7$
   (b) ${}_9C_5$ (e) ${}_{11}C_7$
   (c) ${}_{12}P_4$ (f) ${}_8P_5$
2. Use STAT3 to show that ${}_{10}C_5 = {}_9C_4 + {}_9C_5$.

4

Elementary Probability Concepts with a Prepackaged Program for the Binomial Experiment

4.1 SAMPLE SPACE, SAMPLE POINT, AND EVENT

SAMPLE SPACE

A sample space associated with an experiment or well-defined course of action is the set of all possible outcomes for the experiment.

Examples

1. If we were to select two fenders from a lot of five fenders, then test each of the fenders and label them G (good) or D (defective), the sample space $S = \{ (G,G), (G,D), (D,G), (D,D) \}$.
2. If the names of four different people (A,B,C, or D) were written on separate sheets of paper and placed in a box, and we were to select two names to serve on a committee, the sample space $S = \{ (A,B), (A,C), (A,D), (B,C), (B,D), (C,D) \}$.

SAMPLE POINT

A sample point is an element of a sample space set and represents a possible outcome. A sample point may have any number of components. For example, if we test three dry cell batteries to determine if each is good (g) or defective (d), each sample point would consist of three components, such as (g,g,g) or (d,d,d). If we test only one battery, each sample point would consist of only one component, such as (g) or (d).

EVENT

An event associated with a sample space is any subset of the sample space.

Examples

1. In the experiment where we select and test two fenders and label them G (good) or D (defective), the sample space $S = \{ (G,G), (G,D), (D,G), (D,D) \}$. Consider the following events:

E: Exactly one fender is defective.

$$E = \{ (G,D), (D,G) \}$$

F: The first fender tested is defective.

$$F = \{ (D,G), (D,D) \}$$

$E \cup F$: Exactly one fender is defective, or the first fender tested is defective, or both.

$$E \cup F = \{ (G,D), (D,G), (D,D) \}$$

$E \cap F$: Exactly one fender is defective and the first fender tested is defective.

$$E \cap F = \{ (D,G) \}$$

F': The first fender tested *is not* defective.

$$F' = \{ (G,G), (G,D) \}$$

2. In the experiment where we select a committee of two from 4 persons (A,B,C, or D), the sample space $S = \{ (A,B), (A,C), (A,D), (B,C), (B,D), (C,D) \}$. Consider the following events:

E: Person A is on the committee.

$$E = \{ (A,B), (A,C), (A,D) \}$$

F: Person C is on the committee.

$$F = \{ (A,C), (B,C), (C,D) \}$$

$E \cup F$: Person A *or* person C is on the committee.

$$E \cup F = \{ (A,B), (A,C), (A,D), (B,C), (C,D) \}$$

$E \cap F$: Person A *and* person C are on the committee.

$$E \cap F = \{ (A,C) \}$$

E': Person A *is not* on the committee.

$$E' = \{ (B,C), (B,D), (C,D)\}$$

4.2 ACCEPTABLE ASSIGNMENT OF PROBABILITIES

If $S = \{e_1, e_2, e_3, \ldots, e_n\}$ is a sample space associated with some experiment and we assign each possible outcome (e_j) of the experiment a number called the *probability* of e_j, denoted by $p(e_j)$ where $j = 1,2, \ldots, n$ such that

1. $0 \leq p(e_j) \leq 1$ for $j = 1,2, \ldots, n$ (the probability of each outcome must be between 0 and 1 inclusive) *and*

2. $p(e_1) + p(e_2) + p(e_3) + \cdots + p(e_n) = 1$ (the sum of the probabilities of all the outcomes in S must equal 1), then we are said to have an *acceptable assignment of probabilities.*

Note: Recall that the outcomes of the sample space $e_1, e_2, \ldots, e_n$ are called *sample points.*

4.3 PROBABILITY OF AN EVENT

Given a sample space $S = \{e_1, e_2, e_3, \ldots, e_n\}$ associated with some experiment and some event $E = \{o_1, o_2, o_3, \ldots, o_m\}$ such that $E \subset S$, then the *probability* of event E is denoted by $P(E) = p(o_1) + p(o_2) + p(o_3) + \cdots + p(o_m)$.

Note: If each of the sample points in S has the same probability, that is, $p(e_1) = p(e_2) = \cdots = p(e_n)$, then $P(E) = m/n$ where $m =$ the number of sample points in each E, and $n =$ the number of sample points in the sample space S.

Examples

1. In the experiment where we select two fenders and label them G (good) or D (defective), we saw that

$$S = \{e_1, e_2, e_3, e_4\}$$

where $e_1 = (G,G)$, $e_2 = (G,D)$, $e_3 = (D,G)$, and $e_4 = (D,D)$

Now suppose we are told that the probability of selecting a defective fender on any one trial is 0.1. That is, $p(D) = 0.1$ and $p(G) = 0.9$. The probabilities for each of the sample points in S are as follows:

$$p(e_1) = (0.9)\,(0.9) = 0.81$$
$$p(e_2) = (0.9)\,(0.1) = 0.09$$
$$p(e_3) = (0.1)\,(0.9) = 0.09$$
$$p(e_4) = (0.1)\,(0.1) = 0.01$$

Note: $p(e_1) + p(e_2) + p(e_3) + p(e_4) = 1$. Previously we were given the following events:

$$E = \{\,(G,D),\,(D,G)\,\} = \{e_2,\,e_3\}$$
$$F = \{\,(D,G),\,(D,D)\,\} = \{e_3, e_4\}$$
$$E \cup F = \{\,(G,D),\,(D,G),\,(D,D)\,\} = \{e_2,\,e_3,\,e_4\}$$
$$E \cap F = \{\,(D,G)\,\} = \{e_3\}$$

Therefore,

$$P(E) = p(e_2) + p(e_3) = 0.09 + 0.09 = 0.18$$
$$P(F) = p(e_3) + p(e_4) = 0.09 + 0.01 = 0.10$$
$$P(E \cup F) = p(e_2) + p(e_3) + p(e_4)$$
$$= 0.09 + 0.09 + 0.01 = 0.19$$
$$P(E \cap F) = p(e_3) = 0.09$$

2. In the experiment where we select a committee of two from persons A,B,C, or D, we saw $S = \{e_1, e_2, e_3, e_4, e_5, e_6\}$, where $e_1 = (A,B)$, $e_2 = (A,C)$, $e_3 = (A,D)$, $e_4 = (B,C)$, $e_5 = (B,D)$, $e_6 = (C,D)$.

We assume that each of the sample points in S has the same probability of occurring, and therefore we assign

$$p(e_1) = p(e_2) = p(e_3) = p(e_4) = p(e_5) = p(e_6) = 1/6$$

Previously we were given the following events:

$$E = \{\,(A,B), (A,C), (A,D)\,\} = \{e_1,\,e_2,\,e_3\}$$
$$F = \{\,(A,C), (B,C), (C,D)\,\} = \{e_3,\,e_4,\,e_6\}$$
$$E \cup F\ \{\,(A,B), (A,C), (A,D), (B,C), (C,D)\,\}$$
$$= \{e_1,\,e_2,\,e_3,\,e_4,\,e_6\}$$
$$E \cap F = \{\,(A,C)\,\} = \{e_2\}$$

Since each of the sample points in S have the same probability, we write

$$P(E) = 3/6 = 1/2$$
$$P(F) = 3/6 = 1/2$$
$$P(E \cup F) = 5/6$$
$$P(E \cap F) = 1/6$$

3. A task consists of selecting a committee of five people from five

men and six women. What is the probability that exactly three men will be on the committee of five?

We assume that each of the possible committees has an equal chance of occurring, and we are told that for event E exactly three men are on the committee of five. Therefore,

$$P(E) = \frac{\text{number of elements in } E}{\text{number of elements in sample space } S}$$

$$= \frac{\begin{pmatrix}\text{number of different} \\ \text{ways to select 3 men}\end{pmatrix} \cdot \begin{pmatrix}\text{number of different ways} \\ \text{to select 2 women}\end{pmatrix}}{\text{number of different ways to select committee of 5}}$$

or

$$p(E) = \frac{{}_5C_3 \cdot {}_6C_2}{{}_{11}C_5} = \frac{(10)(15)}{462}$$

$$= \frac{150}{462} = \frac{25}{77}$$

$$\approx 0.32$$

Problem Set 4.1

1. A fair pair of dice are rolled (each die consists of six faces labeled 1,2,3,4,5, and 6, respectively). Events E, F, and G are as follows:

E: The sum of the upper faces read is 5.
F: The sum of the upper faces read is less than 8.
G: The sum of the upper faces read is 7.

(a) List the elements in the sample space S.
(b) Assign probabilities to each of the sample points in S.
(c) List the elements in events E, F, and G.
(d) Find $P(E)$, $P(F)$, and $P(G)$.

2. Three batteries are selected from a large lot of batteries, then tested and labeled A (accepted) or R (rejected). The probability that a battery is accepted on any one try is (0.99) (that is, $p(A) = 0.99$ and $p(R) = 0.01$).

(a) List the eight elements in the sample space S associated with this experiment.
(b) Assign probabilities to each of the sample points in S.

(c) Find the probability that exactly two of the three batteries tested are accepted.

(d) Find the probability that at least two of the three batteries tested are accepted.

3. A TV producer has five different hour-long time slots to be filled with five rerun programs. He has seven comedy reruns, five drama reruns, and eight variety reruns to choose from. If each of the reruns has an equal chance of being chosen,

(a) How many different ways can he fill the five time slots?

(b) What is the probability that he will choose exactly two comedy reruns and exactly two variety reruns?

(c) What is the probability that he will choose five drama reruns?

4. The probability that a rug salesman makes a sale when he calls on a prospective customer is 0.3. If he calls on two prospective customers in one day, what is the probability that he will make at least one sale?

5. Three cards are drawn at random from a standard deck of 52 playing cards. What is the probability of selecting:

(a) Exactly two queens?

(b) At most two queens?

(c) Three hearts?

(d) A heart, a spade, and a club?

6. Ten separate pieces of paper are marked with the digits 0, 1, 2, 3, 4, 5, 6, 7, 8, 9, and placed in a bowl. One piece of paper is drawn for the unit's digit of a five-digit lottery number and then placed back in the bowl. A second piece of paper is drawn for the tens' digit and placed back in the bowl, and the drawing is repeated until the fifth digit for the ten thousands' place of the five-digit lottery number is selected. (Notice. that repetitions of digits are possible.)

(a) How many different five-digit lottery numbers can be drawn?

(b) What is the probability that the lottery number selected is an even number?

(c) What is the probability that the lottery number selected ends in 6?

(d) What is the probability that the lottery number selected begins with a 3 and ends with a 6?

(e) What is the probability that the lottery number selected is 12345?

7. Given a sample space $S = \{e_1, e_2, e_3, e_4, e_5\}$ where $p(e_1) = 0.2$, $p(e_2) = 0.1$, $p(e_3) = 0.2$, $p(e_4) = 0.4$. Find:

(a) $p(e_5)$

(b) The probability of event E, where $E = \{e_2, e_3, e_4\}$

(c) The probability of event F, where $F = \{e_1, e_3, e_5\}$
(d) $P(E \cup F)$
(e) $P(E \cap F)$
(f) $P(E')$
(g) $P(F')$

8. The probability that a salesman makes a sale on any one visit to a prospective customer is 0.3. If he makes three calls on a certain day and we represent a sale by s and no sale by f, we can list the sample space as follows: $S = \{(s,s,s), (s,s,f), (s,f,s), (s,f,f), (f,s,s), (f,s,f), (f,f,s), (f,f,f)\}$.

(a) Assign probabilities to each of the sample points in S.
(b) Find the probability that at most two sales are made.
(c) Find the probability that no sales were made.

4.4 THREE PROBABILITY THEOREMS

If $S = \{e_1, e_2, e_3, \ldots, e_n\}$ is a sample space for some experiment and E and F are events in S, then the following theorems are true.

1. $P(S) = 1$. This theorem is a direct consequence of the fact that in an acceptable assignment of probability the sum of the probabilities of all the sample points in S must be 1.

2. $P(E') = 1 - P(E)$ and $P(E) = 1 - P(E')$. These theorems follow, since $E \cup E' = S$, $E \cap E' = \phi$, and $P(S) = 1$. *Example*: Suppose five fenders are selected at random from a lot of fenders, then tested and labeled G (good) or D (defective). We know that the $p(G) = 0.95$ on any one trial. What is the probability of getting at least one good fender from the five tested? Let E be at least one good fender and E' equal not any, or zero, good fenders $= \{(D,D,D,D,D)\}$. Then

$$
\begin{aligned}
P(E) &= 1 - P(E') \\
P(E) &= 1 - P((D,D,D,D,D,)) \\
P(E) &= 1 - (0.05)\ (0.05)\ (0.05)\ (0.05)\ (0.05) \\
P(E) &= 1 - 0.0000003125 \\
P(E) &= 0.999999685
\end{aligned}
$$

3. $P(E \cup F) = P(E) + P(F) - P(E \cap F)$

Note: $P(E \cup F) = P(E) + P(F)$ if and only if E and F are *mutually exclusive* (that is, $E \cap F = \phi$). If $E \cap F \neq \phi$, then $P(E \cup F) \neq P(E) + P(F)$ because $P(E) + P(F)$ in Fig. 4.1 would add in twice the probability of region 2 in Fig. 4.1.

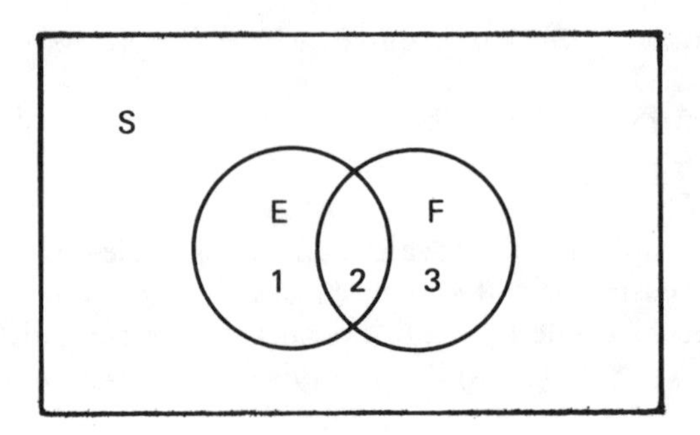

Fig. 4.1

Example

The probability that a person who receives a certain flu vaccine does not contract a virus is 0.8. If three patients are inoculated with the vaccine, what is the probability that at least two of the three will contract a virus, or that the second person inoculated contracts a virus?

Let $p(s) = 0.8$ and $p(f) = 0.2$, where s represents an inoculated person who does not contract a virus; f represents an inoculated person who does contract a virus. Let E = at least two who contract a virus and let F = the second person inoculated who contracts a virus. Then

$$S = \{(s,s,s), (s,s,f), (s,f,s), (s,f,f), (f,s,s), (f,s,f), (f,f,s), (f,f,f)\}$$

and

$$E = \{(s,f,f), (f,s,f), (f,f,s), (f,f,f)\}$$

Therefore,

$$P(E) = 0.032 + 0.032 + 0.032 + 0.008 = 0.104$$

and

$$F = \{(s,f,s), (s,f,f), (f,f,s), (f,f,f)\}$$

Therefore,

$$P(F) = 0.128 + 0.032 + 0.032 + 0.008 = 0.200$$

Now

$$E \cap F = \{(s,f,f), (f,f,s), (f,f,f)\}$$

Therefore,

$$P(E \cap F) = 0.032 + 0.032 + 0.008 = 0.072$$

Then, as in Fig. 4.2,

$$\begin{aligned} P(E \cup F) &= P(E) + P(F) - P(E \cap F) \\ &= 0.104 + 0.200 - 0.072 \\ &= 0.232 \end{aligned}$$

Note: The Venn diagram in Fig. 4.2 clearly shows that we cannot count $P(E \cap F)$ twice to obtain $P(E \cup F)$.

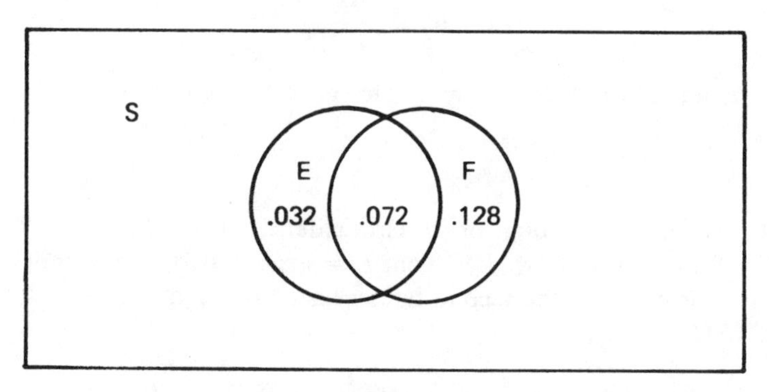

Fig. 4.2

Problem Set 4.2

1. The Hudson Railroad reports that 90% of its trains arrive on time. If we randomly record the arrival of six Hudson Railroad trains, what is the probability that (a) at least one of the six trains will not be on time; (b) exactly one of the six trains will be on time?

2. Given that events E and F in a sample space S have the following probabilities: $P(E) = 0.4$, $P(F) = 0.3$, $P(E \cap F) = 0.1$. Find $P(E \cup F)$.

3. Given that events E and F in a sample space S are mutually exclusive and $P(E) = 0.4$, $P(F) = 0.3$. Find $P(E \cup F)$.

4. A small publishing company publishes two magazines, *Pictorial* and *Today*. One-third of the company's customers subscribe to *Pictorial*, one-sixth subscribe to *Today*, and one-eighteenth subscribe to both magazines. If a customer is selected at random, what is the probability that he/she subscribes to either *Pictorial* magazine or *Today* magazine, or both?

4.5 CONDITIONAL PROBABILITY

If E and F are events in some sample space S, then the *probability* of event E occurring, given event F has *already occurred*, is denoted by

$$P(E|F) = \frac{P(E \cap F)}{P(F)}$$

(We are actually reducing the sample space to event F.) Also, the probability of event F occurring, *given* event E has *already occurred*, is denoted by

$$P(F|E) = \frac{P(F \cap E)}{P(E)}$$

(We are actually reducing our sample space to event E.)

Example

Consider the testing of three fenders, where $P(G) = 0.9$ and $p(D) = 0.1$ on any one try. Let event E = exactly two of the fenders are defective; event F = the second fender tested is defective. Find $P(E|F)$ and $P(F|E)$.

$$S = \{(G,G,G), (G,G,D), (G,D,G), (G,D,D), (D,G,G), (D,G,D), (D,D,G), (D,D,D)\}$$

and

$$E = \{(G,D,D), (D,G,D), (D,D,G)\}$$

Therefore,

$$P(E) = 0.009 + 0.009 + 0.009 = 0.027$$

and

$$F = \{(G,D,G), (G,D,D), (D,D,G), (D,D,D)\}$$

Therefore,

$$P(F) = 0.081 + 0.009 + 0.009 + 0.001 = 0.100$$

Now

$$E \cap F = \{(G,D,D), (D,D,G)\}$$

Therefore,

$$P(E \cap F) = 0.009 + 0.009 = 0.018$$

Then,

$$P(E|F) = \frac{P(E \cap F)}{P(F)} = \frac{0.018}{0.100} = \frac{18}{100} = \frac{9}{50} = 0.18$$

$$P(F|E) = \frac{P(E \cap F)}{P(E)} = \frac{0.018}{0.027} = \frac{18}{27} = \frac{2}{3} \approx 0.67$$

Events E and F are said to be *independent* if and only if $P(E|F) = P(E)$ and $P(F|E) = P(F)$.

Are the events E and F in this example independent? Why or why not?

4.6 BAYES' THEOREM

Bayes' theorem defines the probability of any one of several mutually exclusive events E_n, conditioned by any event D.

Given a sample space S and events $E_1, E_2, \ldots, E_n, D$ in S, such that

$$E_1 \cup E_2 \cup \cdots \cup E_n = S \quad \text{and} \quad E_i \cap E_j = \phi \text{ if } i \neq j$$

then

$$P(E_i|D) = \frac{P(E_i)P(D|E_i)}{P(E_1)P(D|E_1) + P(E_2)P(D|E_2) + \cdots + P(E_n)P(D|E_n)}$$

for any $i = 1,2,3, \ldots, n$.

Example

Machines E_1, E_2, and E_3 produce all of the 20¢ candy bars for the Chewy Chocolate Candy Corporation. Machines E_1, E_2, and E_3 produce 20%, 30%, and 50%, respectively, of the total output of this candy bar. The probability that machine E_1 produces a defective candy bar is 0.02, the probability that machine E_2 produces a defective candy bar is 0.01, and the probability that machine E_3 produces a defective candy bar is 0.03. If one 20¢ candy bar is selected at random and is found to be defective, what is the probability that it was produced by machine E_2?

E_1: Produced by machine E_1.
E_2: Produced by machine E_2.
E_3: Produced by machine E_3.
D: Candy bar is found to be defective.

Note: $E_i \cap E_j = \phi$ if $i \neq j$ and $E_1 \cup E_2 \cup E_3 = S$.

$$P(E_1) = 0.2,\ P(E_2) = 0.3,\ P(E_3) = 0.5$$
$$P(D|E_1) = 0.02,\ P(D|E_2) = 0.01,\ P(D|E_3) = 0.03$$

$$P(E_2|D) = \frac{P(E_2)P(D|E_2)}{P(E_1)P(D|E_1) + P(E_2)P(D|E_2) + P(E_3)P(D|E_3)}$$

$$= \frac{(0.3)(0.01)}{(0.2)(0.02) + (0.3)(0.01) + (0.5)(0.03)}$$

$$= \frac{3}{4 + 3 + 15}$$

Therefore

$$P(E_2|D) = \frac{3}{22} \approx 0.14$$

Sample Problem

For the preceding example, find $P(E_1|D)$ and $P(E_3|D)$.

4.7 PROVING BAYES' THEOREM FOR THREE EVENTS E_1, E_2, E_3

Given a sample space S and events E_1, E_2, E_3, and any event D in S such that $E_i \cap E_j = \phi$ if $i \neq j$, $E_1 \cup E_2 \cup E_3 = S$. Then

$$\begin{aligned} D &= S \cap D \\ &= (E_1 \cup E_2 \cup E_3) \cap D \\ &= (E_1 \cap D) \cup (E_2 \cap D) \cup (E_3 \cap D) \end{aligned}$$

Therefore

$$P(D) = P(E_1 \cap D) + P(E_2 \cap D) + P(E_3 \cap D)$$

Since E_1, E_2, E_3 are pairwise disjoint, and since

$$P(D|E_i) = \frac{P(E_i \cap D)}{P(E_i)}$$

then

$$P(E_i \cap D) = P(E_i)P(D|E_i)$$

Therefore

$$P(D) = P(E_1)P(D|E_1) + P(E_2)P(D|E_2) + P(E_3)P(D|E_3)$$

and

$$P(E_i|D) = \frac{P(E_i)P(D|E_i)}{P(D)}$$

$$= \frac{P(E_i)P(D|E_i)}{P(E_1)P(D|E_1) + P(E_2)P(D|E_2) + P(E_3)P(D|E_3)}$$

Note: Without loss of generality, Bayes' theorem can be proved for $E_1, E_2, \ldots, E_n$, provided that $E_i \cap E_j = \phi$ if $i \neq j$ and $E_1 \cup E_2 \cup \cdots \cup E_n = S$.

Problem Set 4.3

1. If E and F are events in a sample space such that $p(E) = 0.5$, $p(F) = 0.2$, and $p(E \cap F) = 0.1$, find
   (a) The probability of event E occurring, given event F has already occurred;
   (b) The probability of event F occurring, given event E has already occurred.
   (c) Are E and F independent events?

2. There are 35 applicants for supervisory positions in an oil corporation. Some of the applicants have previous supervisory experience and others do not have previous supervisory experience. Some of the applicants are college graduates and some are not. All pertinent information is given in the following chart:

	Previous Supervisory Experience	No Previous Supervisory Experience
College Graduate	13	9
Noncollege Graduate	5	8

Let E = applicants with previous supervisory experience
F = applicants with no previous supervisory experience
G = college graduates
H = noncollege graduates

If an applicant is to be chosen at random and invited for an interview, find the following probabilities:

(a) $P(E \cap G)$ (b) $P(E \cap H)$ (c) $P(F \cap G)$, (d) $P(F \cap H)$
(e) $P(E|G)$ (f) $P(G|E)$ (g) $P(E|H)$ (h) $P(H|E)$
(i) $P(F|G)$ (j) $P(G|F)$ (k) $P(F|H)$ (l) $P(H|F)$

3. A committee of three is to be selected randomly from five possible candidates (A,B,C,D, or E). If K is the event that candidate B

is on the committee chosen and L is the event that candidate C is on the committee chosen, find (a) $P(K|L)$ and (b) $P(L|K)$. (c) Are K and L independent events?

4. Given events E and F are events in a sample space S such that $P(E) = 0.50$, $P(F) = 0.30$, and $P(E \cap F) = 0.15$. Show that E and F are independent events.

5. Five machines produce the total output of a certain product. Machines E_1, E_2, E_3, E_4, and E_5 produce 10%, 20%, 15%, 25%, and 30% of the product, respectively. Machines E_1, E_2, E_3, E_4, and E_5 produce 0.01, 0.01, 0.02, 0.001, and 0.02 defective items, respectively, of their individual outputs. If an item is chosen at random and found to be defective, what is the probability it was (a) produced by machine E_4? (b) produced by machine E_3?

6. In a certain community college, 20% of the first-year students, 15% of the second-year students, and 5% of the nonclassified students are enrolled in a mathematics course. It has been determined that 60% of the school population are first-year students, 30% are second-year students, and 10% are nonclassified students. If a student is selected at random and found to be a second-year student, what is the probability that he is enrolled in a mathematics course?

7. Given E and F are events in a sample space S, $P(E|F) = 0.5$, and $P(F) = 0.2$. Find

(a) $P(E \cap F)$.

(b) If $P(F|E) = 0.4$, what must $P(E)$ equal?

Hint:

$$P(E|F) = \frac{P(E \cap F)}{P(F)} \quad \text{and} \quad P(F|E) = \frac{P(E \cap F)}{P(E)}$$

Therefore

$$P(E \cap F) = P(F)P(E|F) = P(E)P(F|E)$$

4.8 THE BINOMIAL EXPERIMENT

Many experiments result in exactly one of two possible outcomes each time we perform them. For example: A fender is tested and labeled G(good) or D(defective); the recording of a voter's viewpoint as Y(Yes) or N(No) is made on an important issue; the position of a lever on a switch 0(open) or 1(closed) is noted; or P(passing) or F(failing) on an examination is recorded.

A binomial experiment must have the following properties:

1. There are n independent trials of the same experiment.
2. Each trial of the experiment must result in one of two possible outcomes (success or failure).
3. The probability of the two possible outcomes must remain the same in each independent trial of the experiment. (We shall let p = the probability of success on any one trial and q = the probability of failure on any one trial. Note that $p + q = 1$.)

For a specific example of a binomial experiment, let us consider randomly selecting three fenders from a lot of fenders, then testing each of the three fenders selected and labeling them G(good) or D(defective), where the probability of a good fender on any one trial is 0.9. Then

$$S = \{(G,G,G), (G,G,D), (G,D,G), (G,D,D), (D,G,G), (D,G,D), (D,D,G), (D,D,D)\}$$

Now let A_i, where $i = 0,1,2$, or 3, be the event where we obtain exactly i good fenders. Then

$$\begin{aligned}
P(A_0) &= p(D,D,D) \\
&= (0.1)\,(0.1)\,(0.1) \\
&= 1(0.1)^3 \\
P(A_1) &= p(G,D,D) + p(D,G,D) + p(D,D,G) \\
&= (0.9)\,(0.1)\,(0.1) + (0.1)\,(0.9)\,(0.1) + (0.1)\,(0.1)\,(0.9) \\
&= 3(0.9)\,(0.1)^2 \\
P(A_2) &= p(G,G,D) + p(G,D,G) + p(D,G,G) \\
&= (0.9)\,(0.9)\,(0.1) + (0.9)\,(0.1)\,(0.9) + (0.1)\,(0.9)\,(0.9) \\
&= 3(0.9)^2(0.1) \\
P(A_3) &= p(G,G,G) \\
&= (0.9)\,(0.9)\,(0.9) \\
&= 1(0.9)^3
\end{aligned}$$

Note:

$$\begin{aligned}
(0.9 + 0.1)^3 &= {}_3C_3(0.9)^3 + {}_3C_2(0.9)^2(0.1) + {}_3C_1(0.9)\,(0.1)^2 + {}_3C_0(0.1)^3 \\
&= 1(0.9)^3 + 3(0.9)^2(0.1) + 3(0.9)\,(0.1)^2 + 1(0.1)^3
\end{aligned}$$

In general, in three trials of a binomial experiment where p = the probability of success on any one trial and q = the probability of failure on any one trial;

$$\begin{aligned}
P(A_0) &= 1q^3 \\
P(A_1) &= 3pq^2 \\
P(A_2) &= 3p^2q \\
P(A_3) &= 1p^3
\end{aligned}$$

where A_i = exactly i successes, *and*

$$(p + q)^3 = {}_3C_3p^3 + {}_3C_2p^2q + {}_3C_1pq^2 + {}_3C_0q^3$$
$$= 1p^3 + 3p^2q + 3pq^2 + 1q^3$$

By illustrating a number of binomial experiments with n trials of the binomial experiment, we can develop the following theorem:

The probability of r successes in n trials of a binomial experiment where p = the probability of success on any one trial and q = the probability of failure on any one trial is

$$ {}_nC_rp^rq^{n-r} \qquad (p + q = 1, q = 1 - p)$$

Examples

1. Five fenders are selected from a lot of fenders and tested for defects. If the probability of a good fender on any one trial is 0.9, find the probability of obtaining:

(a) Exactly four good fenders. Since we know that $n = 5, p = 0.9$, $q = 1 - 0.9 = 0.1$, $r = 4$, then

$$ {}_5C_4\,(0.9)^4\,(0.1)^1 = 5\,(0.6561)\,(0.1)$$
$$= 0.32805$$
$$\approx 0.33$$

(b) At least four good fenders. Since we know that $n = 5, p = 0.9$, $q = 1 - 0.9 = 0.1$, $r = 4,5$, then

$$P\text{ (at least 4 good)} = P(4\text{ good}) + P(5\text{ good})$$
$$= {}_5C_4(0.9)^4(0.1)^1 + {}_5C_5(0.9)^5(0.1)^0$$
$$= 5(0.6561)\,(0.1) + (1)\,(0.59049)\,(1)$$
$$= 0.32805 + 0.59049$$
$$= 0.91854$$
$$\approx 0.92$$

(c) At most two defective fenders. Since we know that $n = 5$, $p = 0.1$, $q = 0.9$, $r = 0,1,2$, then

$$P\text{ (at most 2 defective)} = P(0\text{ defective}) + P(1\text{ defective}) + P(2\text{ defective})$$
$$= {}_5C_0(0.1)^0(0.9)^5 + {}_5C_1(0.1)^1(0.9)^4 + {}_5C_2(0.1)^2(0.9)^3$$
$$= 0.59049 + 0.32805 + 0.0729$$
$$= 0.99144$$
$$\approx 0.99$$

2. The probability that a student enrolling in a remedial mathematics course will pass the course is 0.7. The only two possible grades are

P(pass), F(fail). If six students enroll in the course, what is the probability that

(a) at least five of the six students will pass the course? Since we know that $n = 6$, $p = 0.7$, $q = 0.3$, $r = 5,6$, then

$$\begin{aligned} P\text{ (at least 5 pass)} &= P(5\text{ pass}) + P(6\text{ pass}) \\ &= {}_6C_5(0.7)^5(0.3)^1 + {}_6C_6(0.7)^6(0.3)^0 \\ &= 6(0.16807)\ (0.3) + (1)\ (0.117649)\ (1) \\ &= 0.302526 + 0.117649 \\ &= 0.420175 \\ &\approx 0.42 \end{aligned}$$

(b) at least one of the six students will pass? Let event E be the event such that at least one of the six students will pass. Then event E' is the event such that none of the six students will pass. [*Remember*: $P(E) = 1 - P(E')$.] Therefore, p(at least one will pass) $= 1 - P$(0 will pass). Since we know that $n = 6$, $p = 0.7$, $q = 0.3$, $r = 0$, then

$$\begin{aligned} P\text{ (at least 1 will pass)} &= 1 - {}_6C_0(0.7)^0(0.3)^6 \\ &= 1 - (1)\ (1)\ (0.000729) \\ &= 1 - 0.000729 \\ &= 0.999271 \end{aligned}$$

Problem Set 4.4

1. The probability that a flu vaccine will be effective on any one person is 0.8. If four people are selected at random and injected with the vaccine, what is the probability that none of the four people will contract the flu?

2. A certain professional basketball player shoots his foul shots with an 0.9 accuracy. What is the probability that he will miss five of the next ten foul shots he attempts?

3. The probability that any one battery produced by the Live Wire Company will be rejected is 0.01. If three batteries are selected at random and then tested and labeled A (accepted) or R (rejected), what is the probability that at least two of the three randomly selected batteries tested will be accepted?

4. A woman professional golfer qualifies for 0.7 of the tournaments in which she desires to participate. What is the probability that she will *not* qualify for two of five randomly selected tournaments she wishes to enter?

5. The Snapshot Company advertises that its flashbulbs work with 0.999 accuracy. If the company's claim is true, what is the probability that all of three randomly selected flashbulbs will be defective?

4.9 A PREPACKAGED PROGRAM FOR THE BINOMIAL EXPERIMENT

Name: STAT4

Description: This program computes the probability of r successes in n independent trials of a binomial experiment.

Instructions: Enter the data beginning in line 150 as follows:

150 DATA N, R, P, D

where:

N = number of independent trials of the binomial experiment

R = number of desired successes

P = probability of success on any one trial of the experiment

D = { 0 if we want to print the desired total binomial probability, or
1 if we want to compute and print a partial sum for a binomial probability.

Note: (1) Line 200 is the END statement. (2) If you want the probability of at least R successes or at most R successes, you must enter more than one set of data. (3) $R \neq 0$.

Sample Problems

1. Find the probability of obtaining exactly five successes in seven trials of a binomial experiment where the probability of success on any one trial is 0.83.

User's Input

```
GET-STAT4
150 DATA 7,5,.83,0
RUN
```

Computer Output

```
#TRIALS        #SUCCESSES       PR. OF SUCC.      BINOMIAL PROB.

 7              5                .83               .23906


#TRIALS        #SUCCESSES       PR. OF SUCC.      BINOMIAL PROB.
```

2. Find the probability of obtaining at least five successes in seven trials of a binomial experiment where the probability of success on any one trial is 0.83.

User's Input

```
GET-STAT4
150 DATA 7,5,.83,1
151 DATA 7,6,.83,1
152 DATA 7,7,.83,0
RUN
```

Computer Output

```
#TRIALS     #SUCCESSES       PR. OF SUCC.         BINOMIAL PROB.

 7           5                .83  PARTIAL SUM= .23906
 7           6                .83  PARTIAL SUM= .628119
 7           7                .83                .899479


#TRIALS     #SUCCESSES       PR. OF SUCC.         BINOMIAL PROB.


LINE     35:    END OF DATA
```

Let us look at the flowchart of STAT4 (Fig. 4.3), which is followed by its coding.

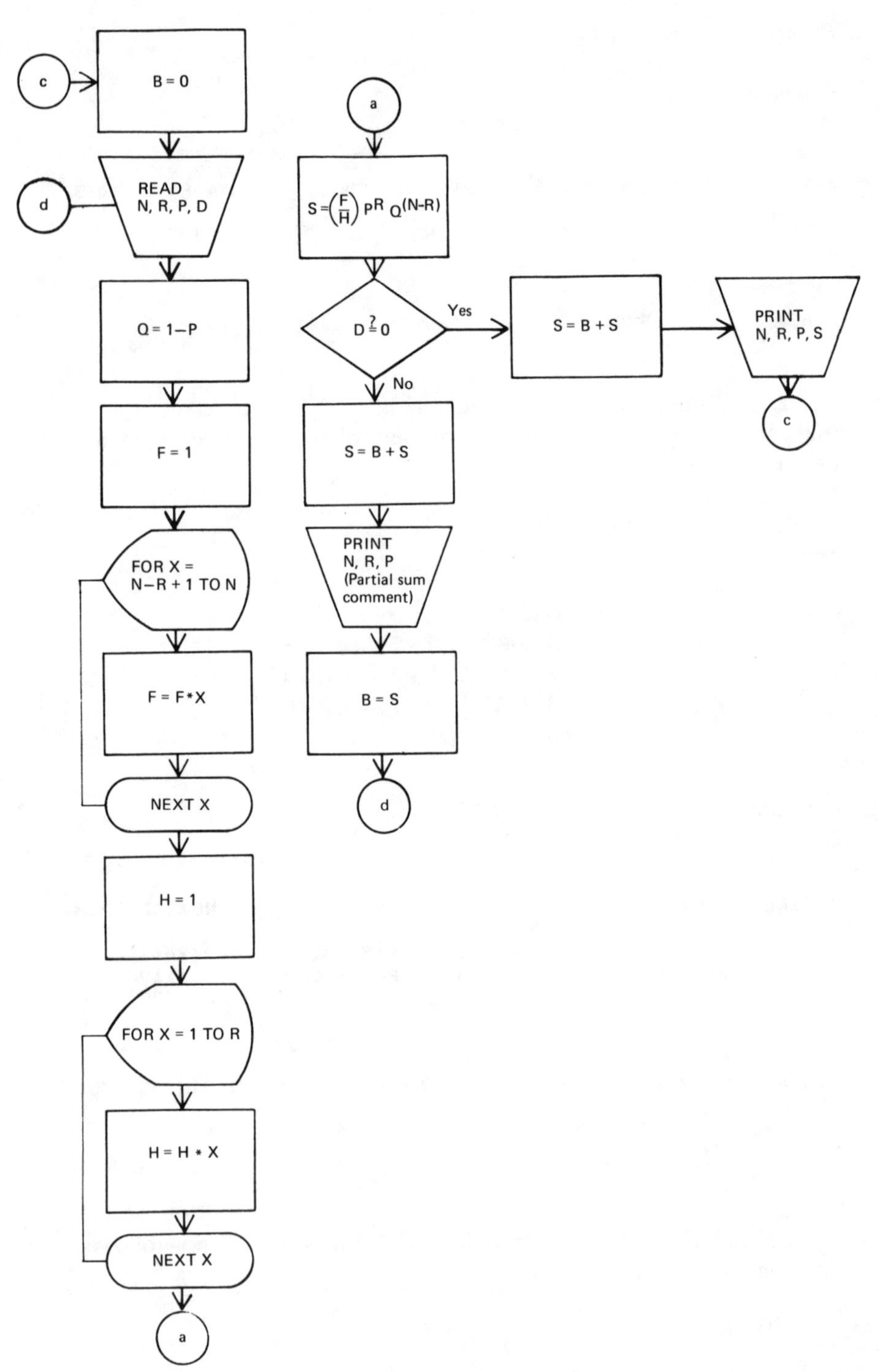

Fig. 4.3. Flowchart 13

The coding of STAT4 is as follows:

```
5 LET B=0
10 PRINT
15 PRINT
20 PRINT
25 PRINT "#TRIALS","#SUCCESSES","PR. OF SUCC.",
         "BINOMIAL PROB."
30 PRINT
35 READ N,R,P,D
40 LET Q = 1-P
45 LET F =1
50 FOR X = N-R+1 TO N
55     LET F = F*X
60 NEXT X
65 LET H = 1
70 FOR X = 1 TO R
75     LET H = H*X
80 NEXT X
85 LET S = F/H*P↑R*Q↑(N-R)
90 IF D = 0 THEN 115
95 LET S = B+S
100 PRINT N,R,P;"PARTIAL SUM="S
105 LET B = S
110 GO TO 35
115 LET S = B+S
120 PRINT N,R,P,S
125 GO TO 5
150 DATA
200 END
```

Problem Set 4.5

Check your manual solutions to each of the five exercises in Problem Set 4.4 by using STAT4 to solve them.

5

Random Variables, Normal Distributions, and Related Prepackaged Programs

5.1 DISCRETE RANDOM VARIABLES

A discrete random variable is a function on the sample space of an experiment which assigns each sample point in a sample space S to exactly one real number. A discrete random variable X also generates a set of ordered pairs $\{(X_i, p(X_i))\}$—called the *probability function*, or *distribution*, of X—where $\{X_i\}$ contains all possible real numbers assigned to sample points in S and $p(X_i)$ represents their associated probabilities. If $\{X_i\}$ contains separate points on a number line, then X is a discrete random variable.

Examples

1. Using the example in Article 4.1, where we randomly select two fenders from a lot of fenders, given that $p(G) = 0.9$ (the probability of a good fender on any one try is 0.9) and $p(D) = 0.1$, let X be the random variable that represents the *number of good fenders* in the sample selected. Then

$$S = \{(G,G), (G,D), (D,G), (D,D)\}$$

Let us assign these four sample points in S to the following numbers:

$$\begin{aligned}(G,G) &\rightarrow \text{number } 2\\ (G,D) &\rightarrow \text{number } 1\\ (D,G) &\rightarrow \text{number } 1\\ (D,D) &\rightarrow \text{number } 0\end{aligned}$$

Possible Values of X	Sample Points Assigned the Value X	Corresponding Probability $p(X)$
0	(D,D)	0.01
1	$(G,D),(D,G)$	$0.09 + 0.09 = 0.18$
2	(G,G)	0.81
		$\Sigma p(X) = 1.00$

We can construct a *probability histogram* for the probability of X, as in Fig. 5.1.

The experiment discussed above is a *binomial experiment* (each trial of the experiment results in exactly one of two possible outcomes, G or D). Also, the probability distribution illustrated in Fig. 5.1 is theoretical. That is, over an infinite number of trials of the experiment we may assume three outcomes:

(1) $$p(0) = 0.01$$

or we obtain 0 good fenders of the 2 fenders tested in 1 of 100 of the trials.

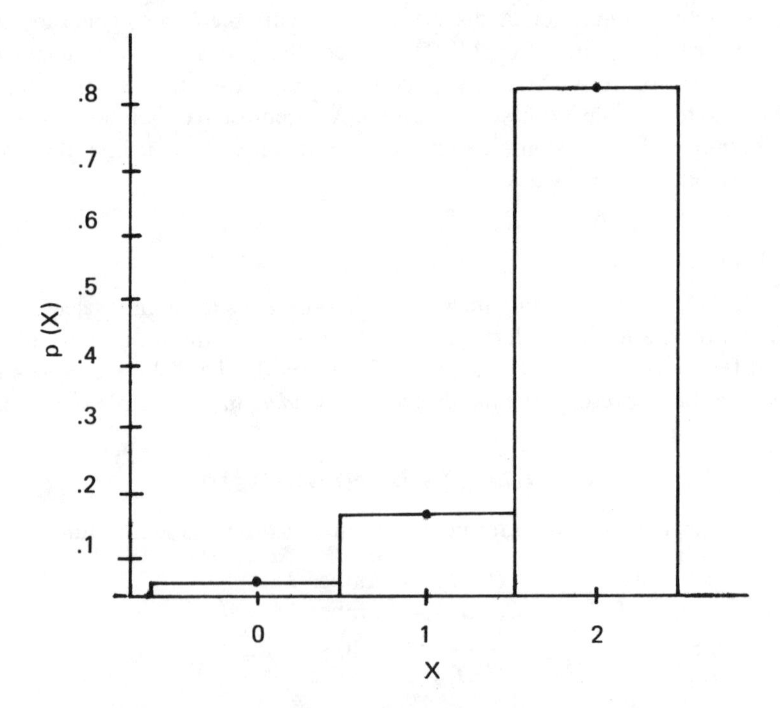

Fig. 5.1. Histogram for probability distribution of X

(2) $$p(1) = 0.18$$

or we obtain 1 good fender of the 2 fenders tested in 18 of 100 of the trials.

(3) $$p(2) = 0.81$$

or we obtain 2 good fenders of the 2 fenders tested in 81 of 100 of the trials.

However, if we try this experiment 100 times (or any finite number of times), we will not necessarily get 0 good fenders 1 time, 1 good fender 18 times, and 2 good fenders 81 times. We will get some variation from this theoretical distribution in repeated finite sampling. Therefore, in Article 5.2 we will discuss the mean, variance, and standard deviation of a discrete random variable X.

2. Each customer leaving the Colonial Diner is asked to state whether the service was G(good), F(fair), or P(poor). In the past, the manager has found $p(G) = 0.7$, $p(F) = 0.2$, and $p(P) = 0.1$. If two customer opinions of the service are selected at random, let Y be the random variable that represents the number of customers in the sample selected who stated the service was poor. Then,

$$\begin{array}{cccccccccc} S = \{ & (G,G), & (G,F), & (G,P), & (F,G), & (F,F), & (F,P), & (P,G), & (P,F), & (P,P) \}. \\ & \downarrow & \downarrow & \downarrow & \downarrow & \downarrow & \downarrow & \downarrow & \downarrow & \downarrow \\ & 0 & 0 & 1 & 0 & 0 & 1 & 1 & 1 & 2 \end{array}$$

Possible Values of Y	Sample Points Assigned the Value Y	$p(Y)$
0	$(G,G),(G,F),(F,G),(F,F)$	$0.49 + 0.14 + 0.14 + 0.04 = 0.81$
1	$(G,P),(F,P),(P,G),(P,F)$	$0.07 + 0.02 + 0.07 + 0.02 = 0.18$
2	(P,P)	0.01
		$\Sigma p(Y) = 1.00$

The histogram in Fig. 5.2 illustrates the probability distribution of Y.

The preceding experiment is *not* a binomial experiment (each trial of the experiment can result in exactly one of three possible outcomes—G, F, or P). Also, the probability distribution illustrated in Fig. 5.2 is theoretical. That is, over an infinite number of trials of the experiment we assume three outcomes:

(1) $$p(0) = 0.81$$

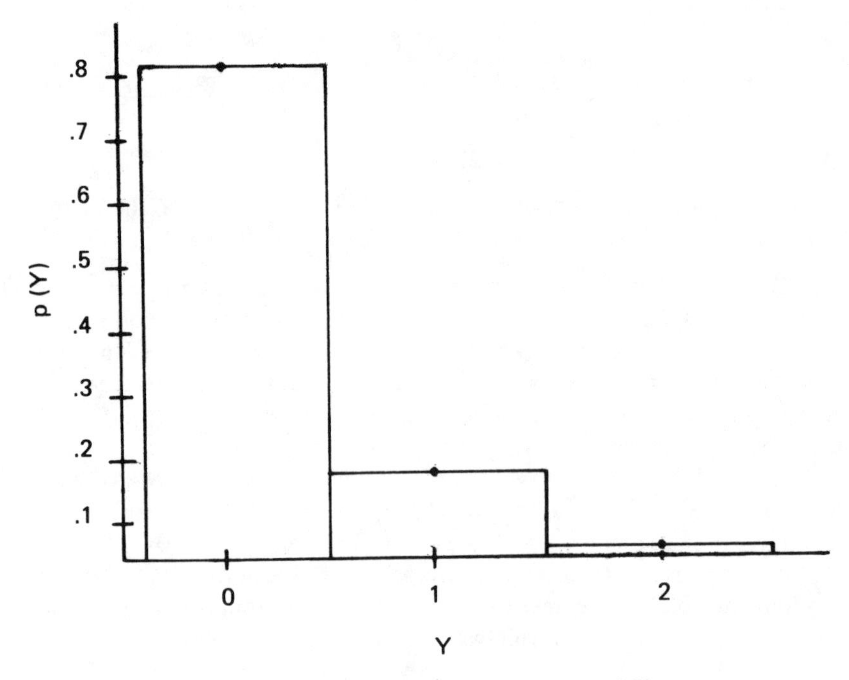

Fig. 5.2. Probability distribution of Y

or we obtain 0 opinions of the 2 opinions selected to state that the service was poor in 81 of 100 of the trials;

(2) $$p(1) = 0.18$$

or we obtain 1 opinion of the 2 opinions selected to state that the service was poor in 18 of 100 of the trials;

(3) $$p(2) = 0.01$$

or we obtain 2 opinions of the 2 opinions selected to state that the service was poor in 1 of 100 of the trials.

However, if we try this experiment 100 times, we will not necessarily get 0 poor service opinions 81 times, 1 poor service opinion 18 times, and 2 good service opinions 1 time. We will get some deviation from this theoretical distribution in repeated finite sampling.

3. Angelo of Angelo's Import Candy House reports that 80% of the customers who enter his store on Valentine's Day buy a box of candy. If three of the customers who enter his store next Valentine's Day are selected at random, let Z be the random variable that represents the number of people in the sample chosen who buy a box of candy. Let s

represent a sold box of candy and n represent a box of candy that was unsold. Then,

$$S = \{(s,s,s),(s,s,n),(s,n,s),(s,n,n),(n,s,s),\\ (n,s,n),(n,n,s),(n,n,n)\}$$

and

$$p(s) = 0.8, \qquad p(n) = 0.2$$

Possible Values of Z	Sample Points Assigned the Value Z	$p(Z)$
0	(n,n,n)	0.008
1	$(s,n,n),(n,s,n),(n,n,s)$	$3(0.032) = 0.096$
2	$(s,s,n),(s,n,s),(n,s,s)$	$3(0.128) = 0.384$
3	(s,s,s)	0.512
		$\Sigma p(Z) = 1.000$

Our probability distribution is represented graphically in Fig. 5.3 as a histogram.

This experiment is a binomial experiment. Why? Also, $p(0) = 0.008$, $p(1) = 0.096$, $p(2) = 0.384$, and $p(3) = 0.512$ is theoretical and will not necessarily happen over a finite number of trials of the experiment; that is, if we try this experiment 1000 times, we will not necessarily get 0 boxes of candy bought 8 times, 1 box of candy bought 96 times, 2 boxes of candy bought 384 times, and 3 boxes of candy bought 512 times.

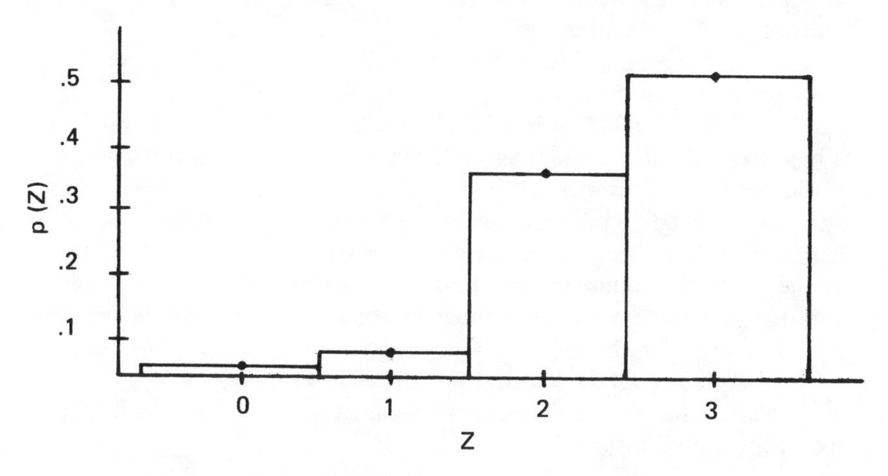

Fig. 5.3. Histogram for the probability distribution of Z

Problem Set 5.1

For each of the following exercises:

(a) Tell whether the experiment is a binomial experiment or not.
(b) List the elements in the sample space S.
(c) Construct a table of values X, $p(X)$ for the probability function of the random variable X.
(d) Draw a histogram for the probability distribution of the random variable X.

1. The Duty Drug Company reports that a new pill to relieve back pain is effective 99 out of 100 times. If two patients are randomly selected from a group of patients with back pain and the new drug is administered, let X be the random variable that represents the number of patients in the sample selected whose back pains are relieved after taking the new drug.
2. A fair die is rolled (a die has six faces with the numbers 1, 2, 3, 4, 5, and 6, each painted on exactly one face). Let X be the random variable that represents the number showing on the upper face.
3. The Today Shirt Company reports that 50% of the shirts it sells are bought by men. If three customers who purchase Today shirts are chosen at random, let X be the random variable that represents the number of female customers in the sample chosen.
4. Memorial High School has determined that six of every ten of its seniors participate in some school-sponsored sport. If two of Memorial High's seniors are selected at random, let X be the random variable that represents the number of seniors in the sample who participate in a school sport.
5. A three-digit lottery number is chosen by randomly selecting one digit at a time (from left to right) from the entire set of digits 0, 1, 2, 3, 4, 5, 6, 7, 8, 9; each digit is painted on a chip and all chips are mixed in a box. Let X be the random variable that represents the number of fives (5) in the lottery number selected.
6. An English professor determines that 95% of the students who enroll in his basic English course receive a grade of C or better. If three of the professor's students are chosen at random, let X be the random variable that represents the number of students in the sample selected who received a grade less than C.
7. A fair coin is tossed four consecutive times. Let X be the random variable that represents the number of heads obtained in the four consecutive tosses.

8. During the past ten years the Slick Hockey Team has achieved the following percentages: $p(\text{win}) = 0.5$, $p(\text{lose}) = 0.4$, $p(\text{tie}) = 0.1$ for any particular game. A Slick Hockey Team fan randomly selects three regular season games to attend. Let X be the random variable that represents the number of tie games the fan sees while attending the three regular season games.

9. The Smooth Taste Beer Company claims that two of every five beer drinkers in Good Time City drink its beer. Two beer drinkers in Good Time City are to be selected at random. Let X be the random variable that represents the number of beer drinkers in the sample of two chosen who do not drink Smooth Taste Beer.

10. One defective radio is inadvertently included in a shipment of four radios to the New Sound Music Shop. A customer randomly selects two radios from the shipment of four radios. Let X be the random variable that represents the number of good radios the customer selects.

5.2 THE MEAN OF A DISCRETE RANDOM VARIABLE

The probability distribution for a discrete random variable X is theoretical, and we assume this distribution will be true over an infinite number of trials of an experiment. However, in a finite number of trials of an experiment a random variable X may deviate from its probability distribution. Before we discuss the deviation of a random variable X, we must first consider its *mean.*

Given the probability distribution for a discrete random variable X, or $\{(X_i, p(X_i))\}$, where $X_1, X_2, \ldots, X_n$ are the real values assigned to sample points in a sample space S under X, and $p(X_1), p(X_2), \ldots, p(X_n)$ are their associated probabilities, then the *mean* of X (or expected value of X), is denoted by

$$\mu = \sum_{i=1}^{n} X_i p(X_i)$$

Examples

Let us consider the three examples discussed in Article 5.1.

1. In the first experiment we randomly selected two fenders from a lot of fenders where $p(G) = 0.9$ and let X be the random variable that represented the number of good fenders in the sample selected; we obtained the following probability distribution:

Possible Values of X	Sample Points Assigned the Value X	$p(X)$
0	(D,D)	0.01
1	(G,D), (D,G)	0.18
2	(G,G)	0.81
		$\Sigma p(X) = 1.00$

The mean (expected value) of X is denoted by $\mu = \Sigma X p(X)$:

$$\begin{aligned}\mu &= 0(0.01) + 1(0.18) + 2(0.81)\\ &= 0.18 + 1.62\\ &= 1.8\end{aligned}$$

Thus, in repeated finite trials of the experiment, the average number of good fenders in the sample of two fenders will likely be 1.8.

2. Each of the two customers leaving the Colonial Diner is asked to state whether the service was G(good), F(fair), or P(poor), and the management has determined $p(G) = 0.7$, $p(F) = 0.2$ and $p(P) = 0.1$. If we let Y be the random variable that represents the number of customers who stated the service was poor, we obtain the following probability distribution:

Possible Values of Y	Sample Points Assigned the Value Y	$p(Y)$
0	(G,G), (G,F), (F,G), (F,F)	0.81
1	(G,P), (F,P), (P,G), (P,F)	0.18
2	(P,P)	0.01
		$\Sigma p(Y) = 1.00$

The mean (expected value) of Y is denoted by

$$\begin{aligned}\mu &= \Sigma Y p(Y)\\ &= 0(0.81) + 1(0.18) + 2(0.01)\\ &= 0.18 + 0.02\\ &= 0.2\end{aligned}$$

Thus, in repeated finite trials of the experiment, the average number of people in the sample of two who stated the service was poor is likely to be 0.2.

3. Angelo of Angelo's Import Candy House reports that 80% of the customers who enter his store on Valentine's Day buy a box of candy. If we randomly select three customers who enter his store next Valentine's Day and let Z be the random variable that represents the

number of people in the sample chosen who buy a box of candy, we obtain the following probability distribution:

Possible Values of Z	Sample Points Assigned the Value Z	$p(Z)$
0	(n,n,n)	0.008
1	(s,n,n), (n,s,n), (n,n,s)	0.096
2	(s,s,n), (s,n,s), (n,s,s)	0.384
3	(s,s,s)	0.512
		$\Sigma p(Y) = 1.000$

The mean (expected value) of Z is denoted by

$$\begin{aligned} \mu &= \Sigma Z p(Z) \\ &= 0(0.008) + 1(0.096) + 2(0.384) + 3(0.512) \\ &= 0.096 + 0.768 + 1.536 \\ &= 2.4 \end{aligned}$$

Thus, in repeated finite trials of the experiment, the average number of people in the sample of three who buy a box of candy is likely to be 2.4.

5.3 THE MEAN OF A BINOMIAL RANDOM VARIABLE

The mean of a binomial random variable X can be computed without developing the probability distribution of X.

Let us consider any binomial random variable for which N = the number of trials of the experiment; p = the probability of success on any one trial; q = the probability of failure on any one trial ($p + q = 1$), and X = the number of successes. Let $N = 1$; then

$$\begin{aligned} \mu &= \Sigma X p(X) \\ &= 0p(0) + 1p(1) \\ &= 0 \cdot q + 1 \cdot p \\ &= p \end{aligned}$$

Let $N = 2$; then

$$\begin{aligned} \mu &= \Sigma X p(X) \\ &= 0p(0) + 1p(1) + 2p(2) \\ &= 0q^2 + 1(2pq) + 2p^2 \\ &= 2pq + 2p^2 \\ &= 2p(q + p) \\ &= 2p \end{aligned}$$

Without loss of generality, we could show that if $N = n$, then $\mu = np$. Again, let us return to the three examples of Article 5.1.

Examples

1. In the first experiment we randomly selected two fenders from a lot of fenders where $p(G) = 0.9$. Let X be the random variable that represents the number of good fenders in the sample selected.

This is a binomial experiment. Why? (Because each trial of the experiment results in exactly one of two possible outcomes, G(good) or D(defective). Thus, we are given

$$n = 2 \text{ (2 trials of the experiment)}$$
$$p = 0.9$$
$$q = 0.1$$

Therefore, since $\mu = np$,

$$\mu = 2(.9) = 1.8$$

Notice that our computed value of μ is the same value for the mean that we obtained by using the general formula $\mu = \Sigma Xp(X)$.

2. The experiment in which each customer leaving the Colonial Diner is asked to state whether the service was G(good), F(fair), or P(poor), and we let Y be the random variable that represents the number of customers who stated the service was poor, is in fact a binomial experiment since service can be classified as "poor" or "not poor." Thus, if $n = 2$, $p = p(\text{poor service}) = 0.1$, and $q = p(\text{not poor service}) = 0.9$; then $\mu = np = 2(0.1) = 0.2$.

3. The experiment involving the random selection of three customers who enter Angelo's Import Candy House on Valentine's Day, given the probability that a customer who enters the store on Valentine's Day buys a box of candy is 0.8, is a binomial experiment. Why? If we let Z be the random variable that represents the number of people in the sample of three who buy a box of candy, we can compute the mean of Z using the binomial mean formula $\mu = np$. In this experiment, $n = 3$, $p = 0.8$, $q = 0.2$. Therefore, $\mu = np = 3(0.8) = 2.4$, which is the same result we obtained using the general formula $\mu = \Sigma Zp(Z)$.

5.4 THE VARIANCE AND STANDARD DEVIATION OF A DISCRETE RANDOM VARIABLE

Given the probability distribution for a discrete random variable, or $\{(X_i), p(X_i)\}$, where $X_1, X_2, \ldots, X_n$ are the real values assigned to sample points in a sample space S under X and $p(X_1), p(X_2), \ldots, p(X_n)$ are their associated probabilities, then the *variance* of X is denoted by

$$\sigma^2 = \sum_{i=1}^{n} (X_i - \mu)^2 p(X_i)$$

where μ is the mean (expected value of X) and the *standard deviation* of X is denoted by $\sigma = \sqrt{\sigma^2}$.

Examples

Again let us consider the same three examples discussed in Article 5.1.

1. The probability distribution for the experiment of randomly selecting two fenders where $p(G) = 0.9$, and X is the random variable that represents the number of good fenders in the sample selected, is as follows:

Possible Values of X	Sample Points Assigned the Value X	$p(X)$
0	(D,D)	0.01
1	$(G,D),(D,G)$	0.18
2	(G,G)	0.81
		$\Sigma p(X) = 1.00$

The *mean* of X was computed as $\mu = 1.8$. We now compute the *variance* of X, denoted by

$$\begin{aligned}\sigma^2 &= \Sigma(X - \mu)^2 p(X) \\ &= (0 - 1.8)^2(0.01) + (1 - 1.8)^2(0.18) + (2 - 1.8)^2(0.81) \\ &= (3.24)(0.01) + (0.64)(0.18) + (0.04)(0.81) \\ &= 0.0324 + 0.1152 + 0.0324 \\ &= 0.18\end{aligned}$$

Then the *standard deviation* of X is computed as

$$\sigma = \sqrt{\sigma^2} = \sqrt{0.18} \approx 0.42$$

2. The probability of distribution for the experiment of randomly selecting three customers and asking them if the service was good, fair, or poor (given, $p(G) = 0.7$, $p(F) = 0.2$, $p(p) = 0.1$), where Y is the random variable that represents the number of customers in the sample selected who stated the service was poor is as follows:

Possible Values of Y	Sample Points Assigned the Value Y	$p(Y)$
0	$(G,G),(G,F),(F,G),(F,F)$	0.81
1	$(G,P),(F,P),(P,G),(P,F)$	0.18
2	(P,P)	0.01
		$\Sigma p(Y) = 1.00$

We computed the mean of Y as $\mu = 0.2$. Then the variance of Y is computed as

$$\begin{aligned}\sigma^2 &= \Sigma(Y - \mu)^2\, p(Y)\\ &= (0 - 0.2)^2(0.81) + (1 - 0.2)^2(0.18) + (2 - 0.2)^2(0.01)\\ &= (0.04)(0.81) + (0.64)(0.18) + (3.24)(0.01)\\ &= 0.0324 + 0.1152 + 0.0324\\ &= 0.18\end{aligned}$$

and the standard deviation of Y is calculated as

$$\sigma = \sqrt{\sigma^2} = \sqrt{0.18} \approx 0.42$$

3. The probability distribution for the experiment of randomly selecting three customers in Angelo's Import Candy House on Valentine's Day (that is, p (customer buys a box of candy) $= 0.8$), where Z is the random variable that represents the number of people in the sample chosen who buy a box of candy, is as follows:

Possible Values of Z	Sample Points Assigned the Value Z	$p(Z)$
0	(n,n,n)	0.008
1	$(s,n,n),(n,s,n),(n,n,s)$	0.096
2	$(s,s,n),(s,n,s),(n,s,s)$	0.384
3	(s,s,s)	0.512
		$\Sigma p(Z) = 1.000$

We have computed the mean of Z as $\mu = 2.4$. Then the variance of Z is computed as

$$\begin{aligned}\sigma^2 &= \Sigma(Z - \mu)^2 p(Z)\\ &= (0 - 2.4)^2(0.008) + (1 - 2.4)^2(0.096) + (2 - 2.4)^2(0.384)\\ &\quad + (3 - 2.4)^2(0.512)\\ &= (5.76)(0.008) + (1.96)(0.96) + (0.16)(0.384) + (0.36)(0.512)\\ &= 0.04608 + 0.18816 + 0.06144 + 0.18432\\ &= 0.48\end{aligned}$$

and the standard deviation of Z is determined as

$$\sigma = \sqrt{\sigma^2} = \sqrt{0.48} \approx 0.69$$

5.5 THE VARIANCE AND STANDARD DEVIATION OF A BINOMIAL RANDOM VARIABLE

The *variance* and *standard deviation* of a binomial random variable X can be computed without developing the probability distribution of X.

Let us consider any binomial random variable for which $N =$ the number of trials of the experiment, p = the probability of success on any one trial, q = the probability of failure on any one trial ($p + q = 1$), and X = the number of successes.

Let $N = 1$; then

$$\mu = np = p$$

and

$$\begin{aligned}\sigma^2 &= \Sigma (X - \mu)^2 p(X) \\ &= (0 - p)^2 p(0) + (1 - p)^2 p(1) \\ &= p^2 q + q^2 p \\ &= pq(p + q) \\ &= pq(1)\end{aligned}$$

or

$$\sigma^2 = pq \quad \text{and} \quad \sigma = \sqrt{pq}$$

Let $N = 2$; then

$$\mu = np = 2p \quad \text{and} \quad \sigma^2 = \Sigma(X - \mu)^2 p(X). \text{ Thus}$$

$$\begin{aligned}\sigma^2 &= (0 - 2p)^2 p(0) + (1 - 2p)^2 p(1) + (2 - 2p)^2 p(2) \\ &= 4p^2(q^2) + (1 - 4p + 4p^2)(2pq) + 4(1 - p)^2(p^2) \\ &= 4p^2q^2 + (1 - 4p + 4p^2)(2pq) + 4q^2p^2 \\ &= 8p^2q^2 + (1 - 4p + 4p^2)(2pq) \\ &= 2pq(4pq + 1 - 4p + 4p^2)\end{aligned}$$

Now substitute $(1 - p)$ for q:

$$\begin{aligned}\sigma^2 &= 2pq(4p(1 - p) + 1 - 4p + 4p^2) \\ &= 2pq(4p - 4p^2 + 1 - 4p + 4p^2) \\ &= 2pq(1)\end{aligned}$$

or

$$\sigma^2 = 2pq \quad \text{and} \quad \sigma = \sqrt{2pq}$$

Without loss of generality we can show that if $N = n$, then $\sigma^2 = npq$ and $\sigma = \sqrt{npq}$.

Examples

Again let us consider our three examples in Article 5.1.

1. The experiment where we randomly select two fenders for which $p(G) = 0.9$, and let X be the random variable that represents the number of good fenders in our sample, is a binomial problem with $n = 2$, $p = 0.9$, $q = 0.1$.

We saw that $\mu = np = 2(0.9) = 1.8$. Now, if

$$\sigma^2 = npq \qquad \text{and} \qquad \sigma = \sqrt{npq}$$

then

$$\sigma^2 = 2(0.9)(0.1) \qquad \text{and} \qquad \sigma = \sqrt{2(0.9)(0.1)}$$

or

$$\sigma^2 = 0.18 \qquad \text{and} \qquad \sigma = \sqrt{0.18} \approx 0.42$$

These are the *same results* we obtained by using the general formulas $\sigma^2 = \Sigma(X - \mu)^2 p(X)$ and $\sigma = \sqrt{\sigma^2}$.

2. The experiment where each customer leaving the Colonial Diner is asked to state whether the service was G(good), F(fair), or P(poor) is in fact a binomial experiment, since service can be classified as "poor" or "not poor." Thus, if $n = 2$, $p = p(\text{poor service}) = 0.1$, and $q = p(\text{not poor service}) = 0.9$; then $\mu = np = 2(0.1) = 0.2$, $\sigma^2 = npq = 2(0.1)(0.9) = 0.18$, and $\sigma = \sqrt{npq} = \sqrt{0.18} \approx 0.42$.

3. The experiment where we randomly select three customers in Angelo's Import Candy House on Valentine's Day, given that p(the customer buys a box of candy) $= 0.8$, is a binomial experiment. Why? If we let Z be the random variable that represents the number of people in the sample who buy a box of candy, we have $N = 3$, $p = 0.8$, $q = 0.2$ and $\mu = np = 3(.8) = 2.4$. Therefore

$$\begin{aligned}\sigma^2 &= npq = 3(0.8)(0.2) \qquad \text{and} \qquad & \sigma &= \sqrt{npq} = \sqrt{3(0.8)(0.2)}\\ &= 0.48 & &= \sqrt{0.48} \approx 0.69\end{aligned}$$

which are the same results we obtained by using the general formulas $\sigma^2 = \Sigma(Z - \mu)^2 p(Z)$ and $\sigma = \sqrt{\sigma^2}$.

Summary

Given a random variable X defined on a sample space S associated with an experiment, with a probability distribution $\{(X_i, p(X_i))\}$, we have determined the following measures.

Measure	In General	If the Experiment is Binomial
The mean of X	$\mu = \sum_{i=1}^{n} X_i p(X_i)$	$\mu = np$
The variance of X	$\sigma^2 = \sum_{i=1}^{n} (X_i - \mu)^2 p(X_i)$	$\sigma^2 = npq$
The standard deviation of X	$\sigma = \sqrt{\sigma^2}$	$\sigma = \sqrt{npq}$

where n = the number of trials, p = probability of success, q = probability of failure.

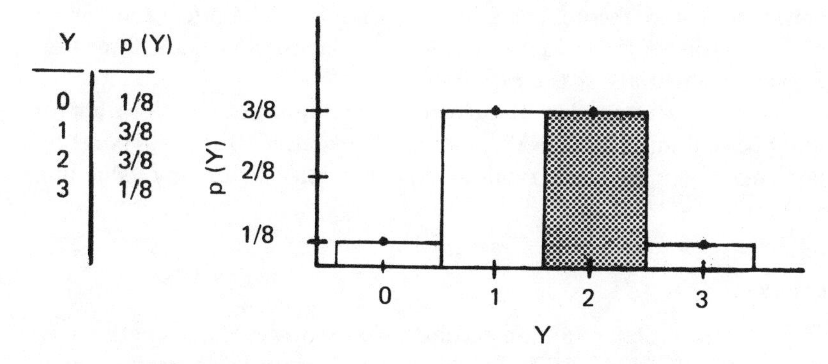

Fig. 5.4. Probability distribution and histogram for Y

Problem Set 5.2

1. For each of the ten exercises in Problem Set 5.1:
   (a) Indicate whether the experiment is a binomial experiment or not.
   (b) Compute the mean of the random variable X.
   (c) Compute the variance of the random variable X.
   (d) Compute the standard deviation of the random variable X.
2. Compute the mean, variance, and standard deviation for the discrete random variable Y whose probability distribution is

Y	$p(Y)$
2	¼
4	½
6	¼

5.6 CONTINUOUS RANDOM VARIABLES

As was previously stated, a random variable X with probability distribution $\{(X_i, p(X_i))\}$—where $X_1, X_2, \ldots, X_n$ represent separate points on a number line—is a discrete random variable. For example, let us look at Fig.5.4, which shows the probability distribution and histogram for a discrete random variable Y, where $Y_1 = 0$, $Y_2 = 1$, $Y_3 = 2$, $Y_4 = 3$ (that is, the random variable Y can take on only the values 0, 1, 2, 3). The shaded area in the histogram for the probability distribution of Y represents $p(2) = 3/8$ (the probability that the random variable Y takes on the value 2 is 3 out of 8). However, we saw that this distribution was theoretical and that in repeated finite sampling the random variable Y would not necessarily take on the value 2 three times in every eight trials of the experiment.

A *continuous random variable* is a random variable X with probability distribution $\{(X, p(X))\}$, where X is defined for *every* point in a particular interval. Many random variables can take on any value in an interval.

Examples

1. Let Y be the random variable that represents the length of time it takes the Memorial High School baseball team to complete a game.
2. Let Z be the random variable that represents the weight of each sixth-grade student in your school district.

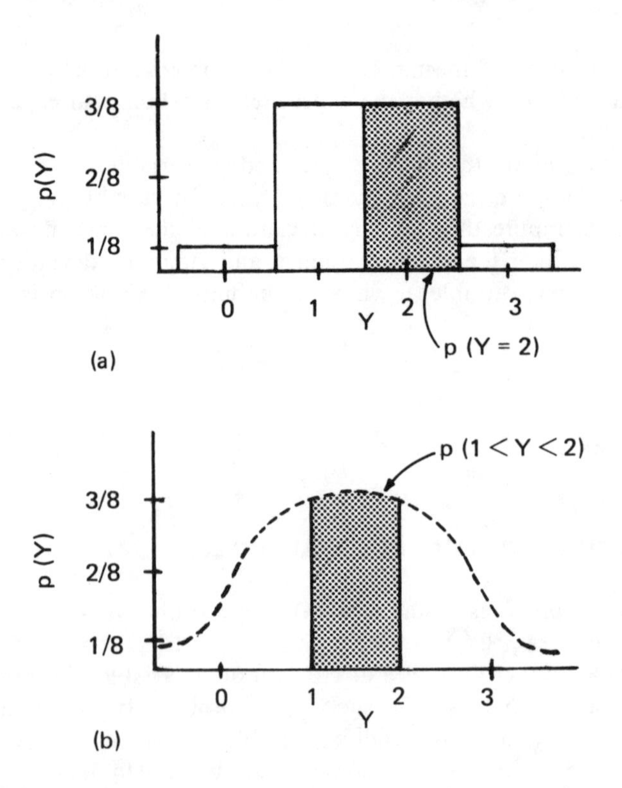

Fig. 5.5. (a) Discrete random variable; (b) continuous random variable

3. Let X be the random variable that represents the amount of juice obtained by manually squeezing three oranges.

In the remainder of this chapter we will be able to convert a *discrete* random variable (Fig. 5.5a) to a continuous random variable [Fig. 5.5(b)] so that we can approximate the probability that the random variable Y will take on any point in a given interval in repeated finite samplings of an experiment.

5.7 NORMAL DISTRIBUTIONS

The most widely used probability distribution in statistics is the *normal probability distribution*, which is generated by the formula

$$p(X) = \frac{1}{\sigma\sqrt{2\pi}} e^{-1/2(x-\mu)^2/\sigma^2}$$

where

$$\begin{aligned} -\infty < X < \infty &= \text{that is, } X \text{ is any real number} \\ \pi &\approx 3.14159 \\ e &\approx 2.71828 \\ \mu &= \text{the } \textit{mean} \text{ of a population} \\ \sigma &= \text{the } \textit{standard deviation} \text{ of a population} \end{aligned}$$

The graphs of normal probability distributions are bell-shaped, but their widths and heights can vary with different population means and standard deviations, as illustrated in Fig. 5.6.

However, all normal probability distributions have the following properties (as shown in Fig. 5.7):

1. The total area (probability) under $p(X) = 1$.
2. The area (probability) is symmetric about μ (the mean).

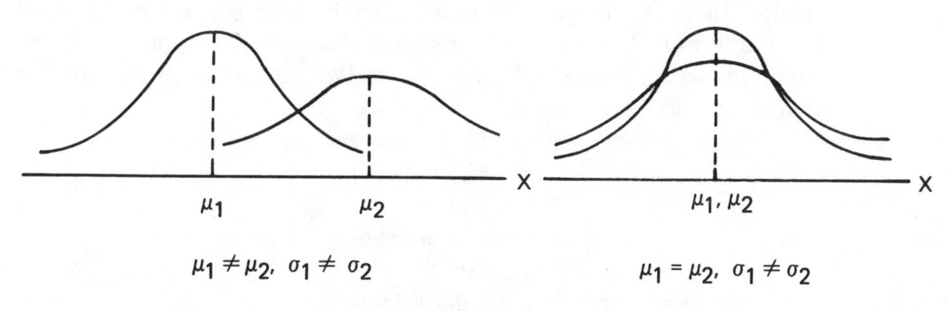

Fig. 5.6

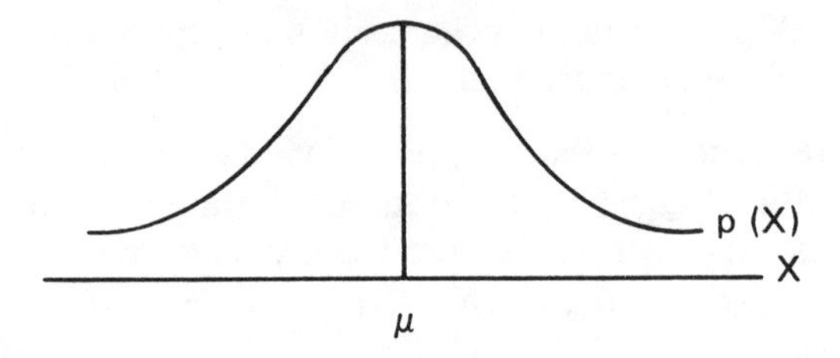

Fig. 5.7

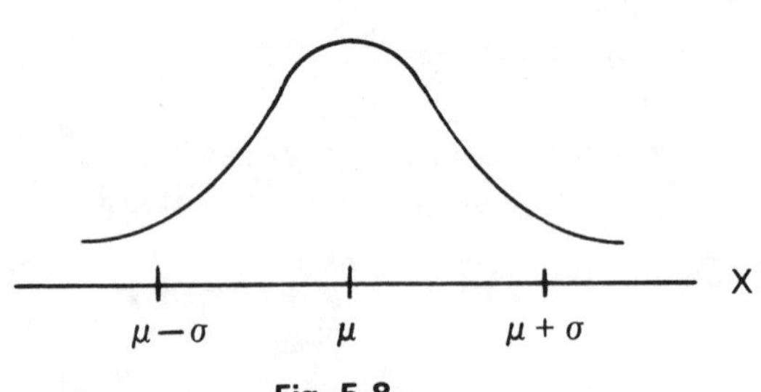

Fig. 5.8

3. $p(X)$ is greatest when $X = \mu$ and the $p(X)$ increases as X gets closer to μ.
4. $p(X)$ approaches 0 as X approaches $+\infty$ or $-\infty$.

5.8 THE STANDARD NORMAL DISTRIBUTION

Any sufficiently large sampling of an experiment with a random variable X defined for it will produce a normal probability distribution. The integral calculus enables us to compute areas (probabilities) under normal curves. Tables are constructed and listed in statistics textbooks for use in calculating specific areas under normal curves. Such a table is discussed in Article 5.9.

We saw previously that normal curves can take on many different shapes. In order to use *only* one area table, we convert our normal distribution (X distribution) to a *standard normal distribution* or Z-distribution, which has $\mu = 0$ and $\sigma = 1$. We do this by means of the following conversion formula:

$$Z = \frac{X - \mu}{\sigma}$$

where X = value assumed by a random variable
μ = population mean
σ = population standard deviation

See Fig. 5.8.

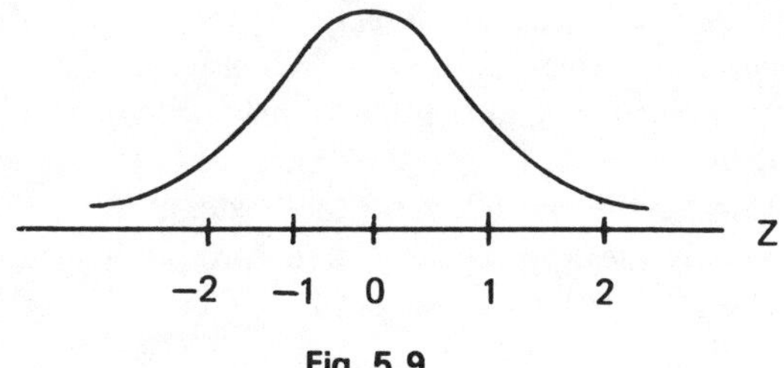

Fig. 5.9

If $Z = (X - \mu)/\sigma$, then

$$Z_{\mu} = \frac{\mu - \mu}{\sigma} = 0$$

$$Z_{\mu+\sigma} = \frac{\mu + \sigma - \mu}{\sigma} = 1$$

$$Z_{\mu-\sigma} = \frac{\mu - \sigma - \mu}{\sigma} = -1$$

$$Z_{\mu+2\sigma} = \frac{\mu + 2\sigma - \mu}{\sigma} = 2$$

$$Z_{\mu-2\sigma} = \frac{\mu - 2\sigma - \mu}{\sigma} = -2$$

etc. . . .

See Fig. 5.9.

5.9 USING A TABLE FOR COMPUTING AREAS (PROBABILITIES) UNDER THE STANDARD NORMAL (*Z*) CURVE

Figure 5.10 depicts the standard normal (Z) curve. From Table F.4 (Appendix) we can locate a value for a Z_i score from 0 to 3.99. In the extreme left column of Table F.4 we read a value of Z_i from 0.0 to 3.9

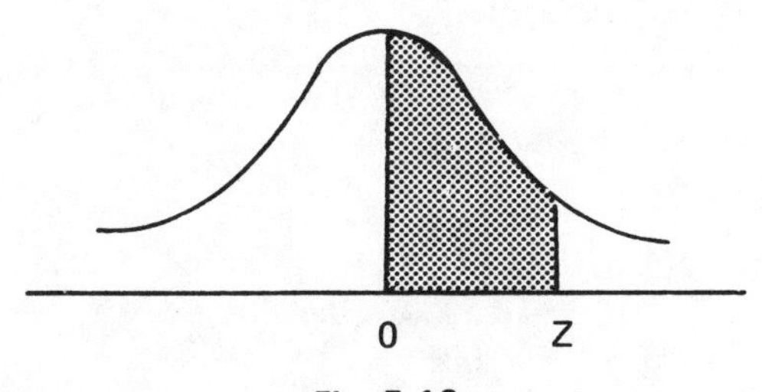

Fig. 5.10

with increments of 0.1. The columns to the right of the Z column allow us to increment Z_i by one-hundredths. When we locate the desired value of Z_i by moving vertically to the tenths' place of Z_i, and horizontally to the hundredths' place of Z_i, this value is the approximate area (probability) under the curve from 0 to Z_i, denoted by $P(0 < Z < Z_i)$. This value also is the approximate area (probability) under the curve from $-Z_i$ to 0, denoted by $p(-Z_i < Z < 0)$.

Sample Problems

Let us consider the following problems in order to become more familiar with the use of Table F.4.

1. What percent of the population in a normal distribution have Z-scores that lie within one standard deviation of the mean? (That is, find $p(-1 < Z < 1.)$

If we locate $Z = 1.00$, we find $p(0 < Z < 1) = 0.3413$; also, $p(-1 < Z < 0) = 0.3413$. Therefore, approximately 68.26% of the population have a Z-score within one standard deviation of the mean.

2. What percent of the population in a normal distribution have Z-scores that lie within two standard deviations of the mean? (That is, find $p(-2 < Z < 2.)$

If we locate $Z = 2.00$, we find $p(0 < Z < 2) = 0.4772$; also, $p(-2 < Z < 0) = 0.4772$. Therefore, 95.44% of the population have a Z-score within two standard deviations of the mean.

3. What percent of the population in a normal distribution have Z-scores that lie within three standard deviations of the mean?

If we locate $Z = 3.00$, we find $p(0 < Z < 3) = 0.4987$. Therefore, 99.74% of the population have a Z-score within three standard deviations of the mean.

Sample problems 1–3 allow us to look more closely at the standard normal curve; see Fig. 5.11.

Note: Only .26% of the population in a normal distribution will have a Z-score that lies beyond three standard deviations of the mean.

4. Find $p(Z > 2.51)$. Refer to Fig. 5.12.

Locate $Z = 2.51$ in Table F.4, and we have

$$p(0 < Z < 2.51) = 0.4940.$$

Therefore,

$$\begin{aligned} p(Z > 2.51) &= p(Z > 0) - p(0 < Z < 2.51) \\ &= 0.5000 - 0.4940 \\ &= 0.006 \end{aligned}$$

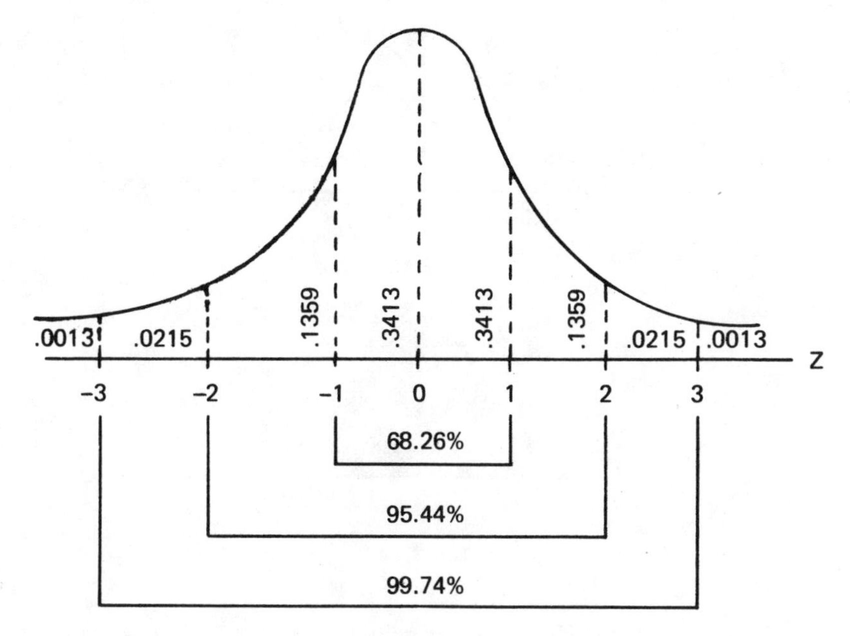

Fig. 5.11

Note: Half of the area is to the left of $Z = 0$ and half of the area is to the right of $Z = 0$.

5. Find $p(Z < -1.07)$. Refer to Fig. 5.13.

$$\begin{aligned} p(Z < -1.07) &= p(Z < 0) - p(-1.07 < Z < 0) \\ &= 0.5000 - 0.3577 \\ &= 0.1423 \end{aligned}$$

6. Find $p(Z > -1.07)$. Refer to Fig. 5.14.

$$\begin{aligned} p(Z > -1.07) &= p(-1.07 < Z < 0) + p(Z > 0) \\ &= 0.3577 + 0.5000 \\ &= 0.8577 \end{aligned}$$

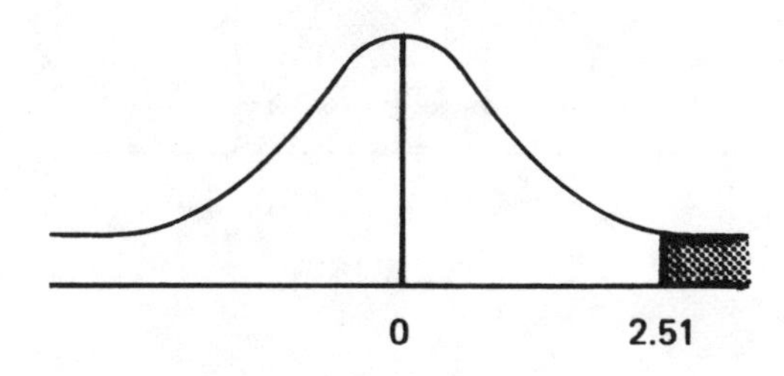

Fig. 5.12

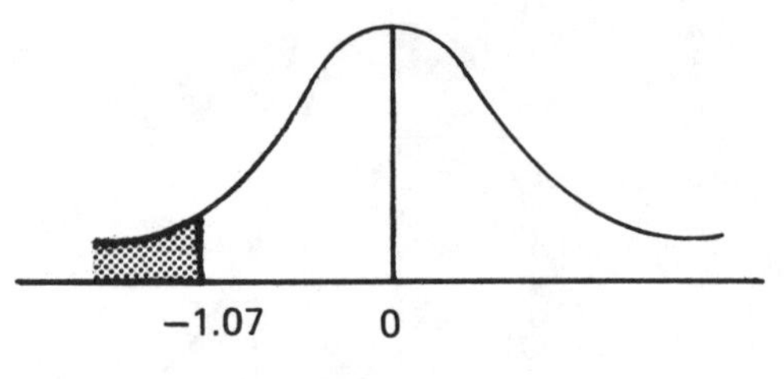

Fig. 5.13

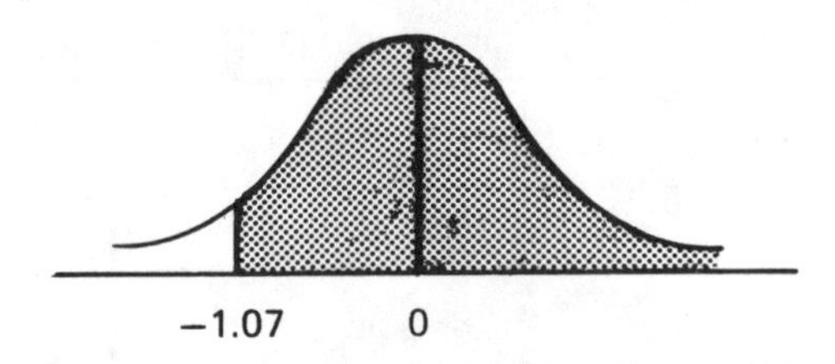

Fig. 5.14

7. If a national IQ test is normally distributed with $\mu = 100$ and $\sigma = 16$, find the probability that a student's raw (X) score will lie between 92 and 112 (that is, find $p(92 < X < 112)$. Refer to Fig. 5.15.

First we must convert our X-scores to Z-scores:

$$Z = \frac{X - \mu}{\sigma}$$

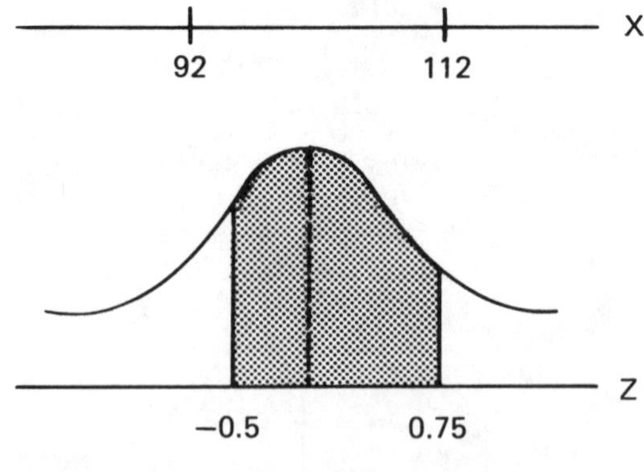

Fig. 5.15

$$Z_{92} = \frac{92 - 100}{16} = \frac{-8}{16} = -\frac{1}{2} = -0.50$$

$$Z_{112} = \frac{112 - 100}{16} = \frac{12}{15} = \frac{3}{4} = 0.75$$

For $p(92 < X < 112)$,

$$\begin{aligned} p(-0.50 < Z < 0.75) &= p(-0.50 < Z < 0) + p(0 < Z < 0.75) \\ &= 0.1915 + 0.2734 \\ &= 0.4649 \end{aligned}$$

that is, 46.49% of the population will have an IQ score between 92 and 112.

8. If scores on a national college entrance exam are approximately normally distributed with $\mu = 500$ and $\sigma = 100$, find the probability that a student's raw (X) score is

(a) Equal to 650 (that is, $p(X = 650)$).

(b) Less than or equal to 577 (that is, $p(X \leq 577)$).

Solution:

(a) By computing Z_{650} and using Table F.4, we could find $p(X > 650)$ or $p(X < 650)$, but we could not find $p(X = 650)$. In order to approximate $p(X) = 650$, we compute Z-scores immediately to the left of Z_{650} and immediately to the right of Z_{650}, and then find the area *between* these two Z-scores. Refer to Fig. 5.16

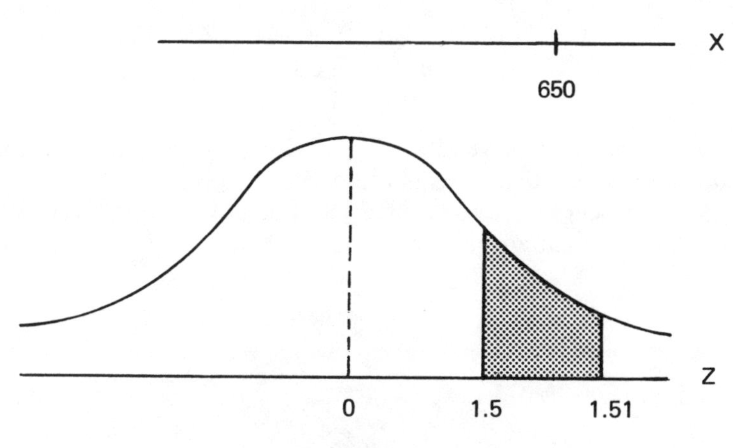

Fig. 5.16

$$Z = \frac{X - \mu}{\sigma}$$

$$Z_{649.5} = \frac{649.5 - 500}{100} = \frac{149.5}{100}$$

$$= 1.495 \approx 1.50$$

$$Z_{650.5} = \frac{650.5 - 500}{100} = \frac{150.5}{100}$$

$$= 1.505 \approx 1.51$$

$$\begin{aligned} p(X = 650) &\approx p(0 < Z < 1.51) - p(0 < Z < 1.50) \\ &= 0.4345 - 0.4332 \\ &= 0.0013 \end{aligned}$$

that is, approximately 0.13% of the population will obtain a college entrance score of 650.

(b) ——+—— X $\quad p(X \leq 577) = ?$
577

In order to include $p(X) = 577$ in our approximation, we will compute

$$Z_{577.5} = \frac{577.5 - 500}{100} = \frac{77.5}{100} = 0.775 \approx 0.78$$

Refer to Fig. 5.17.

$$\begin{aligned} p(X \leq 577) &\approx p(Z < 0.78) \\ &= p(Z < 0) + p(0 < Z < 0.78) \\ &= 0.5000 + 0.2823 \\ &= 0.7823 \end{aligned}$$

that is, approximately 78.23% of the population will have a college entrance score less than or equal to 577.

9. Consider Sample Problem 8, where we were given that a national college entrance exam has $\mu = 500$ and $\sigma = 100$. Suppose we

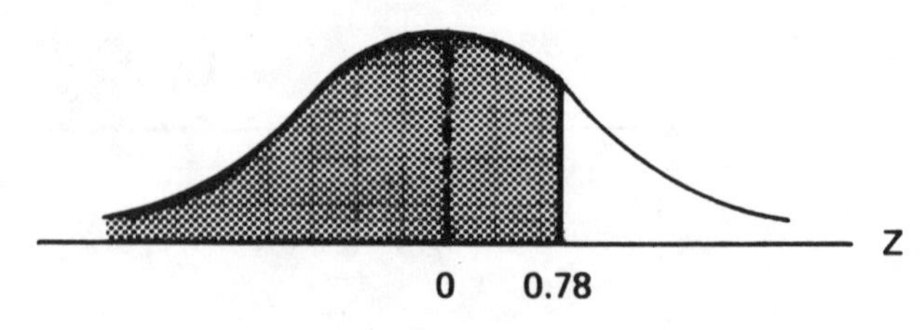

Fig. 5.17

were asked to find a raw score X such that only 10% of the population obtain a score greater than X (that is, find an X such that $p(X_i > X) = 0.1000$).

Solution. We are looking for a raw score X, but we must use our *standard* normal curve (Z-scores) to find it. Note: $Z = (X - \mu)/\sigma$, which implies

$$Z\sigma = X - \mu$$
$$\mu + Z\sigma = X$$

or

$$X = \mu + Z\sigma$$

We know $\mu = 500$ and $\sigma = 100$; now we must find a Z such that $p(Z_i > Z) = 0.1000$. Refer to Fig. 5.18

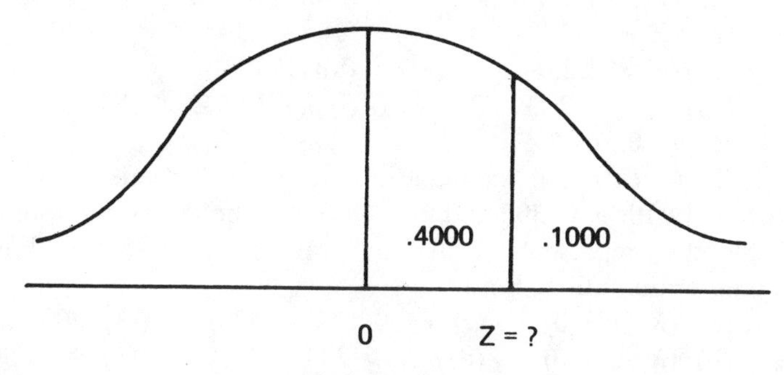

Fig. 5.18

If $p(Z_i > Z) = 0.1000$, then $p(0 < Z_i < Z) = 0.4000$. In Table F.4 we find $Z = 1.28$ such that $p(0 < Z_i < 1.28) = 0.3997$, which we will use as our approximation for the value of Z we are looking for. Refer to Fig. 5.19.

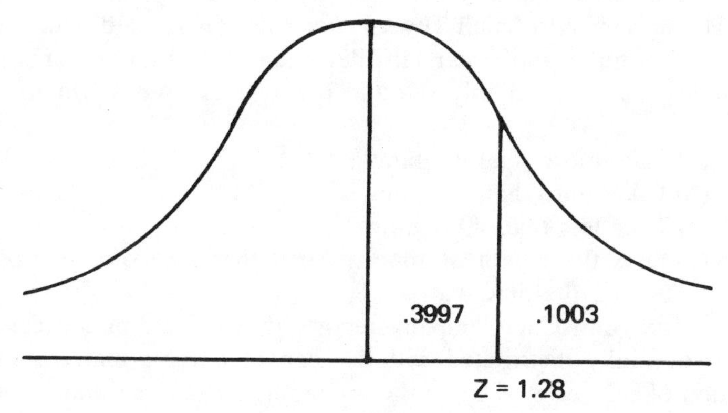

Fig. 5.19

Therefore, if $X = \mu + Z\sigma$, then

$$\begin{aligned} X &= 500 + (1.28)(100) \\ &= 500 + 128 \\ &= 628 \end{aligned}$$

that is, approximately 10% of the population will have a college entrance score greater than 628.

Problem Set 5.3

1. Using Table F.4 of standard normal Z-scores, find

(a) $p(Z > 1.25)$
(b) $p(Z < -2.1)$
(c) $p(Z < 1.25)$
(d) $p(Z > -2.1)$
(e) $p(-0.51 < Z < 0.79)$
(f) $p(1 < Z < 1.5)$
(g) $p(-1.3 < Z < -0.55)$
(h) $p(Z < 0)$
(i) $p(-3 < Z < 3)$
(j) $p(Z > 3)$

2. The entrance examination at State University is normally distributed with $\mu = 300$ and $\sigma = 60$. Let X be the continuous random variable that represents a student's raw score on State University's entrance examination. Find

(a) $p(X > 350)$
(b) $p(X \leq 360)$
(c) $p(180 < X < 420)$
(d) $p(X \geq 200)$
(e) $p(200 \leq X \leq 420)$
(f) $p(X \leq 300)$

3. In Problem 2, where $\mu = 300$ and $\sigma = 60$ for State University's entrance examination, find a student's raw score X such that:

(a) 25% of the population obtains a score greater than X.
(b) 69.5% of the population obtains a score less than X.

Note: If $Z = (X - \mu)/\sigma$, then $X = \mu + Z\sigma$.

4. The Arkansas On Time Bus Company reports that its daily trip from Home Town to Small Town is normally distributed with a mean time of 45 minutes and a standard deviation of 3 minutes. What is the probability that a randomly selected trip from Home Town to Small Town will

(a) Take more than 45 minutes?
(b) Take more than 50 minutes?
(c) Take less than 40 minutes?
(d) Find the length of time X such that 30% of the trips will exceed the time X.

5. The Mini Car Company reports that its mid-priced car's life span is normally distributed with a mean of 6 years and a standard deviation of 1.5 years. What is the probability that a randomly selected mid-priced Mini Car will have a life span

(a) Greater than or equal to 7.5 years?
(b) Between 4 and 7.5 years?

6. The State Track Association reports that the time for the 100-yard dash at its annual track meet is normally distributed with a mean of 11.25 seconds and a standard deviation of 1.05 seconds. What is the probability that a randomly selected participant in the annual State-sponsored 100-yard dash will register a time of:
   (a) Less than 10 seconds?
   (b) More than 12 seconds?
   (c) Between 10 seconds and 12 seconds?
   (d) At least 9.8 seconds?
   (e) Less than 9.5 seconds?

5.10 USING THE NORMAL CURVE TO APPROXIMATE BINOMIAL PROBABILITIES

In Chapter 4 we saw that the probability of r successes in n independent trials of a binomial experiment, where p = the probability of success on any one trial and q = the probability of failure on any one trial, can be computed as follows: ${}_nC_r p^r q^{n-r}$, where $q = 1 - p$. For example, if ten fenders are selected at random from a lot of fenders and tested for defects and the probability of a good fender on any one trial is 0.9, find the probability of obtaining exactly six good fenders.

Exact Probability

If X is the random variable that represents the number of good fenders selected, then

$$\begin{aligned} P(X = 6) &= {}_{10}C_6(0.9)^6(0.1)^4 \\ &= (210)\,(0.531441)\,(0.0001) \\ &\approx 0.0112 \end{aligned}$$

Normal Curve Approximation

Let X be the random variable that represents the number of good fenders selected. We know that the mean and standard deviations of a binomial random variable are

$$\mu = np \qquad \text{and} \qquad \sigma = \sqrt{npq}$$

In the sample problem above,

$$\begin{aligned} \mu &= 10(0.9) \\ &= 9 \end{aligned} \qquad \text{and} \qquad \begin{aligned} \sigma &= \sqrt{10.(0.9)(0.1)} \\ &\approx 0.95 \end{aligned}$$

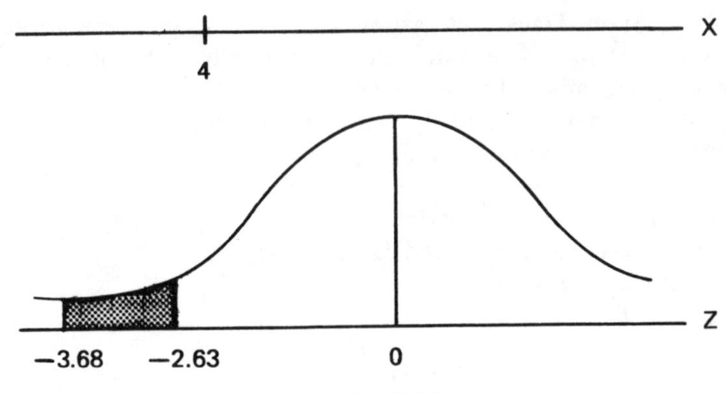

Fig. 5.20

To approximate $p(X = 6)$, using the standard normal curve areas in Table F.4, we compute $Z = X - \mu/\sigma$:

$$Z_{5.5} = \frac{5.5 - 9}{0.95} \qquad \text{and} \qquad Z_{6.5} = \frac{6.5 - 9}{0.95}$$

$$= \frac{-3.5}{0.95} \qquad\qquad = \frac{-2.5}{0.95}$$

$$\approx -3.68 \qquad\qquad \approx -2.63$$

Refer to Fig. 5.20.

Therefore,

$$\begin{aligned} p(X = 6) &= p(-3.68 < Z < 0) - p(-2.63 < Z < 0) \\ &= 0.4999 - 0.4957 \\ &= 0.0042 \end{aligned}$$

Although the normal approximation is close to the exact area, one might question the use of the normal approximation if the exact probability is readily available. However, in a very large sampling of a binomial random variable, the exact probability may be very tedious to compute. Consider the following variation of the sample problem about randomly selecting and testing of fenders.

Sample Problem

Suppose we were to randomly select 40 fenders from a lot of fenders and test them for defects, where $p = 0.9$. If we let X be the random variable that represents the number of good fenders selected, we can find the probability of obtaining exactly 30 good fenders; that is, $p(X = 30)$.

Then

1. The exact probability would be extremely tedious to compute:

$p(X = 30) = {}_{40}C_{30}(0.9)^{30}(0.1)^{10}$

2. The normal approximation of $p(X = 30)$ would be computed as follows:

$$\mu = np = (40)(0.9) = 36 \qquad \sigma = \sqrt{npq} = \sqrt{40(0.9)(0.1)} \approx 1.9$$

$$Z_{29.5} = \frac{29.5 - 36}{1.9} = \frac{-6.5}{1.9} \approx -3.42 \quad \text{and} \quad Z_{30.5} = \frac{30.5 - 36}{1.9} = \frac{-5.5}{1.9} \approx -2.89$$

Refer to Fig. 5.21

Therefore,

$$p(X = 30) \approx p(-3.42 < Z < 0) - p(-2.9 < Z < 0) = 0.4997 - 0.4981 = 0.0016$$

Now let us consider some sample problems of large binomial random sampling for which we will use the normal curve to compute an approximation for the desired binomial probabilities.

Sample Problems

1. In the preceding experiment we randomly selected 40 fenders from a lot of fenders and tested them for defects, where $p = 0.9$. If we let

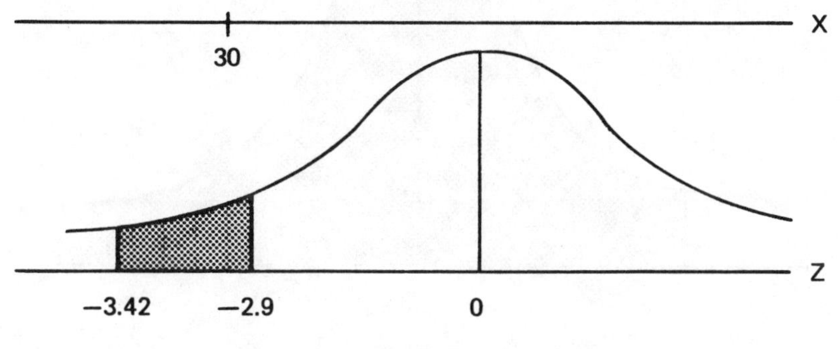

Fig. 5.21

X be the random variable that represents the number of good fenders selected, we can find the normal curve approximation før the probability of

(a) Obtaining at least 30 good fenders; that is, $p(X \geq 30) = p(X = 30) + p(X = 31) + \cdots + p(X = 40)$.

$$\mu = np = 40(0.9) = 36$$
$$\sigma = \sqrt{npq} = \sqrt{40(0.9)(0.1)} \approx 1.9$$
$$Z_{29.5} = \frac{29.5 - 36}{1.9} \approx -3.42$$

Refer to Fig. 5.22.

Therefore,

$$\begin{aligned} p(X \geq 30) &\approx p(-3.42 < Z < 0) + p(Z > 0) \\ &= 0.4997 + 0.5000 \\ &= 0.9997 \end{aligned}$$

(b) Obtaining at most 30 good fenders; that is. $p(X \leq 30) = p(X = 30) + p(X = 29) + \cdots + p(X = 1) + p(X = 0)$.

$$\mu = np = 40(0.9) = 36$$
$$\sigma = \sqrt{npq} = \sqrt{40(0.9)(0.1)} \approx 1.9$$
$$Z_{30.5} = \frac{30.5 - 36}{1.9} \approx -2.89$$

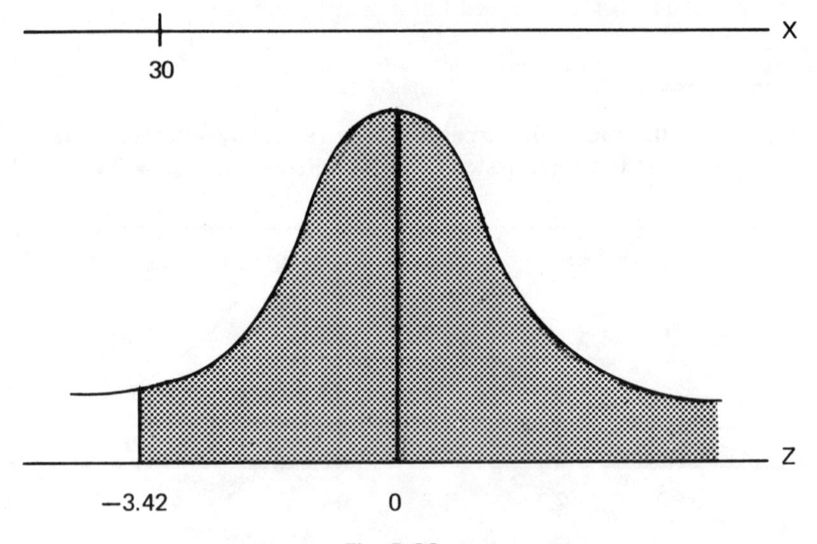

Fig. 5.22

Refer to Fig. 5.23.

Then

$$p(X \le 30) \approx p(Z < 0) - p(-2.89 < Z < 0)$$
$$= 0.5000 - 0.4981$$
$$= 0.0019$$

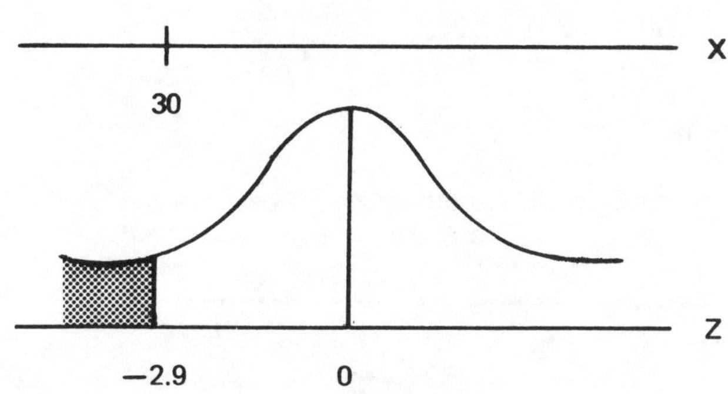

Fig. 5.23

2. The probability that a student enrolling in a remedial mathematics course will pass the course is 0.7 (the only two possible grades are P(pass) and F(fail)). If 700 students enroll in the remedial course, and we let X be the random variable that represents the number of students who fail, find the normal curve approximation to the probability that

(a) Between 200 and 225 students will fail the course (that is, if we let p = the probability of failing = 0.3 and q = the probability of passing = 0.7). Find $p(200 < X < 225)$.

$$\mu = np = 700(0.3) = 210$$
$$\sigma = \sqrt{npq} = \sqrt{700(0.3)(0.7)} = \sqrt{147} \approx 12.124$$

Since the binomial distribution represents a discrete random variable, we must use a *continuity correction* when approximating binomial probabilities via the standard normal distribution.

$$Z_{199.5} = \frac{199.5 - 210}{12.124} = \frac{-10.5}{12.124} \approx -0.87.$$

$$Z_{225.5} = \frac{225.5 - 210}{12.124} = \frac{15.5}{12.124} \approx 1.29$$

Refer to Fig. 5.24

Then,

$$p(200 < X < 225) \approx p(-0.87 < Z < 0) + p(0 < Z < 1.29)$$
$$= 0.3078 + 0.4015$$
$$= 0.7093$$

(b) Less than 230 students will fail the course; that is, find $p(X < 230)$, where $p = 0.3$ and $q = 0.7$.

$$\mu = (700)(0.3) = 210 \qquad \sigma = \sqrt{700(0.3)(0.7)} \approx 12.124.$$

Using the continuity correction, we obtain

$$Z_{230.5} = \frac{230.5 - 210}{12.124} = \frac{20.5}{12.124} \approx 1.69$$

Refer to Fig. 5.25.

Then

$$p(X < 230) \approx p(0 < Z < 1.69) + p(Z < 0)$$
$$= 0.4545 + 0.5000 = 0.9545$$

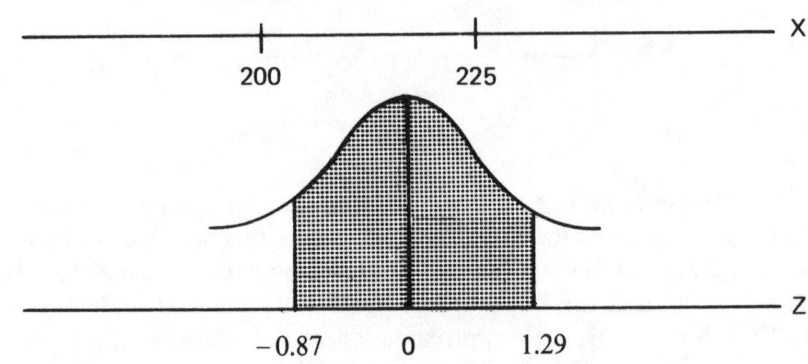

Fig. 5.24

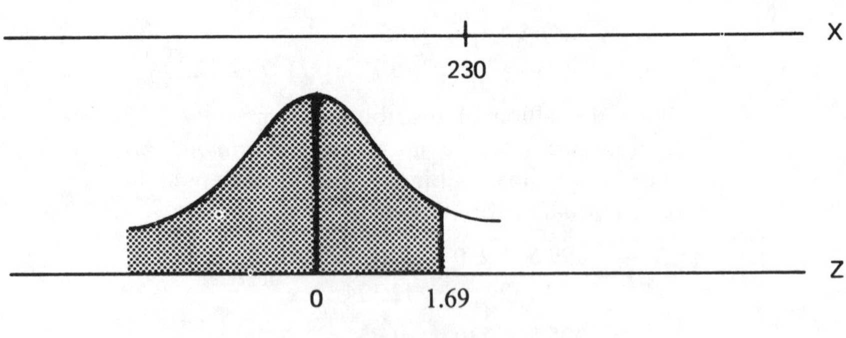

Fig. 5.25

Problem Set 5.4

Use the *standard normal curve* to approximate the desired binomial probabilities in each of the following exercises.

1. The probability that a flu vaccine will be effective on any one person is 0.8. If 200 people are selected at random and injected with the

vaccine, what is the probability that less than 20 people will contract the flu?

2. A certain professional basketball player shoots his foul shots with a 0.6 accuracy. What is the probability that he will make at least 35 of the 40 randomly selected foul shots he attempts?

3. The probability that any one battery produced by the Live Wire Company is rejected is 0.01. If 100 batteries are selected at random and then tested and labeled A(accepted) or R(rejected), what is the probability that at least two batteries are rejected but at most four batteries are rejected (that is, $p(2 \leq X \leq 4) = p(X = 2) + p(X = 3) + p(X = 4)$)?

4. A certain woman professional golfer qualifies for five of ten of the tournaments in which she desires to participate. What is the probability that she will qualify for exactly 35 of the 50 randomly selected tournaments she wishes to enter?

5. The Snap Shot Company advertises that its flashbulbs work with 0.999 accuracy. If its claim is true, what is the probability that at most 4 of 1000 randomly selected flashbulbs are defective?

6. A marksman claims he hits the bull's-eye on six of every ten shots he fires. If his claim is true, what is the probability that he will not hit the bull's-eye on 6 of his next 15 shots?

5.11 A PREPACKAGED PROGRAM FOR COMPUTING THE Z-SCORE/SCORES WHEN USING THE NORMAL CURVE TO APPROXIMATE BINOMIAL PROBABILITIES

Name: STAT5

Description: This program computes the Z-score/scores and tells which area to compute when using the normal curve to approximate binomial probabilities.

Instructions: Enter the data beginning in line 150 as follows:

```
150  DATA N,P,R1, R2, X
```

where

N = number of independent trials in the binomial experiment

P = probability of success on any one trial

R_1 and R_2 = desired number of successes if we have a lower limit and upper limit; otherwise, $R_1 = R_2$

$$X = \begin{cases} 0 \text{ if we want exactly } R_1 = R_2 \text{ successes;} \\ 1 \text{ if we want less than } R_1 = R_2 \text{ successes;} \\ 2 \text{ if we want more than } R_1 = R_2 \text{ successes;} \\ 3 \text{ if we want at most } R_1 = R_2 \text{ successes;} \\ 4 \text{ if we want at least } R_1 = R_2 \text{ successes;} \\ 5 \text{ if we want between } R_1 \text{ and } R_2 \text{ successes;} \\ 6 \text{ if we want between } R_1 \text{ and } R_2 \text{ inclusive successes.} \end{cases}$$

Note: Line 200 is the END statement.

Sample Problem

If we randomly select and test ten fenders from a lot of fenders, where the probability of selecting a good fender on any one try is 0.7, we use the normal curve approximation to find the probability of obtaining one of seven outcomes, as follows:

1. Exactly five good fenders: ($P(X = 5)$).

User's Input

```
GET-STAT5
150 DATA 10,.7,5,5,0
RUN
```

Computer Output

```
FIND AREA BETWEEN Z = -1.72516 AND Z = -1.0351
```

User's Calculations

$$p(-1.73 < Z < -1.04) = 0.4582 - 0.3508 = 0.1074$$

2. Less than five good fenders: ($P(X < 5)$).

User's Input

```
GET-STAT5
150 DATA 10,.7,5,5,1
RUN
```

Computer Output

```
FIND THE AREA TO THE LEFT OF Z = -1.38013
```

User's Calculations

$$p(Z < -1.38) = 0.5000 - 0.4162 = 0.0838$$

3. More than five good fenders: ($P(X > 5)$).

User's Input

```
GET-STAT5
150 DATA 10,.7,5,5,2
RUN
```

Computer Output

```
FIND THE AREA TO THE RIGHT OF Z = -1.38013
```

User's Calculations

$$p(Z > -1.38) = 0.5000 + 0.4162 = 0.9162$$

4. At most five good fenders: ($p(X \leq 5)$).

User's Input

```
GET-STAT5
150 DATA 10,.7,5,5,3
RUN
```

Computer Output

```
FIND THE AREA TO THE LEFT OF Z = -1.0351
```

User's Calculations

$$p(Z < -1.04) = 0.5000 - 0.3508 = 0.1492$$

5. At least five good fenders: ($P(X \geq 5)$).

User's Input

```
GET-STAT5
150 DATA 10,.7,5,5,4
RUN
```

Computer Output

```
FIND THE AREA TO THE RIGHT OF Z = -1.72516
```

User's Calculations

$$p(Z > -1.73) = 0.5000 + 0.4582 = 0.9582$$

6. Between four and nine good fenders: (($P(4 < X < 9)$).

User's Input

```
GET-STAT5
150 DATA 10,.7,4,9,5
RUN
```

Computer Output

FIND AREA BETWEEN Z = −2.0702 AND Z = 1.38013

User's Calculations

$$p(-2.07 < Z < 1.38) = 0.4808 + 0.4162 = 0.8970$$

7. Between four and nine inclusive good fenders: ($P(4 \leq X \leq 9)$).

User's Input

```
GET-STAT5
150 DATA 10,.7,4,9,6
RUN
```

Computer Output

FIND AREA BETWEEN Z = −2.41623 AND Z = 1.72516

User's Calculations

$$p(-2.42 < Z < 1.73) = 0.4922 + 0.4582 = 0.9504$$

Let us look at the flowcharting of STAT5 (Fig. 5.26).
The coding of STAT5 is as follows:

```
5 READ N,P,R1,R2,X
10 LET M=N*P
15 LET D=SQR(N*P*(1-P))
20 IF X=0 THEN 70
25 IF X=1 THEN 85
30 IF X=2 THEN 100
35 IF X=3 THEN115
40 IF X=4 THEN125
45 IF X=5 THEN 135
50 LET Z1=((R1-.5)-M)/D
55 LET Z2=((R2+.5)-M)/D
60 PRINT "FIND AREA BETWEEN Z="Z1;"AND Z="Z2
65 GO TO 5
70 LET Z1=((R1-.5)-M)/D
75 LET Z2=((R1+.5)-M)/D
80 GO TO 60
85 LET Z=(R1-M)/D
90 PRINT "FIND THE AREA TO THE LEFT OF Z="Z
95 GO TO 5
100 LET Z=(R1-M)/D
105 PRINT "FIND THE AREA TO THE RIGHT OF Z="Z
110 GO TO 5
```

```
115 LET Z=((R1+.5)-M)/D
120 GO TO 90
125 LET Z=((R1-.5)-M)/D
130 GO TO 105
135 LET Z1=(R1-M)/D
140 LET Z2=(R2-M)/D
145 GO TO 60
150 DATA
200 END
```

Problem Set 5.5

Test your solutions to each of the six problems in Problem Set 5.4 by using STAT5 to solve them.

5.12 NORMALIZING RAW TEST SCORES VIA A PREPACKAGED COMPUTER PROGRAM

How many times in the past have you or a fellow classmate asked a teacher to "curve" the results of a test? The proper terminology is normalizing test results. To normalize a raw test score X, one must compute a normal score $N = \mu + Z\sigma$, where $Z = (X - \bar{X})/s$, μ = population mean (or an estimate of the population mean), σ = population standard deviation (or an estimate of the population standard deviation).

For example, suppose the final examination grades in Statistics I at New College have yielded a mean $\mu = 75$ and a standard deviation $\sigma = 15$ during the past 20 semesters. Suppose two sections of Statistics I at New College achieve the following results on their final examination during the past semester:

Section I				Section II			
X_i	X_i^2	Z_i	N_i	X_i	X_i^2	Z_i	N_i
50	2500	−0.74	63.9	60	3600	−1.15	57.75
40	1600	−1.38	54.3	95	9025	1.36	95.4
75	5625	0.88	88.2	85	7225	0.65	84.75
82	6724	1.32	94.8	80	6400	0.28	79.2
60	3600	−0.09	73.65	60	3600	−1.15	57.75
$\Sigma X = 307$	$\Sigma X^2 = 20{,}049$			$\Sigma X = 380$	$\Sigma X^2 = 29{,}850$		

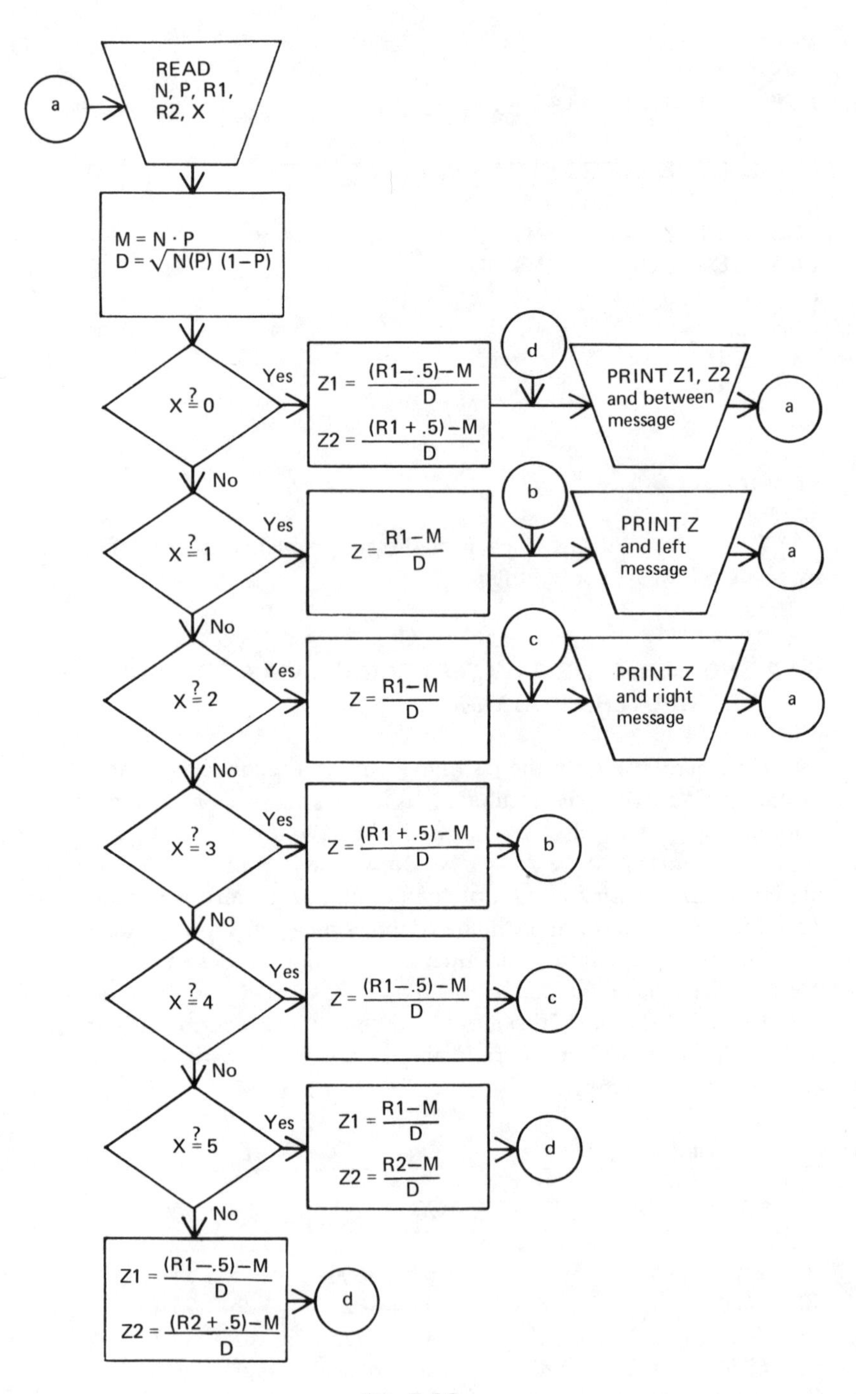
a
READ
N, P, R1,
R2, X
M = N · P
D = √N(P) (1−P)
X ?= 0
Yes
No
Z1 = (R1−.5)−M / D
Z2 = (R1 + .5)−M / D
d
PRINT Z1, Z2
and between
message
a
X ?= 1
Yes
No
Z = R1−M / D
b
PRINT Z
and left
message
a
X ?= 2
Yes
No
Z = R1−M / D
c
PRINT Z
and right
message
a
X ?= 3
Yes
No
Z = (R1 + .5)−M / D
b
X ?= 4
Yes
No
Z = (R1−.5)−M / D
c
X ?= 5
Yes
No
Z1 = R1−M / D
Z2 = R2−M / D
d
Z1 = (R1−.5)−M / D
Z2 = (R2 + .5)−M / D
d

Fig. 5.26

Using these data:

$$\overline{X}_1 = \frac{307}{5} = 61.4 \qquad \overline{X}_2 = \frac{380}{5} = 76$$

$$s_1 = \sqrt{\frac{\Sigma X^2 - (\Sigma X)^2/n}{n}} \qquad s_2 = \sqrt{\frac{29{,}850 - (380)^2/5}{5}}$$

$$= \sqrt{\frac{20{,}049 - (307)^2/5}{5}} \qquad = \sqrt{\frac{29{,}850 - 28{,}880}{5}}$$

$$= \sqrt{\frac{20{,}049 - 18{,}849.8}{5}} \qquad = \sqrt{194}$$

$$= \sqrt{239.84} \qquad s_2 \approx 13.93$$

$$s_1 \approx 15.49$$

Note:

$$Z_i = \frac{X_i - \overline{X}}{s}, \qquad N_i = \mu + Z_i\sigma$$

Now, using the population of 20 semesters of Statistics I, we obtain $N_i = 75 + Z_i(15)$.

For Section I, each normal score is higher than its corresponding raw score, while for Section II only one normal score is higher than its corresponding raw score.

Name: STAT6

Description: This program normalizes a set of n raw scores, where $n \leq 30$.

Instructions: Enter the data beginning in line 150 as follows:

```
150  DATA N,X1, X2, . . ., Xn, U, D
```

where

N = number of raw scores

$X_1, X_2, \ldots, X_n$ = actual raw scores

U = mean or approximated mean of the population

D = standard deviation or approximated standard deviation of the population

Note: Line 200 is the END statement.

Sample Problem

User's Input

```
GET-STAT6

150 DATA 5,50,40,75,82,60,75,15
RUN
```

Computer Output

```
X                    Z                    N
 50                  -.736111              63.9583
 40                  -1.38182              54.2727
 75                   .878168              88.1725
 82                   1.33017              94.9525
 60                  -9.03992E-02          73.644

X                    Z                    N

LINE    10:   END OF DATA

TIME 0.0 SECS.
```

User's Input

```
GET-STAT6

150 DATA 5,60,95,85,80,60,75,15
RUN
```

Computer Output

```
X                    Z                    N
 60                  -1.14873              57.769
 95                   1.36412              95.4618
 85                   .646162              84.6924
```

```
 80                .287183              79.3077
 60               -1.14873              57.769

X                  Z                    N

LINE    10:    END OF DATA

TIME 0.0 SECS.
```

Let us look at the flowcharting of STAT6 (Fig. 5.27).
The coding of STAT6 is as follows:

```
2 DIM X(30), Z(30), N(30)
5 PRINT "X","Z","N"
10 READ N
15 LET S1=0
20 LET S2=0
25 FOR I=1 TO N
30      READ X(I)
35      LET S1=S1+X(I)
40      LET S2=S2+(X(I))↑2
45 NEXT I
50 LET M=S1/N
55 LET S=SQR((S2-S1↑2/N)/N)
60 READ U,D
65 FOR I=1 TO N
70      LET Z(I)=(X(I)-M)/S
75      LET N(I)=U+Z(I)*D
80 PRINT X(I),Z(I),N(I)
85 NEXT I
90 PRINT
95 GO TO 5
150 DATA
200 END
```

Problem Set 5.6

Use STAT6 to normalize the raw-score test results for each of the following two sections of statistics, given population mean = 70 and population standard deviation = 10.

Section I	Section II
X_i	X_i
64	90
76	67
62	85
60	67
65	65
65	75
74	82
70	78
55	84
61	70
78	98
87	74
51	48
30	90
64	61
80	78
84	95
73	70
46	66
58	100
37	70
48	
91	
70	
70	
33	
65	

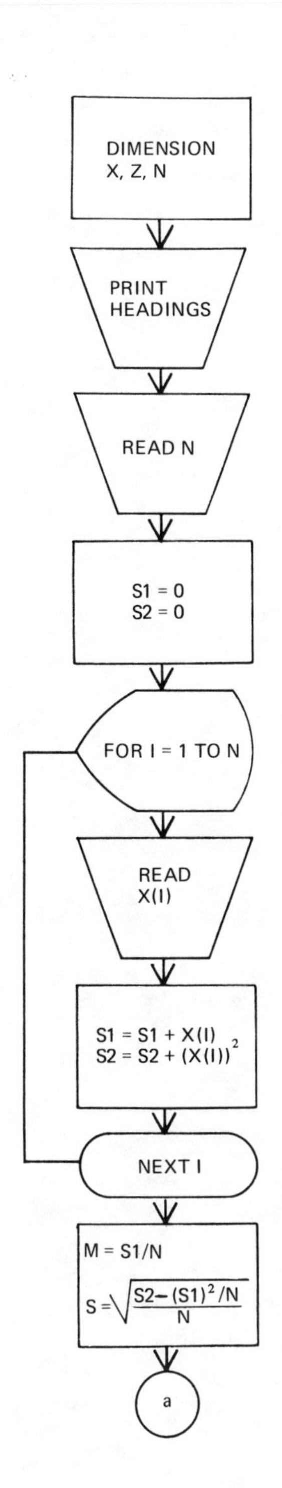
DIMENSION
X, Z, N
PRINT
HEADINGS
READ N
S1 = 0
S2 = 0
FOR I = 1 TO N
READ
X(I)
S1 = S1 + X(I)
S2 = S2 + (X(I))²
NEXT I
M = S1/N
S = √((S2 − (S1)²/N)/N)
a

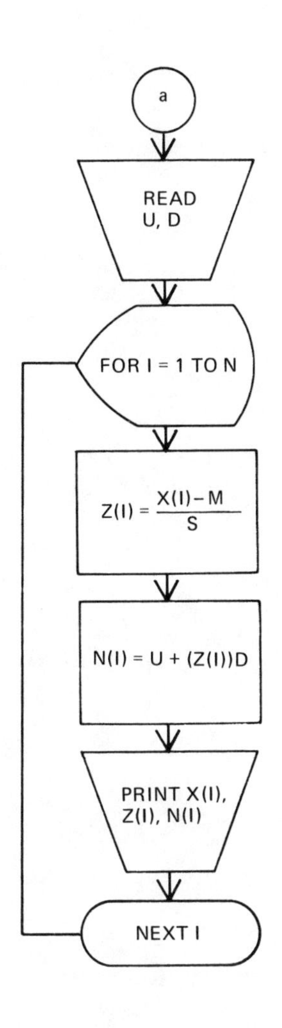
a
READ
U, D
FOR I = 1 TO N
Z(I) = (X(I) − M)/S
N(I) = U + (Z(I))D
PRINT X(I),
Z(I), N(I)
NEXT I

Fig. 5.27

6

Hypothesis Testing and Related Prepackaged Computer Programs

6.1 INFERENTIAL STATISTICS

In Chapter 2 we mentioned that inferential statistics allows us to make predictions or decisions about a population that are based on information obtained from a sample as part of the population. For example, if we wanted to make a prediction about a national election, the population would consist of all people who are registered voters in the United States. We might attempt to make our prediction based on information obtained by random sampling of 1000 registered voters from each state of the union.

If we wanted to decide whether one method of teaching statistics yielded better results than another, it would be obviously impractical to use the total population of all students enrolled in statistics courses, but we could randomly select 200 students enrolled in a statistics course, then randomly place 100 of them in sections teaching statistics by method 1, and place the remaining 100 students in sections teaching statistics by method 2. Then by administering a pretest and post test to each student at the beginning and end of the course, we might be able to decide whether one method yields better results than the other.

In inferential statistics we want to estimate some *measures of the population*, called *parameters*, which are based on *measures contained in a sample*, called statistics. The mean and standard deviation *of a population* will be denoted by the symbols μ and σ, respectively, and the mean and standard deviation of a sample will be denoted by the symbols $\bar{X}$ and s, respectively.

As we discussed earlier, the tools of inferential statistics are descriptive statistics and elementary probability. Most of the time we are interested in the population parameters, such as μ and σ, but we usually must estimate them from the sample statistics, such as $\bar{X}$ and s. It is extremely important that our sample be randomly selected; that is, that every element of the population has an *equal chance* of being included in the sample. Thus the selection of random samples should in some way be left to chance by using, for example, a form of gambling device (dice, slips of paper, etc.) or specially designed tables of *random numbers*.

6.2 THE CENTRAL LIMIT THEOREM AND CONFIDENCE INTERVALS

Suppose we are *given* a population with μ = mean and σ = standard deviation. Then if we *randomly select* a certain number of samples, each of size n, where $n \geq 30$, from the population, and compute the mean = $\bar{X}$ and standard deviation = s *for each* sample, we can construct a frequency distribution of the sample means, and obtain a distribution that, according to the Central Limit Theorem:

1. Is approximately normal.
2. Has a mean $\mu_{\bar{x}} = \mu$ and a standard deviation $\sigma_{\bar{x}} = \sigma/\sqrt{n}$, where n is the size of each sample.

Interval Estimation. Random sample means tend to be normally distributed around the mean of the universe. It can therefore be ascertained from the normal curve table (Table F.4, Appendix) that in the case of a perfectly normal sampling distribution 68.27 percent of the sample means will fall within the one σ (standard deviation or standard error) range, 95.45 percent within the two σ range, and 99.73 percent within the three σ range. This implies that the probability that a given sample mean would fall no further from the universe mean than one, two, or three standard errors is approximately 0.683, 0.955, and 0.997, respectively. In statistical applications, one of the most frequently used ranges is the 95% *confidence interval*, which corresponds to a range of 1.96 standard errors.

If one can be 95 percent confident that a sample mean lies no further from the universe mean than 1.96 standard errors in either direction, then he may have equal confidence that the universe mean is no further away from the sample mean. Accordingly, a universe mean based upon a single given sample mean can be estimated in the form of a confidence interval for $\mu = \bar{X} \pm Z\sigma_{\bar{x}}$ (see Fig. 6.1). In this equation μ connotes the universe or population mean, $\bar{X}$ the sample mean, $\sigma_{\bar{x}}$ (which

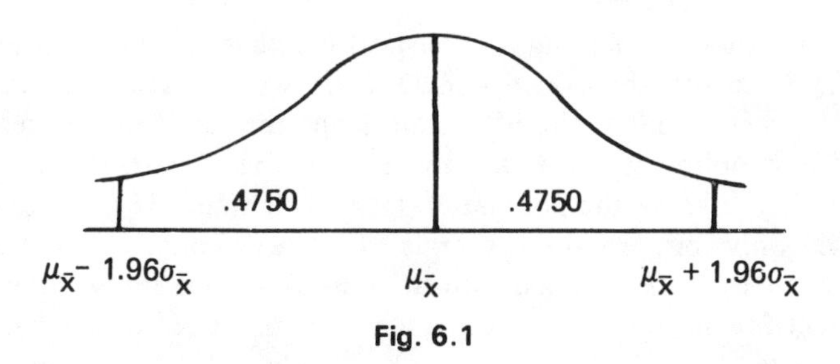

Fig. 6.1

is equal to $\sigma/\sqrt{n}$) the standard deviation of the sample distribution or the standard error of the mean, and Z the number of standard errors.

In the case of a 95 percent confidence interval, for example, Z equals 1.96, as can be ascertained from a normal curve table. When a 95 percent confidence interval is used to estimate the mean of the universe, the *level of confidence* associated with this estimate is also 95 percent. This means that with repeated sampling of a given size n, in the long run 95 out of 100 such confidence intervals actually include the universe mean, while in five cases the universe mean lies outside the estimating range.

Examples

1. A college entrance exam is normally distributed with $\sigma = 100$. A random sample of 36 students is selected from high school seniors who take the college entrance exam. What is the 95% confidence interval for the population mean of the college entrance examination, if $\bar{X}$ is found to be 500?

We know $\bar{X} = 500$, $\sigma = 100$, and $n = 36$; therefore, the 95% confidence interval is

$$\bar{X} - 1.96\sigma_{\bar{x}} \text{ to } \bar{X} + 1.96\sigma_{\bar{x}} \text{ inclusive}$$

or

$$500 - 1.96(100/\sqrt{36}) \text{ to } 500 + 1.96(100/\sqrt{36}) \text{ inclusive}$$

or

$$500 - 32.67 \text{ to } 500 + 32.67 \text{ inclusive}$$

or

$$467.33 \le \mu \le 532.67$$

2. A random sampling of 100 sport cars costing less than $10,000 shows a mean life span of $\bar{X} = 6.32$ years with a standard deviation $s = 2.31$ years. What is the 95% confidence interval for the mean life span of a sampling of sport cars costing less than $10,000?

We do *not* know the population mean and standard deviation, and we have only one random sample. In the equation $\sigma_{\bar{x}} = \sigma/n$ we approximate $\sigma_{\bar{x}}$ by replacing σ with the standard deviation of sample (s). To correct for an inherent bias in such an estimate of σ, $\sqrt{n}$ has been replaced by $\sqrt{n-1}$. Thus, $\hat{\sigma}_{\bar{x}}$ (est.) $= s/\sqrt{n-1}$. Therefore, to approximate the 95% confidence interval, we write

$$\bar{X} = 6.32$$

and compute

$$\hat{\sigma}_{\bar{x}} = \frac{s}{\sqrt{n-1}} = \frac{2.31}{\sqrt{99}} = \frac{2.31}{9.95} = 0.232$$

The 95% confidence interval is then

$$\bar{X} - 1.96\hat{\sigma}_{\bar{x}} \leq \mu \leq \bar{X} + 1.96\hat{\sigma}_{\bar{x}}$$

or

$$6.32 - 1.96(0.232) \leq \mu \leq 6.32 + 1.96(0.232)$$

or

$$5.88 \leq \mu \leq 6.77$$

Now, if we were to choose a certain number of random samples (each of size n) from a given population and construct a frequency distribution of the sample means, with mean $= \mu_{\bar{x}}$ and standard deviation $= \sigma_{\bar{x}}$, we might be interested in finding an interval from $\bar{X} - Z\sigma_{\bar{x}}$ to $\bar{X} + Z\sigma_{\bar{x}}$ inclusive, such that if one of the randomly selected samples has a mean $= \bar{X}$, then $\bar{X} - Z\sigma_{\bar{x}} \leq \mu \leq \bar{X} + Z\sigma_{\bar{x}}$ for 99% of the time (that is, the mean of the population will lie in this interval 99% of the time).

From Table F.4 (Appendix) we can see that $p(-2.576 < Z < +2.576) = 0.9900$. Therefore, in a random sampling of the population if one sample has a mean $= \bar{X}$, then

$$\bar{X} - 2.576\sigma_{\bar{x}} \leq \mu \leq \bar{X} + 2.576\sigma_{\bar{x}}$$

where $\sigma_{\bar{x}} = \sigma/\sqrt{n}$ 99% of the time. Refer to Fig. 6.2. The interval $\bar{X} - 2.576\sigma_{\bar{x}}$ to $\bar{X} + 2.576\sigma_{\bar{x}}$ inclusive is called the *99% confidence interval* for the population mean.

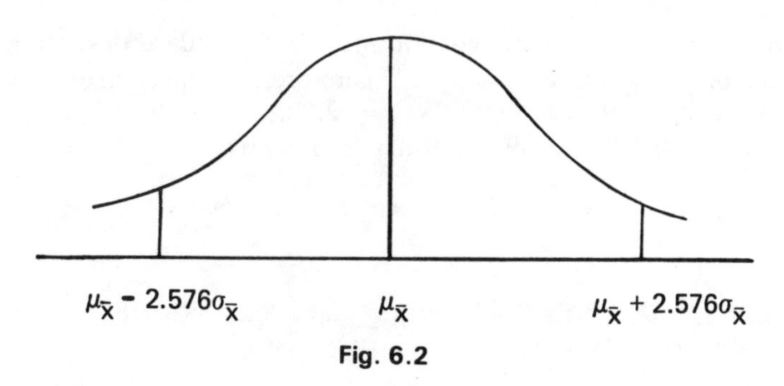

Fig. 6.2

Examples

1. A national intelligence test is normally distributed with $\sigma = 16$. A random sample of 36 students is selected from students completing the intelligence test. What is the 99% confidence interval for the population mean of this national intelligence test, if $\bar{X} = 100$?

We know $\bar{X} = 100$, $\sigma = 16$, and $n = 36$; therefore the 99% confidence interval is

$$\bar{X} - 2.576\sigma_{\bar{x}} \text{ to } \bar{X} + 2.576\sigma_{\bar{x}} \text{ inclusive}$$

or

$$100 - 2.576(16/\sqrt{36}) \text{ to } 100 + 2.576(16/\sqrt{36}) \text{ inclusive}$$

or

$$100 - 6.87 \text{ to } 100 + 6.87 \text{ inclusive}$$

or

$$93.13 \leq \mu \leq 106.87$$

2. A random sampling of 125 one-mile harness horse races at Goshen Race Track had a mean time of 2.02 minutes for the mile race, with a standard deviation of 0.31 minute. What is the 99% confidence interval for the mean time of one-mile horse races at Goshen Race Track?

We do not know the population mean and standard deviation, and we have only one random sample. Therefore, to approximate the 99% confidence interval, we know $\bar{X} = 2.02$, and $\sigma_{\bar{x}} = s/\sqrt{n-1} = 0.31/\sqrt{124} \approx 0.028$. The 99% confidence interval is

$$\bar{X} - 2.576\hat{\sigma}_{\bar{x}} \leq \mu \leq \bar{X} + 2.576\sigma_{\bar{x}}$$

or

$$2.02 - 2.576(0.028) \leq \mu \leq 2.02 + 2.576(0.028)$$

or

$$1.95 \leq \mu \leq 2.09$$

Problem Set 6.1

1. We saw that if any random sample of a population has a mean $= \bar{X}$, then

$$\bar{X} - 1.96\sigma_{\bar{x}} \leq \mu \leq \bar{X} + 1.96\sigma_{\bar{x}}$$

is the 95% confidence interval for the mean of the population, and

$$\bar{X} - 2.575\sigma_{\bar{x}} \leq \mu \leq \bar{X} + 2.575\sigma_{\bar{x}}$$

is the 99% confidence interval for the mean of the population.

(a) Now find the Z-score such that $\bar{X} - Z\sigma_{\bar{x}} \leq \mu \leq \bar{X} + Z\sigma_{\bar{x}}$ is the 90% level of confidence for the mean of the population.

(b) Also find a Z-score such that $\bar{X} - Z\sigma_{\bar{x}} \leq \mu \leq \bar{X} + Z\sigma_{\bar{x}}$ is the 99.9% level of confidence for the mean of the population.

2. The Smooth Riding Tire Company states that the life span of its radial tires are normally distributed with a standard deviation of 4500 miles. What is the 95% level of confidence for the population mean of Smooth Riding radial tires, if $\bar{X} = 38{,}000$ miles and $n = 100$?

3. A random sampling of 300 freshmen at Jersey College had a mean grade-point average of 2.23, with a standard deviation of 0.48. Find the 95% confidence interval for the freshman mean grade-point average at this college.

4. A random sampling of 100 patients at Hope Hospital showed they had a mean amount of glucose (sugar) in their blood of 105.6 mg/100 ml with a standard deviation of 3.8 mg/100 ml. Find the 99%

confidence interval for the mean amount of glucose in the blood of patients at Hope Hospital.

5. A medical report states that the mean age of 100 women who have a normal childbirth delivery is 23.8 years with a standard deviation of 6.1 years. Find the 99% confidence interval for the mean age of women who had normal childbirths.

6. A supervisor for the Monroe Department of Sanitation reports that a random sampling of 100 daily loads of garbage for a three-man crew shows a mean load of 2600 pounds with a standard deviation of 620 pounds. Find the 95% confidence interval for the mean number of pounds of garbage carried daily by a crew of three men.

7. Given a random sample of 100 drawn from a given population, with the sample mean = 70 and a standard deviation = 10, find the following confidence intervals for the population:

(a) 95% confidence interval
(b) 99% confidence interval
(c) 90% confidence interval
(d) 86.12% confidence interval
(e) 98.22% confidence interval
(f) 99.80% confidence interval

6.3 A PREPACKAGED PROGRAM FOR COMPUTING CONFIDENCE INTERVALS FOR THE MEAN OF A POPULATION

Name: STAT7

Description: This program computes confidence intervals for the mean of a population.

Instructions: Enter the data beginning in line 150 as follows:

```
150 DATA M,D,N
```

where

M = the mean of a sample = $\bar{X}$ of size n

D = if $D = \sigma/\sqrt{n}$ or $D = s/\sqrt{n}$, where s = the standard deviation of a sample and n is the size of the sample

N = 1 if we want the 95% confidence interval
2 if we want the 99% confidence interval
3 if we want the 99.9% confidence interval

Note: Line 200 is the END statement.

Sample Problem

User's Input

```
GET-STAT7

150 DATA 2.02,.028,1
151 DATA 2.02,.028,2
152 DATA 2.02,.028,3
RUN
```

Computer Output

```
THE 95 PERCENT CONFIDENCE INTERVAL
WITH M= 2.02      AND SD= 2.80000E-02    IS:
 1.96512      TO 2.07488      INCLUSIVE.

THE 99 PERCENT CONFIDENCE INTERVAL
WITH M= 2.02      AND SD= 2.80000E-02    IS:
 1.94787      TO 2.09213      INCLUSIVE.

THE 99.9 PERCENT CONFIDENCE INTERVAL
WITH M= 2.02      AND SD= 2.80000E-02    IS:
 1.92788      TO 2.11212      INCLUSIVE.


LINE    10:   END OF DATA

TIME 0.0 SECS.
```

Let us look at the flowcharting of STAT7 (Fig. 6.3).
The coding of STAT7 is as follows:

```
5 PRINT
10 READ M,D,N
15 IF N=1 THEN 55
20 IF N=2 THEN 75
25 LET L1=M-3.29*D
30 LET L2=M+3.29*D
```

```
35 PRINT "THE 99.9 PERCENT CONFIDENCE INTERVAL"
40 PRINT "WITH M="M;"AND SD="D;"IS:"
45 PRINT L1; "TO" L2; "INCLUSIVE."
50 GO TO 5
55 LET L1=M-1.96*D
60 LET L2=M+1.96*D
65 PRINT "THE 95 PERCENT CONFIDENCE INTERVAL"
70 GO TO 40
75 LET L1=M-2.576*D
80 LET L2=M+2.576*D
85 PRINT "THE 99 PERCENT CONFIDENCE INTERVAL"
90 GO TO 40
150 DATA
200 END
```

Problem Set 6.2

Use STAT7 to check your manual calculations for Problems 2 through 6 in Problem Set 6.1.

6.4 TESTING STATISTICAL HYPOTHESES: PARAMETERS KNOWN

Scores on national college entrance examinations are normally distributed with $\mu = 500$ and $\sigma = 100$. A random sample of 144 Memorial High School students shows $\overline{X} = 518$ on this entrance examination. You assume that the results of the sampling chosen will indicate that Memorial High School students, on the average, scored higher than those students in the national distribution. Is your assumption or hypothesis valid? Obviously, we must develop a means to *test* your hypothesis.

Motivated Hypothesis and Null Hypothesis

In order to solve the problem, first let us state two mutually exclusive hypotheses: (1) a *null hypothesis* (H_o) which specifies hypothesized values for one or more of the population parameters. A null hypothesis is always in terms of some parameter of the population study. (2) The *motivated* or *alternative* hypothesis (H_m) which asserts that the population parameter is some value other than the one hypothesized. The motivated hypothesis may be *directional* or *non-directional*. When H_m only asserts that the population parameter is different from the one hypothesized in H_o, it is referred to as a non-directional or *two-tail* hypothesis. In our example above, however, H_m is directional or *one-tail*. In this instance, in addition to asserting that the population parameter is different from the one hypothesized, we assert the direction of that difference.

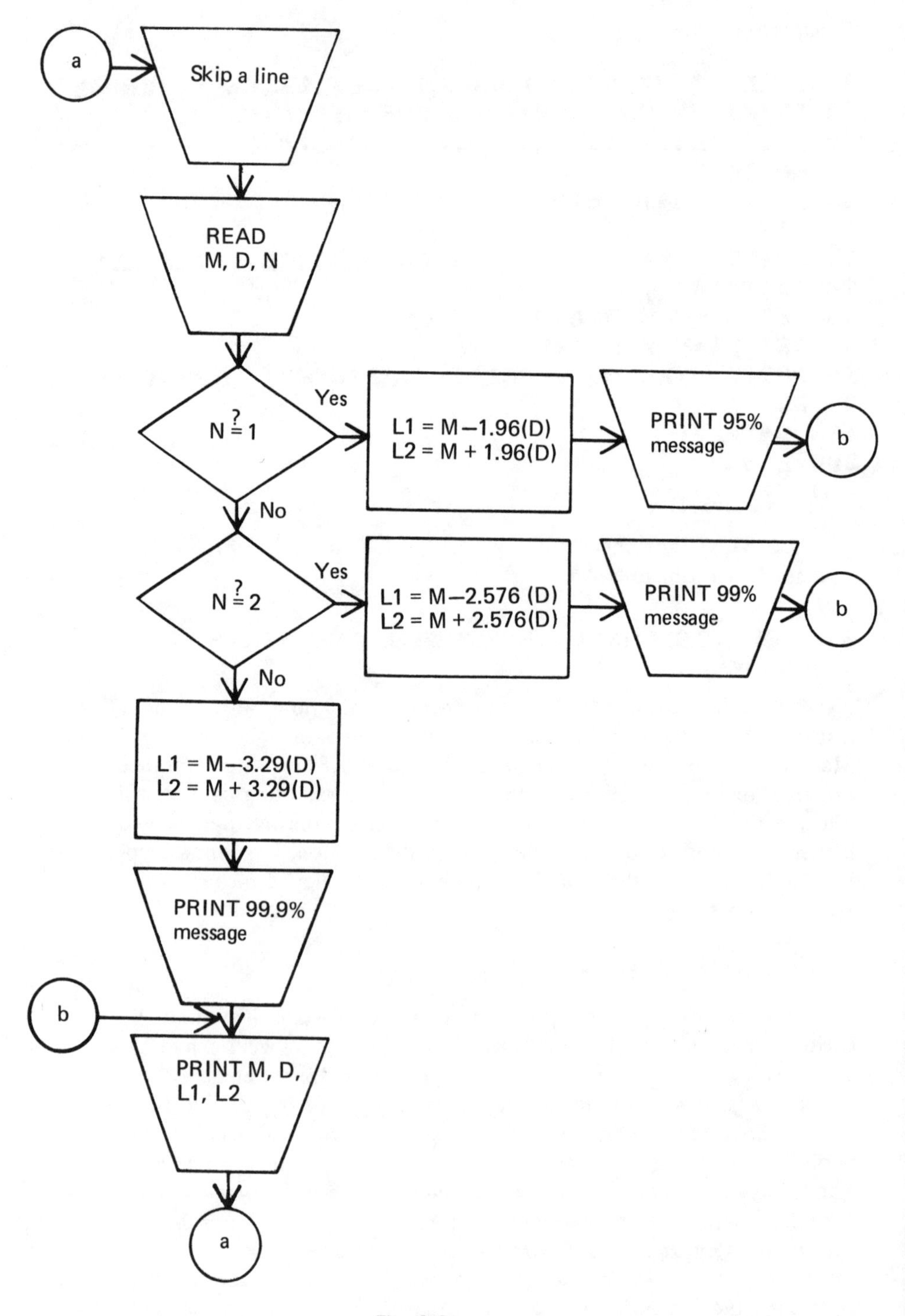
a
Skip a line
READ
M, D, N
N ≟ 1
Yes
L1 = M−1.96(D)
L2 = M + 1.96(D)
PRINT 95%
message
b
No
N ≟ 2
Yes
L1 = M−2.576 (D)
L2 = M + 2.576(D)
PRINT 99%
message
b
No
L1 = M−3.29(D)
L2 = M + 3.29(D)
PRINT 99.9%
message
b
PRINT M, D,
L1, L2
a

Fig. 6.3

In our example, given a sample mean of 518 for $N = 144$, is it reasonable to assume that the sample was drawn from the given population whose mean is 500? Or was it drawn from another population or universe?

Thus we state

(Null hypothesis) H_o: The mean of the population (μ) from which the sample was drawn equals 500; that is, $\mu = \mu_o = 500$.

(Motivated hypothesis) H_m: The mean of the population from which the sample was drawn is greater than 500; that is, $\mu_o > \mu$. Note that H_m is directional; consequently a one-tail test of significance will be employed.

The Observed Z-score

To test our motivated hypothesis, we compute an observed Z-score for the difference between a sample mean and the population mean. The observed Z-score for the difference between the sample's mean and the population's mean is

$$Z = \frac{\bar{X} - \mu}{\sigma/\sqrt{n}}$$

where $n \geq 30$.

In the sampling of 144 Memorial High School students' results on the college entrance examination, $\mu = 500$, $\sigma = 100$, $\bar{X} = 518$, $n = 144$. Therefore,

$$Z = \frac{518 - 500}{100/\sqrt{144}} = \frac{18}{100/12} = 2.16$$

Levels of Significance—Critical Z-values

In order to show whether our assumption that H_o, or the null hypothesis, is true or is warranted, we set a *level of significance*, denoted by an area of the normal curve such that the probability of our observed Z-score falling in this region *due to chance alone* is predetermined.

For example, if we set our level of significance at the 0.05 level while trying to prove that a sample mean is greater than the population mean, we want a critical Z-score such that the probability that our observed Z-score will be greater than Z (due to chance alone) is 0.05. That is, if H_m: $\mu_o > \mu$, then we want a critical Z such that $p(Z_i > Z) = 0.05$. Refer to Fig. 6.4.

Therefore, if we want to show H_m: $\mu_o > \mu$ by assuming H_o: $\mu_o \leq \mu$ is true, our critical $Z = 1.645$. That is, the probability that we will get an observed Z greater than 1.645 (due to chance alone) is 5/100.

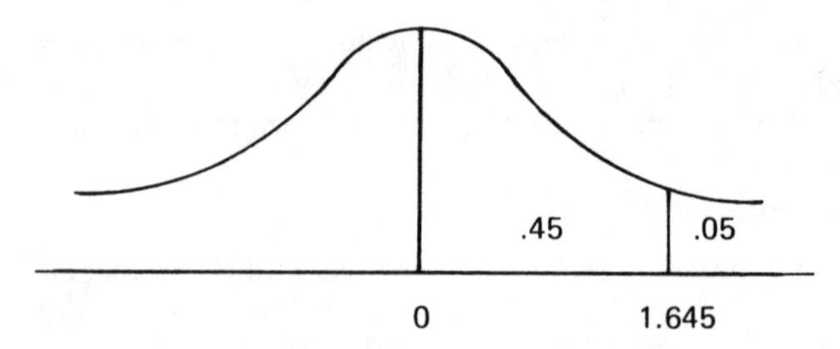

Fig. 6.4. $p(Z_i > 1.645) = 0.05$

For the 144 Memorial High School student results on the college entrance examination, we have

$$\mu = 500, \quad \sigma = 100, \quad \bar{X} = 518, \quad n = 144$$
$$H_m: \mu_o > \mu, \quad H_o: \mu_o = \mu$$

$$\text{Observed } Z = \frac{\bar{X} - \mu}{\sigma/\sqrt{n}} = \frac{518 - 500}{100/\sqrt{144}} = 2.16$$

$$\text{Critical } Z \text{ (0.05 level of significance)} = 1.645$$
$$\text{Observed } Z > \text{critical } Z, \text{ or } 2.16 > 1.645$$

Therefore, we can reject H_o at the 0.05 level of significance.

Note: If we wanted to show $H_m: \mu_o < \mu$ at the 0.05 level of significance, our critical $Z = -.1645$. Refer to Fig. 6.5. If we wanted to show that $H_o: \mu_o = \mu$ or $H_m: \mu_o \neq \mu$ at the 0.05 level of significance, our critical Z-scores would be $Z = \pm 1.96$. Refer to Fig. 6.6.

If we set our level of significance at the 0.01 level while trying to prove that our sample mean is greater than our population mean, we want a critical Z-score such that the probability that our observed Z_i score will be greater than Z (due to chance alone) is 0.01. That is, if H_m: $\mu_o > \mu$, we want a critical Z such that $p(Z_i > Z) = 0.01$. Refer to Fig. 6.7.

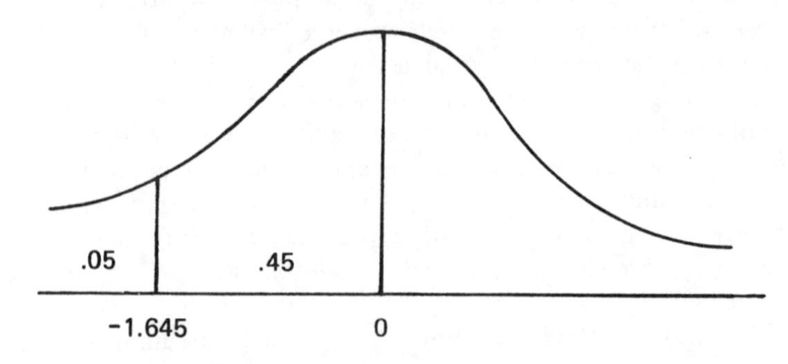

Fig. 6.5. $p(Z_i < -1.645) = 0.5$

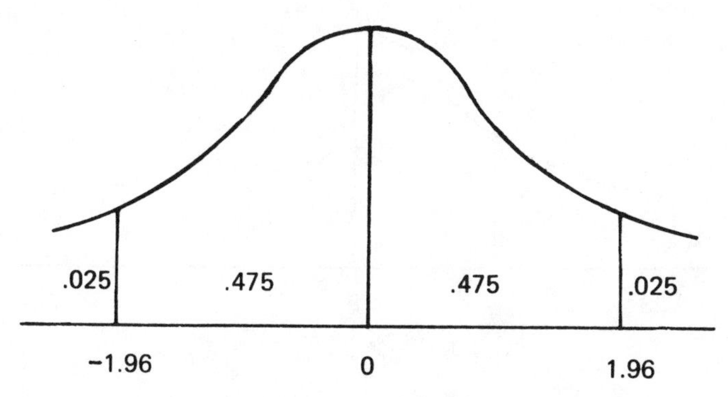

Fig. 6.6. $[p(Z_i < -1.96) \text{ or } p(Z_i > 1.96)] = 0.025 + 0.025 = 0.5$

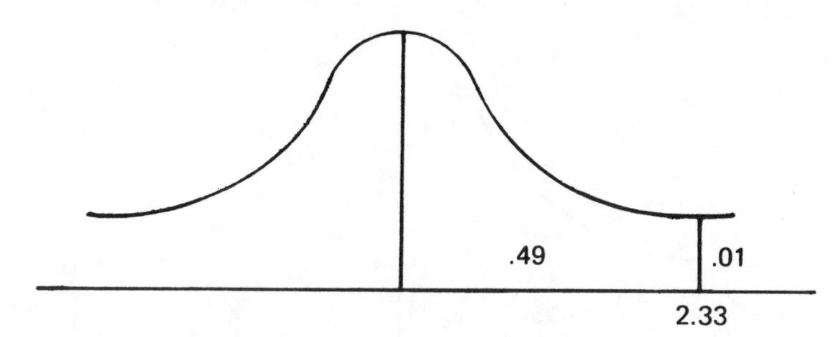

Fig. 6.7. $p(Z_i > 2.33) = 0.01$

Therefore, if we want to show H_m: $\mu_o > \mu$ by assuming that H_o: $\mu_o = \mu$ is true, our critical value $Z = 2.33$. That is, the probability that we will get an observed Z greater than 2.33 (due to chance alone) is 1/100.

For the 144 Memorial High School student results on the college entrance examination, we have

$$\mu = 500, \quad \sigma = 100, \quad \overline{X} = 518, \quad n = 144$$
$$H_m: \mu_o > \mu, \quad H_o: \mu_o = \mu$$
$$\text{Observed } Z = 2.16$$
$$\text{Critical } Z \text{ (0.01 level of significance)} = 2.33$$
$$\text{Observed } Z \not> \text{critical } Z \quad \text{or} \quad 2.16 \not> 2.33$$

Therefore we cannot reject H_o at the 0.01 level of significance. Thus, there is no significant difference in the mean scores of Memorial High students and students in the national distribution on the college entrance examination.

Note: If we wanted to show H_m: $\overline{X} < \mu$ at the 0.01 level of significance, our critical $Z = -2.33$. Refer to Fig. 6.8. If we wanted to

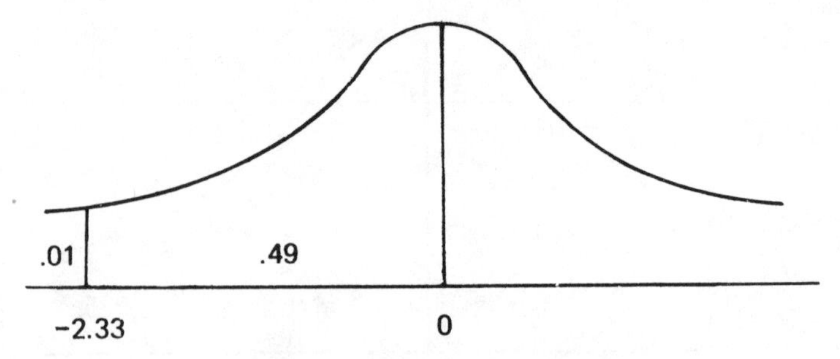

Fig. 6.8. $p(Z_i < -2.33) = 0.01$

show that H_o: $\mu_o = \mu$ or H_m: $\mu_o \neq \mu$ at the 0.01 level of significance, our critical Z-scores would be $Z = \pm 2.58$. Refer to Fig. 6.9.

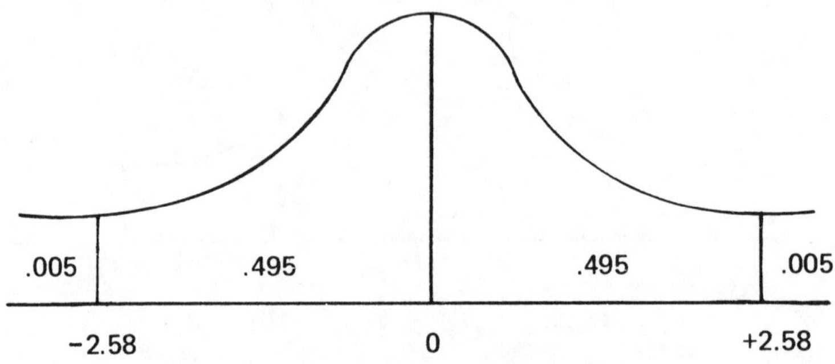

Fig. 6.9. $p(Z_i < -2.58)$ or $p(Z_i > 2.58) = 0.005 + 0.005 = 0.01$

6.5 ONE-TAIL TEST VERSUS TWO-TAIL TEST

Motivated hypotheses such as H_m: $\mu_o > \mu$ or H_m: $\mu_o < \mu$ employ *one-tail tests* because we are testing to see if our observed Z_i-score lies in *one region* of the normal curve. Refer to Fig. 6.10.

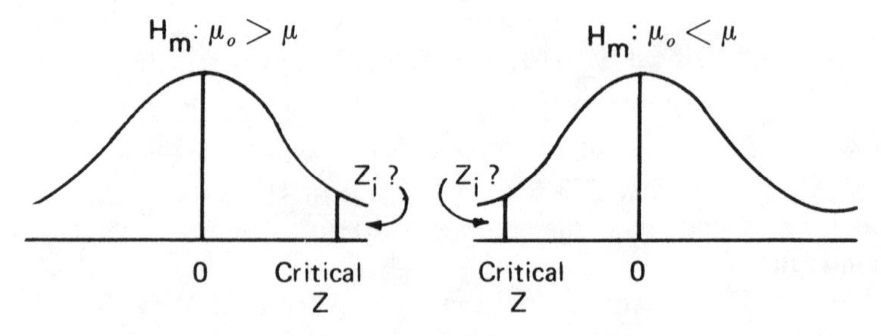

Fig. 6.10

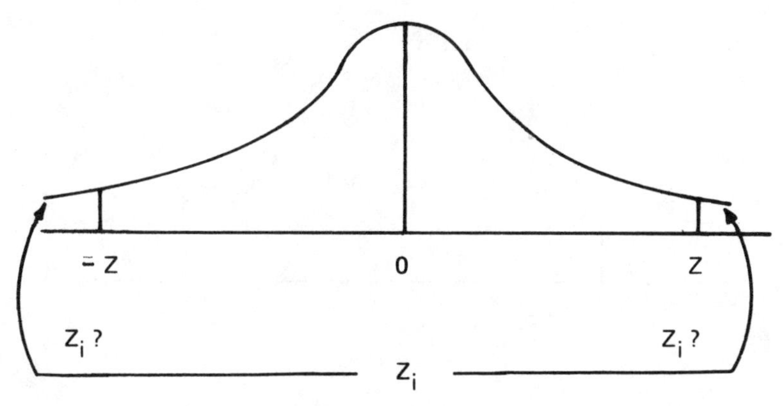

Fig. 6.11

Motivated hypotheses such as H_m: $\mu_o \neq \mu$ employ *two-tail tests* because we are testing to see if our observed Z_i-score lies in *any of two regions* of the normal curve. See Fig. 6.11.

6.6 STATISTICAL HYPOTHESES: PARAMETERS UNKNOWN

In Chapter 2 we learned to compute the *standard deviation* for a sample by using one of the following equivalent formulas:

$$s = \sqrt{\frac{\sum_{i=1}^{n} (X_i - \bar{X})^2}{n}} \quad \text{or} \quad s = \sqrt{\frac{\sum_{i=1}^{n} X_i^2 - \left(\left(\sum_{i=1}^{n} X_i\right)^2\right)/n}{n}}$$

It can be shown that for small samples where n is less than 30, the preceding formulas usually *underestimate* the population's standard deviation. In order to obtain a more accurate estimate of the population's standard deviation, statisticians use the following *unbiased* standard deviation formula to estimate the population's standard deviation:

$$s' = \sqrt{\frac{\sum_{i=1}^{n} (X_i - \bar{X})^2}{n - 1}} \quad \text{or} \quad s' = \sqrt{\frac{n}{n-1}}\,(s)$$

When testing hypotheses for samples of size $n < 30$, we use the *Student t-*, or t-distribution (see Table F.5, Appendix), for our critical *t*-scores. The *t*-distribution takes into account the underestimation of the standard deviation for small samples, and is therefore used instead of the

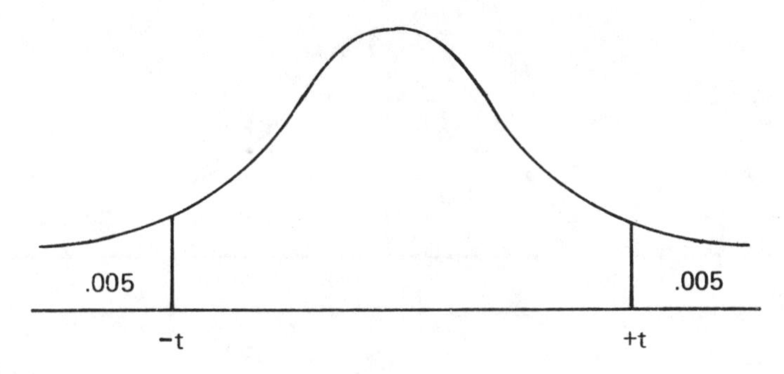

Fig. 6.12

normal distribution when $n < 30$. The df (degrees of freedom) column is part of the correction factor, where df = $n - 1$ and n is the size of the sample.

Therefore, if we wanted to show H_m: $\mu_o > \mu$ at the 0.05 level of confidence where

1. $n = 21$, the df = $n - 1 = 20$, and Table F.5 shows our critical $t = 1.725$.

2. $n = 29$, then df = 29 − 1 = 28, and Table F.5 shows our critical $t = 1.701$.

Note: When df $\geq$ 30, our critical $t = 1.645$, which is equivalent to a critical $Z = 1.645$ at the 0.05 level of significance for a one-tail test.

If we wanted to show $H_0{:}\mu_0 = \mu$ or $H_m{:}\mu_0 \neq \mu$ at the 0.05 level of significance, we would look under the 0.025 level of significance in Table F.5 because we want $[p(t_i < t)$ or $p(t_i > t)] = 0.025 + 0.025 = 0.05$. Therefore, for a two-tail test at the 0.05 level of significance, if

1. $n = 21$, then df = 20 and Table F.5 shows that our critical t-scores would be $t = \pm 2.086$.

2. $n = 29$, then df = 28, and Table F.5 shows that our critical t-scores would be $t = \pm 2.048$.

Note: When df $\geq$ 30 our critical $t = 1.96$, which is equivalent to a critical $Z = 1.96$ at the 0.05 level of significance for a two-tail test. Refer to Fig. 6.12.

When looking for critical t-scores at the 0.01 level for a two-tail test, look under the 0.005 column in Table F.5.

Observed T-score

In testing for differences between a sample mean and a population mean where the size of the sample $n < 30$, then the observed t-score is computed as follows:

$$t = \frac{\bar{X} - \mu}{s'/\sqrt{n}}$$

where $s' = \sqrt{n/(n-1)}(s)$, $n < 30$, df $= n - 1$.

6.7 TESTING HYPOTHESES CONCERNING THE DIFFERENCE BETWEEN A SAMPLE MEAN AND A POPULATION MEAN

The following algorithm (step-by-step procedure) can be used when testing hypotheses concerning a sample mean and a population mean:

1. State the motivated hypothesis (H_m) and the null hypothesis (H_o).
2. Decide whether you are using a one-tail test or a two-tail test.
3. Decide whether to use a critical Z-score ($n \geq 30$) or a critical t-score ($n < 30$).
4. Find the critical Z-score (Table F.5) or the critical t-score (Table F.5) for the desired level of significance.
5. Compute the observed Z-score $= (\bar{X} - \mu)/(\sigma/\sqrt{n})$ or the observed t-score $= (\bar{X} - \mu)/(s'/\sqrt{n})$, where $s' = \sqrt{n/(n-1)}(s)$.
6. State your conclusion (that is, that H_o *can* be rejected or that H_o *cannot* be rejected).

Sample Problems

1. Scores on a national intelligence test are normally distributed with $\mu = 100$. A random sampling of 21 students at Home Town High School reveals that they obtained $\bar{X} = 95$ and $s = 25$ on the intelligence test. Assume that this random sampling indicates that, on the average, the Home Town High students score below the national average on the intelligence test.

Test your motivated hypothesis at the 0.05 level of significance. Given: $\mu = 100$, $\bar{X} = 95$, $s = 25$ (σ not given).

(a) H_o: $\mu = \mu_o = 100$.
H_m: $\mu_o < \mu$.

(b) One-tail test.

(c) Use t-score because $n < 30$.

(d) Find critical t-score, using Table F.5, at 0.05 level of significance and df $= 21 - 1 = 20$. Therefore, critical $t = -1.725$. Refer to Fig. 6.13.

(e) Observed $t_i = \dfrac{\bar{X} - \mu}{s'/\sqrt{n}} = \dfrac{95 - 100}{25.62/\sqrt{21}} = \dfrac{-5}{25.62/4.58} \approx -0.89$

$$\left(s' = \sqrt{\frac{n}{n-1}}(s) = \sqrt{\frac{21}{20}}(25) = 25.62\right)$$

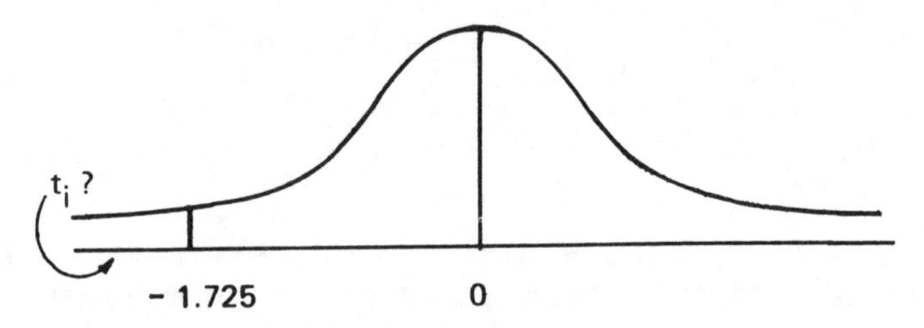

Fig. 6.13

(f) Conclusion:

$$\text{Observed } t_i \not< \text{critical } t$$
$$-0.89 \not< -1.725$$

Therefore, we cannot reject H_o at the 0.05 level of significance. That is, there is no significant difference in the mean scores of the sample and national population.

2. A national medical survey shows that the mean systolic (rhythmic contracted) blood pressure for ten-year-old children is 108.5 mm Hg with a standard deviation of 8 mm Hg. A random sample of 150 ten-year-old students in Busy Town, U.S.A., has a mean blood pressure of 110.4 mm Hg. You assume this sample indicates, on the average, that Busy Town ten-year-olds have blood pressures higher than the national average.

Test your hypothesis at the 0.01 level of significance.

$$\mu = 108.5, \qquad \sigma = 8, \qquad \overline{X} = 110.4, \qquad n = 150.$$

(a) H_m: $\mu_o > \mu$, $\quad H_o$: $\mu_o = \mu$.
(b) One-tail test.
(c) Use Z-score because $n \geq 30$.
(d) Find critical Z at 0.01 level of significance as 2.33. Refer to Fig. 6.14.
(e) Observed $Z_i = \dfrac{\overline{X} - \mu}{\sigma/\sqrt{n}} = \dfrac{110.4 - 108.5}{8.0/\sqrt{150}} = \dfrac{1.9}{8.0/12.25}$

$$\approx 2.91$$

(f) Conclusion:

$$\text{Observed } Z_i > \text{critical } Z$$
$$2.91 > 2.33$$

Therefore, we can reject H_o at the 0.01 level of significance. Hence, the mean blood pressure of Busy Town ten-year-olds is significantly greater than the national population mean.

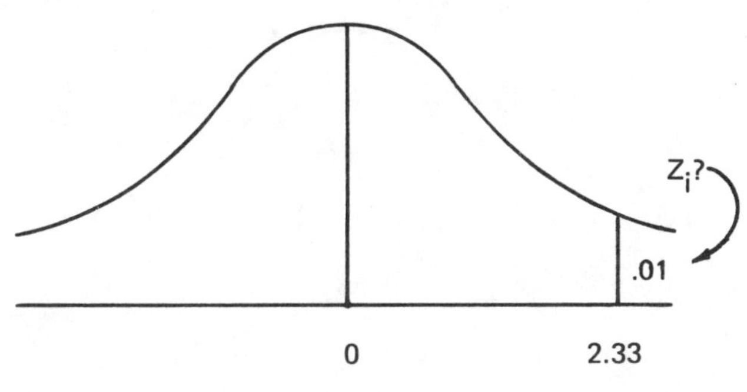

Fig. 6.14

3. A certain automobile manufacturer reports that its most expensive car has a mean life span of seven years. A random sampling of 29 of these expensive cars shows a mean life span of 5.9 years, with a standard deviation of 3.3 years. You suppose that this sample of 29 cars is not representative of the population employed for the manufacturer's claim. Test your hypotheses at the 0.05 level of significance.

$$\mu = 7, \qquad \bar{X} = 5.9, \qquad s = 3.3, \qquad n = 29$$

$$s' = \sqrt{\frac{n}{n-1}}\,(s) = \sqrt{\frac{29}{28}}\,(3.3) \approx 3.36$$

(a) H_m: $\mu_o \neq \mu$, $\qquad H_o$: $\mu_o = \mu$
(b) Two-tail test.
(c) Use t-scores because $n < 30$.
(d) Find critical t-scores at 0.05 level of significance, where df $= 29 - 1 = 28$ (look under 0.025 column in Table F.5); $t = \pm 2.048$. Refer to Fig. 6.15.
(e) Observed $t_i = \dfrac{\bar{X} - \mu}{s'/\sqrt{n}} = \dfrac{5.9 - 7}{3.36/\sqrt{29}} = \dfrac{-1.1}{3.36/5.39}$

$$\approx -1.76$$

(f) Conclusion:

$$\text{Observed } t_i \not< \text{critical } t$$
$$-1.76 \not< -2.048$$

Therefore, we cannot reject H_o at the 0.05 level of significance. Thus, there is no significant difference between the manufacturer's claimed population mean life span in years of the car and the sample mean life span of the car.

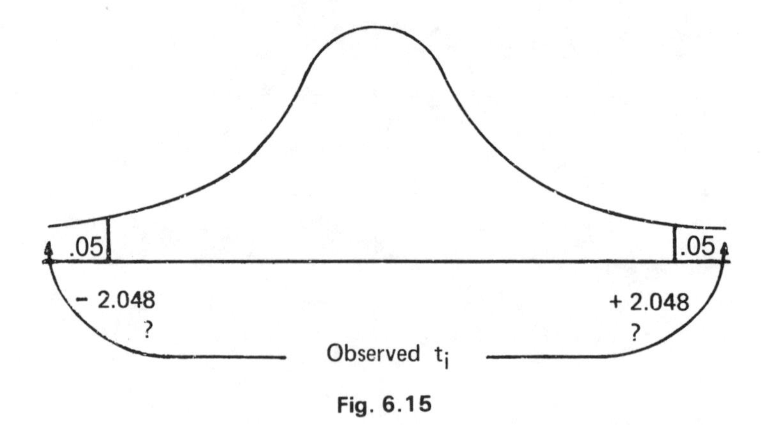

Fig. 6.15

6.8 A PREPACKAGED PROGRAM FOR COMPUTING OBSERVED Z-SCORE OR OBSERVED t-SCORE FOR DIFFERENCE BETWEEN A SAMPLE MEAN AND A POPULATION MEAN

Name: STAT8

Description: This program computes observed Z-score or observed t-score for differences between a sample mean and a population mean.

Instructions: Enter the data beginning in line 150 as follows:

```
150 DATA M1, M2, D, N
```

where

$M1$ = sample mean
$M2$ = population mean
D = population's standard deviation (σ) or sample's standard deviation (s)
N = size of the sample

Note: Line 200 is the END statement.

Sample Problem

User's Input

```
GET-STAT8

150 DATA 110.4, 108.5, 8, 150
151 DATA 5.9, 7, 3.3, 29
RUN
```

Computer Output

```
OBSERVED Z= 2.90876
OBSERVED T=-1.76383

LINE      5:    END OF DATA

TIME 0.0 SECS.
```

Let us look at the flowcharting of STAT8.

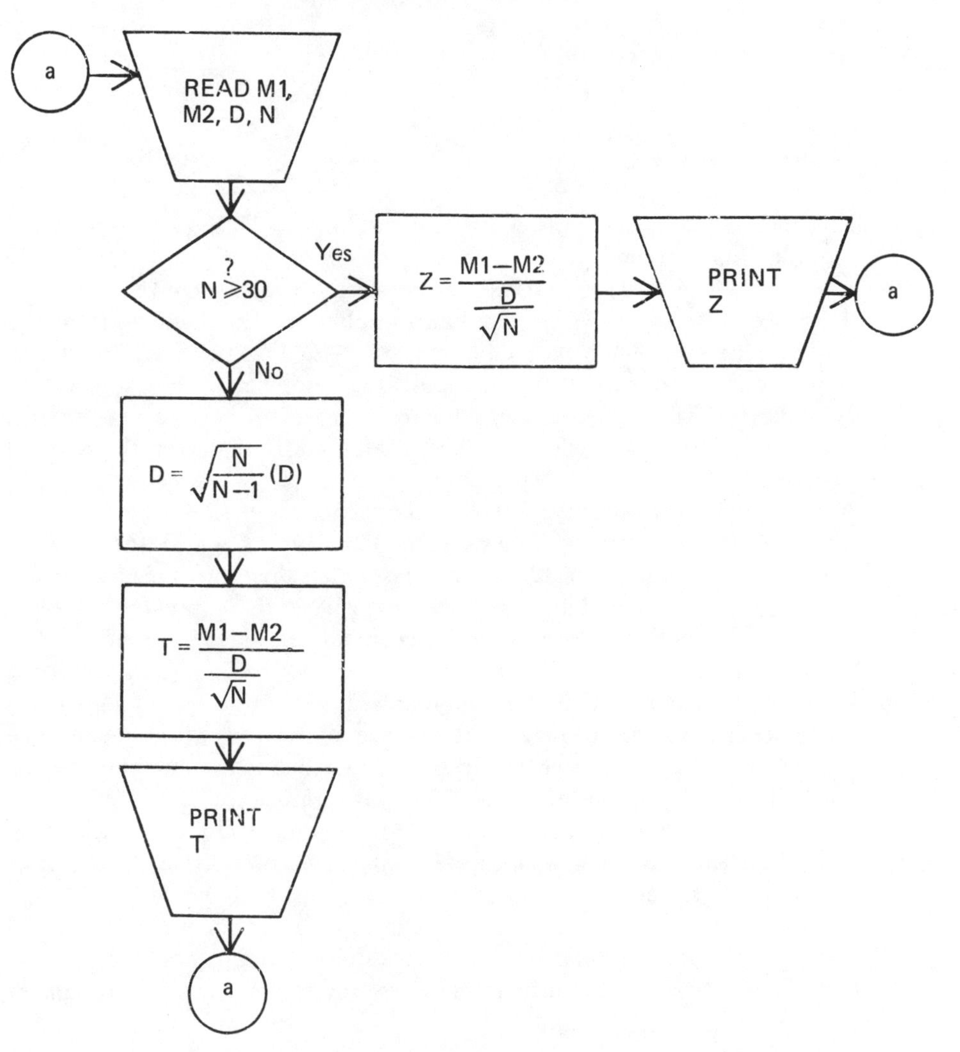

Fig. 6.16. Flowchart 17

The coding of STAT8 is as follows:

```
5 READ M1,M2,D,N
10 IF N>=30 THEN 35
15 LET D=SQR(N/(N-1))*D
20 LET T=(M1-M2)/(D/SQR(N))
25 PRINT "OBSERVED T="T
30 GO TO 5
35 LET Z=(M1-M2)/(D/SQR(N))
40 PRINT "OBSERVED Z="Z
45 GO TO 5
150 DATA
200 END
```

Problem Set 6.3

Use the algorithm in Article 6.7 and/or STAT8 as tools to solve the following problems.

1. The college entrance exam used by State University has a $\mu = 400$ and $\sigma = 100$. A random sampling of 125 students from the northern part of the state determines $\overline{X} = 420$ on the State University entrance examination. You suppose the sampling of 125 students to indicate that, on the average, students from the northern part of the state perform better on the entrance examination than do the general population.
   (a) Test your hypothesis at the 0.05 level of significance.
   (b) Test your hypothesis at the 0.01 level of significance.

2. A national health survey reports that the mean age of a woman who has a normal childbirth delivery is 23.8 years. A random sampling of 25 patients at Sweet Charity Hospital shows the mean age and standard deviation for normal deliveries to be $\overline{X} = 22.4$ years and $s = 7$ years. You assume the information obtained in the sample of 25 patients indicates, on the average, that the age of women who have normal deliveries at Sweet Charity Hospital is less than the national average. Test your hypothesis at the 0.05 level of significance.

3. A certain city reports that a three-man crew in its department of sanitation collects a daily mean of 2600 pounds of garbage. A supervisor randomly selects 28 three-man crews and finds that their daily collection of garbage weighs 2500 pounds, on the average, with a standard deviation of 625 pounds. Assume this random sampling of 28 three-man crews indicates that this sample does not support the department of

sanitation report. Test your hypothesis at the 0.05 level of significance.

4. The Business Training Institute reports that its graduates earn a mean salary of $12,000, with a standard deviation of $3,000. A random sampling of 169 Business Training Institute graduates shows their salaries to have a $\overline{X}$ = $12,300. You hypothesize that, on the average, the mean of the sample of 169 salaries is higher than the mean salary of the general population of Business Training Institute graduates. Test your hypothesis at the 0.10 level of confidence. Refer to Fig. 6.17. (Hint: In a one-tail test of H_m: $\mu_o > \mu$, the critical Z for the 0.10 level of confidence would be $Z = 1.282$.)

6.9 MORE ON CONFIDENCE INTERVALS

If we find a significant difference between a sample mean and population mean when testing a hypothesis at a certain level of confidence, then we should ask ourselves: Why do we have a significant difference? Confidence intervals for the mean of a sample will help us decide *why* we have a significant difference. Confidence intervals are computed from our critical Z- or critical t-values. For example,

Table 6.1 Critical Z and t values

Z	t
$\dfrac{\overline{X} - \mu}{\sigma/\sqrt{n}} = \pm Z$	$\dfrac{\overline{X} - \mu}{s'/\sqrt{n}} = \pm t$
	where $s' = \sqrt{\dfrac{n}{n-1}}(s)$
or $\overline{X} - \mu = \pm Z\left(\dfrac{\sigma}{\sqrt{n}}\right)$	or $\overline{X} - \mu = \pm t\left(\dfrac{s'}{\sqrt{n}}\right)$
or $\mu = \overline{X} \pm Z\left(\dfrac{\sigma}{\sqrt{n}}\right)$	or $\mu = \overline{X} \pm t\left(\dfrac{s'}{\sqrt{n}}\right)$
That is,	That is,
$\overline{X} - Z\left(\dfrac{\sigma}{\sqrt{n}}\right) \leq \mu \leq \overline{X} + Z\left(\dfrac{\sigma}{\sqrt{n}}\right)$	$\overline{X} - t\left(\dfrac{s'}{\sqrt{n}}\right) \leq \mu \leq \overline{X} + t\left(\dfrac{s'}{\sqrt{n}}\right)$

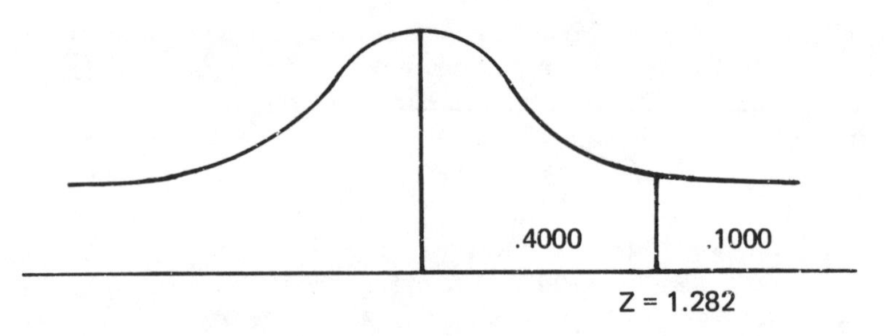

Fig. 6.17

A *large* confidence interval might suggest that we obtained a significant difference because our sample was too small or perhaps our sample was not random. A *small* confidence interval suggests that we obtained a significant difference because we really do have a significant difference between the sample's mean and the population's mean.

Consider the following examples, discussed previously, in which we tested for significant differences between a sample mean and a population mean. Notice that we have now changed the level of significance in these 2 examples.

Examples

1. A national medical survey shows that the mean systolic blood pressure for ten-year-old children is 108.5 mm Hg, with a standard deviation of 8 mm Hg. A random sampling of 150 ten-year-old pupils in Busy Town, U.S.A., has a mean blood pressure of 110.4. Assume this sample indicates that, on the average, Busy Town ten-year-olds have blood pressures that are higher than the national average. Let us test this hypothesis at the 0.05 level of significance..

$$\mu = 108.6, \qquad \sigma = 8, \qquad \overline{X} = 110.4, \qquad n = 150$$

(a) H_m: $\mu_o > \mu$, $\quad H_o$: $\mu_o = \mu$.

(b) One-tail test.

(c) Use Z-score because $n \geq 30$.

(d) Find critical Z at 0.05 level of significance as 1.65.

(e) Observed $Z_i = \dfrac{\overline{X} - \mu}{\sigma/\sqrt{n}} = \dfrac{110.4 - 108.5}{8.0/\sqrt{150}} \approx 2.91$

(f) Conclusion:

Observed $Z_i >$ critical Z

2.91 $>$ 1.65

Therefore, we can reject H_o at the 0.05 level of significance. That is, the mean blood pressure of Busy Town ten-year-olds is significantly greater than the national population mean. Note that previously we rejected the null hypotheses of this problem at the .01 level of significance. Now we have also rejected H_o at the .05 level of significance.

2. A certain automobile manufacturer reports that its most expensive car has a mean life span of seven years. A random sampling of 29 of these expensive cars shows a mean life span of 5.9 years, with a standard deviation of 3.3 years. You hypothesize that this sample of 29 cars is not representative of the population employed for the manufacturer's claim. Test your hypothesis at the 0.01 level of significance.

$$\mu = 7, \qquad \overline{X} = 5.9, \qquad s = 3.3, \qquad n = 29$$

$$s' = \sqrt{\frac{n}{n-1}}(s) = \sqrt{\frac{29}{28}}(3.3) \approx 3.36$$

(a) H_m: $\mu_o \neq \mu$, $\quad H_o$: $\mu_o = \mu$.
(b) Two-tail test.
(c) Use t-scores because $n < 30$.
(d) Find critical t-scores at the 0.01 level of significance, where df $= 29 - 1 = 28$ (see Table F.5); $t = \pm 2.763$.
(e)

$$\text{Observed } t_i = \frac{\overline{X} - \mu}{s'/\sqrt{n}} = \frac{5.9 - 7}{3.36/\sqrt{29}} \approx -1.76$$

(f) Conclusion:

$$\text{Observed } t_i \not< \text{critical } t$$
$$-1.76 \not< -2.763$$

Therefore, we cannot reject H_o at the 0.01 level of significance. That is, there is no significant difference between the manufacturer's claimed population mean life span of the car and the sample mean life span of the car.

Notice that previously we did not reject the null hypothesis of this problem at the .05 level of significance. Now we cannot reject H_o at the 0.01 level of significance.

Problem Set 6.4

Test the hypotheses of exercises 2 and 3 in problem set 6.3 at the .01 level of significance.

6.10 TESTING FOR SIGNIFICANT DIFFERENCES BETWEEN TWO SAMPLE MEANS

Before we begin testing for significant differences between two sample means, let us review the algorithm (step-by-step procedure) for hypothesis testing, which was discussed in Article 6.7.

1. State the motivated hypothesis (H_m) and the null hypothesis (H_o).
2. Decide whether you are using a one-tail test or a two-tail test.
3. Decide whether to use a critical Z-score ($n \geq 30$) or a critical t-score ($n < 30$).
4. Find the critical Z-score (Table F.5, Appendix) or the critical t-score (Table F.5, Appendix) for the desired level of significance.
5. Compute the observed Z_i score or the observed t_i score.
6. State your conclusion (that is, either we can reject H_o or we cannot reject H_o).

The formulas for computing the observed Z-score or observed t-score for the difference between *two sample means* are *not* the same as the formulas used for computing the observed Z-score or observed t-score for the difference between *a sample mean and a population mean.* See Table 6.2.

Table 6.2 Comparative formulas for observed *Z*- and *t*-scores.

Observed Z-, t-scores for differences between a sample mean ($\bar{X}$) and a population mean (μ)	Observed Z-, t-scores for differences between two sample means ($\bar{Y}_1$ and $\bar{Y}_2$)
$Z_{obs} = \dfrac{\bar{X} - \mu}{\sigma/\sqrt{n}}$ where $n \geq 30$.	$Z_{obs} = \dfrac{\bar{Y}_1 - \bar{Y}_2}{\sqrt{(s_1{}^2/n_1) + (s_2{}^2/n_2)}}$ where $n_1 \geq 30$ and $n_2 \geq 30$.
$t_{obs} = \dfrac{\bar{X} - \mu}{s'/\sqrt{n}}$ where $n < 30$, $s' = \sqrt{n/n - 1}\,(s)$, and df $= n - 1$.	$t_{obs} = \dfrac{\bar{Y}_1 - \bar{Y}_2}{\sqrt{((n_1)s_1^2 + (n_2)s_2^2)/(n_1 + n_2 - 2)}(\sqrt{(1/n_1) + (1/n_2)})}$ where $n_1 < 30$ or $n_2 < 30$, df $= n_1 + n_2 - 2$, and $s^2 = \Sigma(Y - \bar{Y})^2/n$.

Examples

1. Five students are taught elementary statistics, using a computer-assisted approach; and five students are taught elementary statistics, using a traditional approach. Both sections are given the same final examination and obtain the following results:

Section Y_1 (computer-assisted): 81, 92, 50, 61, 88
Section Y_2 (traditional): 80, 73, 52, 65, 89

Assume that the computer-assisted approach yields better final examination results than does the traditional approach. Test your hypothesis at the 0.05 level of significance.

Solution. First we must compute some necessary statistics. See Table 6.3.

Table 6.3 Values of Y_1 and Y_2.

Y_1	Y_1^2	Y_2	Y_2^2
81	6561	80	6400
92	8464	73	5329
50	2500	52	2704
61	3721	65	4225
88	7744	89	7921
$\Sigma Y_1 = 372$	$\Sigma Y_1^2 = 28{,}990$	$\Sigma Y_2 = 359$	$\Sigma Y_2^2 = 26{,}579$

Referring to Table 6.2.

$$\overline{Y}_1 = \frac{\Sigma Y_1}{n_1} = \frac{372}{5} = 74.4 \qquad \overline{Y}_2 = \frac{\Sigma Y_2}{n_2} = \frac{359}{5} = 71.8$$

$$s_1^2 = \frac{\Sigma Y_1^2 - ((\Sigma Y_1)^2/n_1)}{n_1} \qquad s_2^2 = \frac{\Sigma Y_2^2 - ((\Sigma Y_2)^2/n_2)}{n_2}$$

where

$$s_1^2 = \frac{28{,}990 - ((372)^2/5)}{5} = 262.64$$

and

$$s_2^2 = \frac{26{,}579 - ((359)^2/5)}{5} = 160.56$$

Now let us use our six-step algorithm (Art. 6.7) for hypothesis testing:

(a) H_m: $\mu_1 > \mu_2$, H_o: $\mu_1 = \mu_2$

(b) One-tail test.

(c) Use critical t-score because the samples are very small ($n_1 = n_2 = 5$, $n_1 < 30$, and $n_2 < 30$).

(d) Compute the critical t-score at the 0.05 level of significance, where df $= n_1 + n_2 - 2 = 5 + 5 - 2 = 8$. Look under the 0.05 column in Table F.5 (Appendix) to find $t = 1.86$.

(e) Observed

$$t_i = \frac{\overline{Y}_1 - \overline{Y}_2}{\sqrt{((n_1)s_1^2 + (n_2)s_2^2)/(n_1 + n_2 - 2)}(\sqrt{(1/n_1) + (1/n_2)})}$$

$$= \frac{74.4 - 71.8}{\sqrt{((5)(262.64) + (5)(160.56))/(5 + 5 - 2)}(\sqrt{(1/5) + (1/5)})}$$

$$= \frac{2.6}{(\sqrt{264.5})(\sqrt{0.4})} \approx \frac{2.6}{(16.26)(0.63)}$$

$$\approx 0.25$$

(f) Conclusion:

$$\begin{array}{lcl} \text{Observed } t_i & \not> & \text{critical } t \\ 0.25 & \not> & 1.860 \end{array}$$

Therefore, we cannot reject H_o at the 0.05 level of significance. That is, there is no significant difference between the mean scores of sections Y_1 and Y_2.

2. A random sampling of 150 ten-year-old students in Busy Town, U.S.A., discloses a mean systolic blood pressure of 110.4 mm Hg with a standard deviation of 8.8 mm Hg. Another random sampling of 174 ten-year-old students in Home Town shows a mean systolic blood pressure of 108.7 mm Hg with a standard deviation of 7.91 mm Hg. You assume these two samples indicate that, on the average, the blood pressures of ten-year-old Home Town students are significantly lower than the blood pressures of ten-year-old Busy Town students. Test your hypothesis at the 0.05 level of significance.

Solution. Let Y_1 represent the Home Town sample and Y_2 represent the Busy Town sample:

$$\begin{array}{ll} \bar{Y}_1 = 108.7 & \bar{Y}_2 = 110.4 \\ s_1 = 7.91 & s_2 = 8.8 \\ n_1 = 174 & n_2 = 150 \end{array}$$

(a) H_m: $\mu_1 < \mu_2$, H_o: $\mu_1 = \mu_2$

(b) One-tail test.

(c) Use critical Z-score because the samples are large ($n_1 = 174$ and $n_2 = 150$, $n_1 \geq 30$ and $n_2 \geq 30$).

(d) Find the critical Z-score at the 0.05 level of significance as -1.645. Refer to Fig. 6.18.

(e) Observed $Z_i = \dfrac{\bar{Y}_1 - \bar{Y}_2}{\sqrt{(s_1^2/n_1) + (s_2^2/n_2)}}$

$$= \frac{108.7 - 110.4}{\sqrt{((7.91)^2)/174) + ((8.8)^2/150)}}$$

$$= \frac{-1.7}{\sqrt{(62.5681)/174 + (77.44/150)}}$$

$$\approx \frac{-1.7}{\sqrt{0.88}}$$

$$\approx \frac{-1.7}{0.94}$$

$$\approx -1.82$$

(f) Conclusion:

$$\begin{array}{ccc} \text{Observed } Z_i & < & \text{critical } Z \\ -1.82 & < & -1.645 \end{array}$$

Therefore, we can reject H_o at the 0.05 level of significance. That is, the mean blood pressure of Home Town ten-year-olds is significantly less than the mean blood pressure of Busy Town ten-year-olds.

3. In Example 2, what would be our conclusion if we had set our significant level at the 0.01 level?

$$H_m: \mu_1 < \mu_2, \qquad H_o: \mu_1 = \mu_2$$
$$\text{Observed } Z_i \approx -1.82$$

However, our critical Z at the 0.01 level of significance is −2.33 (see Fig. 6.19). Thus, our *conclusion* would be that

$$\begin{array}{ccc} \text{Observed } Z_i & \not< & \text{critical } Z \\ -1.82 & \not< & -2.33 \end{array}$$

Therefore, we *cannot* reject H_o at the 0.01 level of significance. That is, the mean blood pressure of Busy Town ten-year-olds is not significantly greater than the mean blood pressure of Home Town ten-year-olds.

4. A random sample of 35 Live Wire 49¢ batteries had a mean lifetime of 230 hours, with a standard deviation of 22 hours. A second random sample of 37 Live Wire 49¢ batteries had a mean lifetime of 250

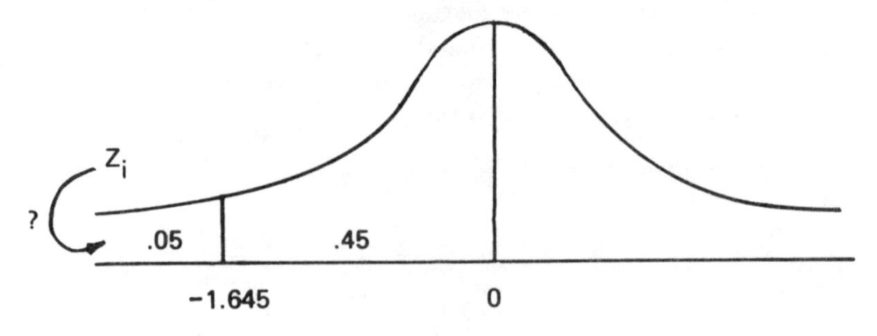

Fig. 6.18

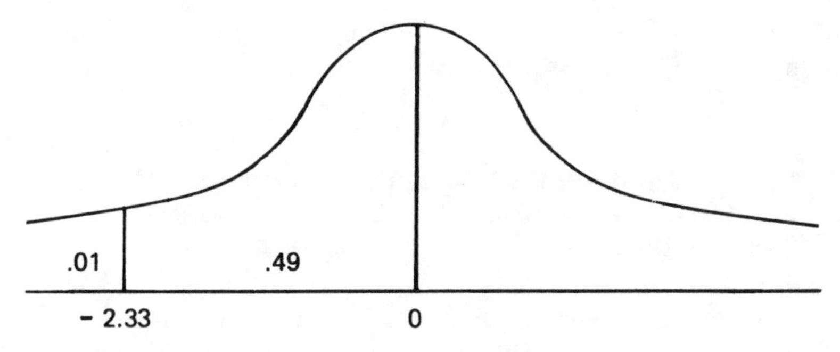

Fig. 6.19

hours, with a standard deviation of 25 hours. Assume a significant difference between these two samples. Test your hypothesis at the 0.01 level of significance, given

$$\bar{Y}_1 = 230 \qquad s_1 = 22 \qquad n_1 = 35$$
$$\bar{Y}_2 = 250 \qquad s_2 = 25 \qquad n_2 = 37$$

Solution

(a) $H_m: \mu_1 \neq \mu_2$, $\quad H_o: \mu_1 = \mu_2$.

(b) Two tail test.

(c) Use critical Z-scores because $n_1 \geq 30$ and $n_2 \geq 30$.

(d) Compute the critical Z-scores at the 0.01 level of significance for a two-tail test as $Z = \pm 2.58$.

(e) Observed $Z_i = \dfrac{\bar{Y}_1 - \bar{Y}_2}{\sqrt{(s_1^2/n_1) + (s_2^2/n_2)}}$

$$= \frac{230 - 250}{\sqrt{((22)^2/35) + ((25)^2/37)}}$$

$$= \frac{-20}{\sqrt{13.828571 + 16.891891}}$$

$$= \frac{-20}{\sqrt{30.720462}}$$

$$\approx \frac{-20}{5.54}$$

$$\approx -3.61$$

(f) Conclusion:

$$\text{Observed } Z_i < \text{critical } Z$$
$$-3.61 < -2.58$$

Therefore, we can reject H_o at the 0.01 level of significance. That is, the mean lifetime of batteries in the first sample is significantly less than the mean lifetime of batteries in the second sample.

6.11 A PREPACKAGED PROGRAM FOR COMPUTING THE OBSERVED Z-SCORE OR OBSERVED t-SCORE FOR DIFFERENCE BETWEEN TWO SAMPLE MEANS

Name: STAT9

Description: This program computes the observed Z-score or observed t-score for differences between two sample means.

Instructions: Enter the data beginning in line 150 as follows:

```
150 DATA M1, M2, V1, V2, N1, N2
```

where

$M1$ = first sample's mean
$M2$ = second sample's mean
$V1$ = first sample's variance
$V2$ = second sample's variance
$N1$ = size of first sample
$N2$ = size of second sample

Note: Line 200 is the END statement.

Sample Problem

User's Input

```
GET-STAT9

150 DATA 108.7,110.4,62.5681,77.44,174,150
151 DATA 74.4,71.8,262.64,160.56,5,5
RUN
```

Computer Output

```
OBSERVED Z=-1.81649
OBSERVED T= .252772

LINE     5:   END OF DATA

TIME 0.0 SECS.
```

We now look at the flowcharting of STAT9 (Fig. 6.20).

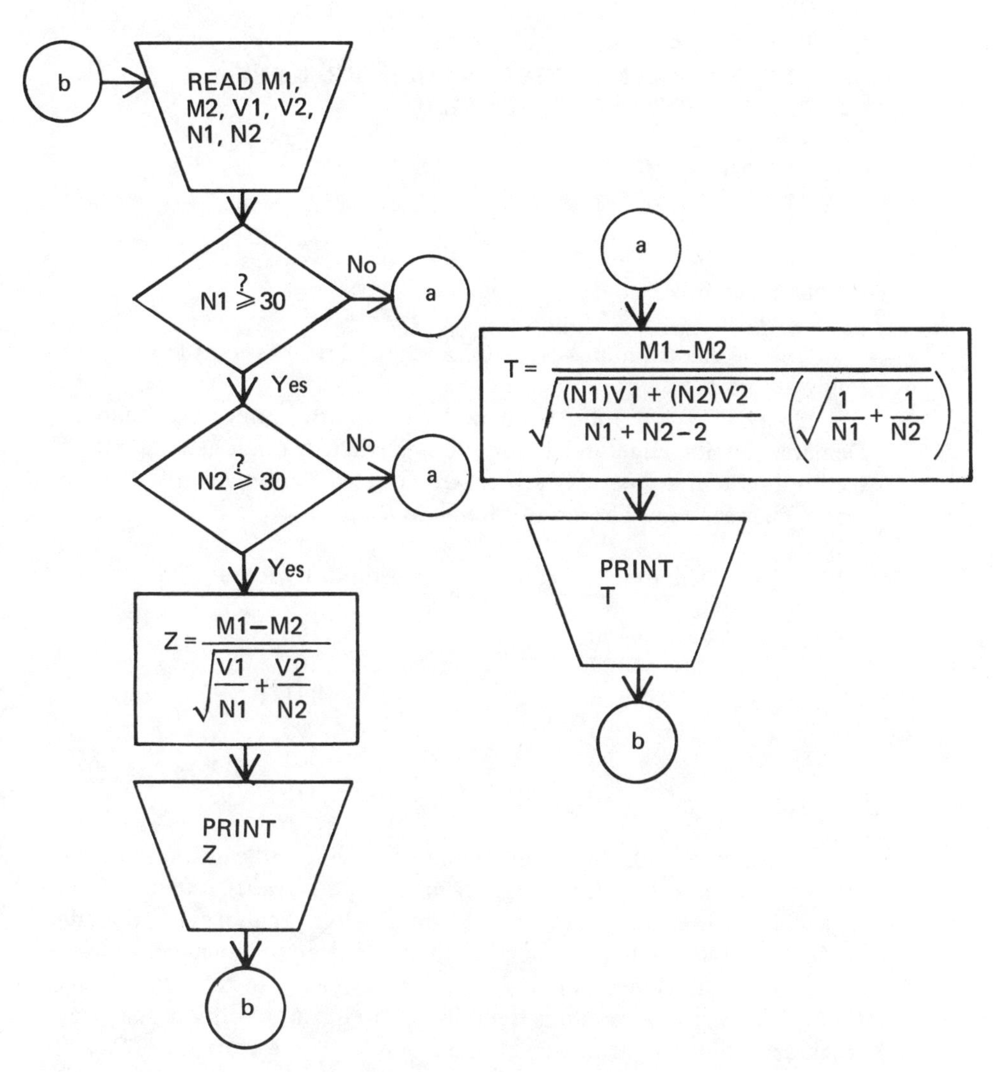

Fig. 6.20. Flowchart 18

The coding of STAT9 is as follows:

```
5 READ M1,M2,V1,V2,N1,N2
10 IF N1>=30 THEN 30
14 LET D=SQR(((N1)*V1+(N2)*V2)/(N1+N2-2))*SQR(1/N1+1/N2)
15 LET T=(M1-M2)/D
20 PRINT "OBSERVED T="T
25 GO TO 5
30 IF N2>=30 THEN 40
```

```
35 GO TO 14
40 LET Z=(M1-M2)/SQR(V1/N1+V2/N2)
45 PRINT "OBSERVED Z="Z
50 GO TO 5
150 DATA
200 END
```

Problem Set 6.5

Use the algorithm in Article 6.7 and/or STAT9 as tools to solve the following problems.

1. The City of Hoboken, New Jersey, has two junior high schools, Demarest Junior High and Brandt Junior High. A random sampling of six ninth-grade student IQ scores are selected from each of the two junior high schools. The results are as follows:

Demarest IQ scores	Brandt IQ scores
100	100
107	105
110	111
95	90
90	92
120	115

You assume that the results of the sampling indicate that, on the average, Demarest Junior High School's ninth-grade students have higher IQ scores than do Brandt Junior High School's ninth-grade students. Test your hypothesis at the 0.05 level of significance.

2. Two machines, A and B, produce 15¢ candy bars for the Cavity Candy Company. A random sampling of 100 candy bars produced by machine A shows a mean weight of 3.2 ounces with a variance of 0.0289 ounce. A random sampling of 90 candy bars produced by machine B determines a mean weight of 3.15 ounces with a variance of 0.0225 ounce. You suppose these samples indicate that, on the average, machine B produces candy bars of less weight than those produced by machine A. Test your hypothesis at the 0.05 level of significance.

3. Test the hypothesis in Problem 2 at the 0.01 level of significance.

4. Given the following information for two random samples:

Sample 1: $\overline{Y}_1 = 50$ $s_1^2 = 100$ $n_1 = 35$
Sample 2: $\overline{Y}_2 = 46.174$ $s_2^2 = 95$ $n_2 = 40$

(a) Test the motivated hypothesis H_m: $\mu_1 \neq \mu_2$ at the 0.05 level of significance.
(b) Test the motivated hypothesis H_m: $\mu_1 \neq \mu_2$ at the 0.01 level of significance.

5. Machines A, B, and C manufacture the same type of pipe for the Wet Plumbing Company. The following table indicates a random sampling of pipes and their mean diameters made by machines A, B, and C.

Machine	Number in Sample	Mean Diameter, Inches	Variance of Diameter
A	35	2.03	0.001
B	40	2.032	0.015
C	35	2.038	0.030

Test the following motivated hypotheses at the 0.05 level of significance:
(a) H_m: $\mu_A < \mu_B$
(b) H_m: $\mu_C > \mu_B$
(c) H_m: $\mu_A \neq \mu_B$
(d) H_m: $\mu_A \neq \mu_C$

Now repeat each of the tests above at the 0.01 level of significance.

6. Given the following information for two random samples:

Sample 1: $\overline{Y}_1 = 75$ $s_1^2 = 100$ $n_1 = 10$
Sample 2: $\overline{Y}_2 = 80$ $s_2^2 = 68$ $n_2 = 8$

Test the motivated hypothesis H_m: $\mu_1 < \mu_2$ at the 0.01 level of significance.

6.12 HYPOTHESIS TESTING OF PROPORTIONS FOR LARGE SAMPLINGS OF A BINOMIAL EXPERIMENT

In Article 5.10 we used the standard normal distribution to approximate binomial probabilities when the number of trials of a binomial experiment was large.

In computing our $Z = (X - \mu)/\sigma$ score when approximating binomial probabilities, you may recall the following: X represents the number of successes in n trials of a binomial experiment, $\mu = np$, $\sigma = \sqrt{np(1 - p)}$, and p is the theoretical probability of success for any one trial of a binomial experiment.

In a large sampling of a binomial experiment, where $n \geq 100$, we may test hypotheses that state there are significant differences between the sample proportion of successes (X/n) and the theoretical proportion of successes (p), where our observed

$$Z = \frac{(X/n) - p}{\sqrt{(p(1-p))/n}} = \frac{X - pn}{\sqrt{p(1-p)n}}$$

Sample Problems

1. A pharmaceutical company claims to have developed a flu vaccine that is more than 85% effective. In order to test the flu vaccine, a random sampling of 500 people are inoculated with the vaccine. Of the 500 people inoculated, 435 did not contract the flu. Test the claim of the pharmaceutical company at the 0.05 level of significance.

Solution $(n = 500, X = 435, p = 0.85)$

(a) $H_m: p > 0.85$
$H_o: p = 0.85$

(b) One-tail test.

(c) Use Z-score because $n = 500 \geq 30$.

(d) Critical Z at the 0.05 level of significance is 1.645 (see Fig. 6.21).

(e) Observed $Z_i = \dfrac{X - pn}{\sqrt{p(1-p)n}}$

$$= \frac{435 - (0.85)(500)}{\sqrt{(0.85)(0.15)(500)}}$$

$$\approx \frac{10}{7.98}$$

or

$Z_i \approx 1.25$

(f) Conclusion:

Observed $Z_i \not> $ critical Z
$1.25 \not> 1.645$

Therefore, we cannot reject H_o at the 0.05 level of significance. Thus, there is no significant difference in the population proportion of

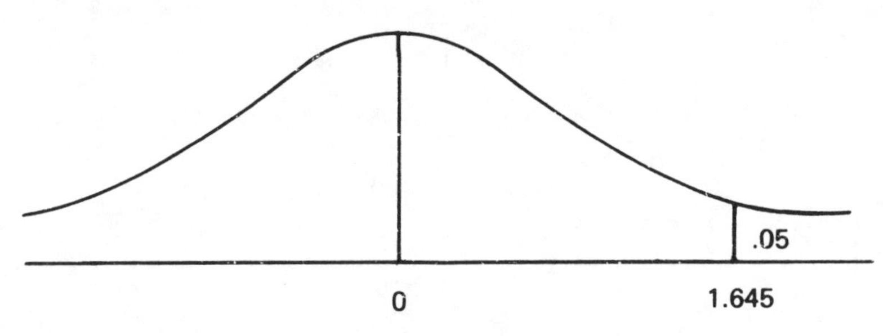

Fig. 6.21

successes as claimed by the company and the sample proportion of successes. Thus, the company's claim is probably false.

2. A candidate for president of the student council at Longfellow College stated that 70% of the student body wanted to change from a two-semester to a three-semester school year. Six hundred fifty students were randomly selected and asked if they would prefer a three-semester schedule to a two-semester schedule. Five hundred of these students said they preferred the three-semester schedule. Test the candidate's claim at the 0.01 level of significance.

Solution ($n = 650$, $X = 500$, $p = 0.7$)

(a) $H_m\colon p \neq 0.7 \qquad H_o\colon p = 0.7$

(b) Two-tail test.

(c) Use Z-score because $n = 650 \geq 30$.

(d) Critical Z-scores at the 0.01 level of significance are ± 2.575 (see Fig. 6.22).

(e) Observed $Z_i = \dfrac{X - pn}{\sqrt{p(1-p)n}} = \dfrac{500 - 0.7(650)}{\sqrt{(0.7)(0.3)(650)}}$

or

$$Z_i = \frac{45}{\sqrt{136.5}} \approx \frac{45}{11.68} \approx 3.85$$

(f) Conclusion:

$$\text{Observed } Z > 2.575$$
$$3.85 > 2.575$$

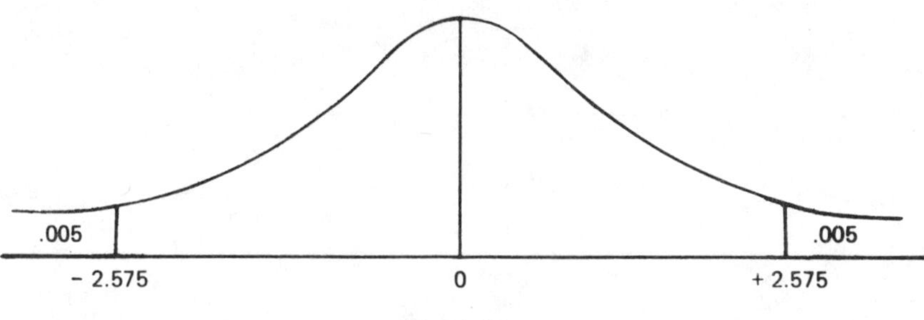

Fig. 6.22

Therefore, we can reject H_o. Thus, there is a significant difference in the proportion of the student body who preferred change as claimed by the candidate and the sample proportion of students who preferred change. The indication is that more than 70% of the population preferred such a change.

6.13 A PREPACKAGED PROGRAM FOR COMPUTING THE OBSERVED Z-SCORE AND 95% CONFIDENCE INTERVAL IN TESTING PROPORTIONS FOR A BINOMIAL EXPERIMENT

Name: BINOM

Description: This program computes the observed Z-score and 95% confidence interval for tests of proportions in a binomial experiment.

Instructions: Enter the data beginning in line 150 as follows:

```
150 DATA N, X, P
```

where

$N \geq 100$ = number of trials of a binomial experiment

X = number of successes in N trials

P = theoretical probability of success on any one trial

Note: Line 200 is the END statement.

Sample Problem

User's Input

```
GET-BINOM

150 DATA 500,435,.85
RUN
```

Computer Output

```
OBSERVED Z= 1.25245
THE 95% CONFIDENCE INTERVAL IS:
 .840522    TO .899478

LINE     5:   END OF DATA

TIME 0.0 SECS.
```

Let us look at the flowcharting of BINOM (Fig. 6.23).
The coding of BINOM is as follows:

```
5 READ N, X, P
10 LET P2 = X/N
15 LET D = SQR (P2*(1-P2)/N)
20 LET Z = (X-P*N)/SQR(P*(1-P)*N)
25 LET L1 = P2 - 1.96*D
30 LET L2 = P2 + 1.96*D
35 PRINT "OBSERVED Z="Z
40 PRINT "THE 95% CONFIDENCE INTERVAL IS:"
45 PRINT L1;"TO";L2
50 GO TO 5
150 DATA
200 END
```

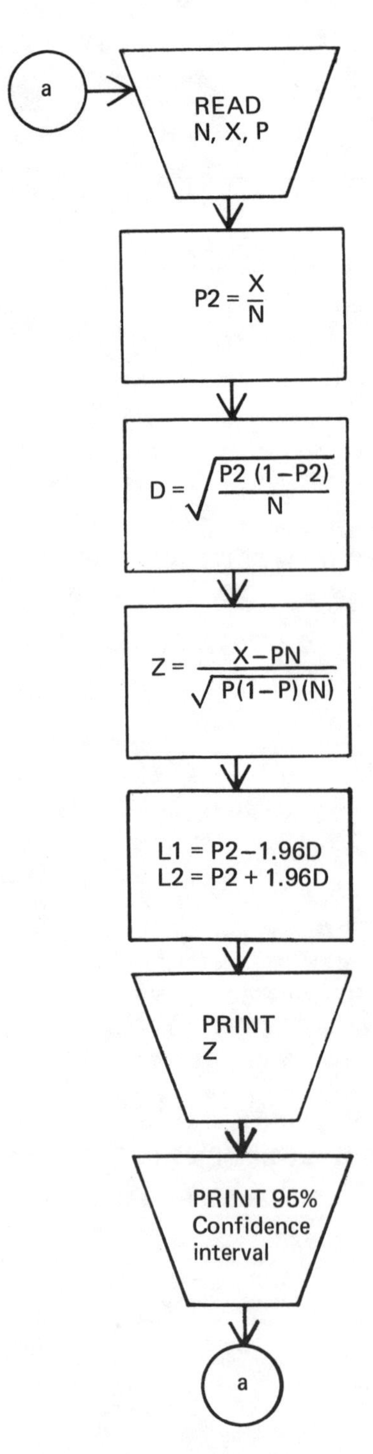
a
READ
N, X, P
P2 = X/N
D = √(P2 (1−P2)/N)
Z = (X−PN)/√(P(1−P)(N))
L1 = P2−1.96D
L2 = P2 + 1.96D
PRINT
Z
PRINT 95%
Confidence
interval
a

Fig. 6.23

Problem Set 6.6

Use the algorithm for hypothesis testing and the prepacked program BINOM as tools to solve the following problems.

1. The Live Wire Company states that no more than 1% of the batteries they produce are defective. Four hundred Live Wire batteries are randomly selected and tested for defects. Five of the batteries tested proved to be defective. Test the Live Wire Company's claim at the 0.05 level of significance.
2. A certain professional basketball player claims he shoots his foul shots with 0.82 accuracy. We randomly select and record the results of 125 of his foul attempts and find that 109 of them are successful. Test his claim at the 0.01 level of significance.
3. A candidate for the United States Senate states that 75% of the voters in his state favor some form of gasoline rationing. Two hundred thirty-five of 320 randomly selected voters in the candidate's state indicate that they do favor some form of gasoline rationing. Test the candidate's claim at the 0.05 level of significance.
4. A foreign automobile manufacturer claims that 65% of the people who own the car he manufactures are college graduates. A random sampling of 135 people who own the manufacturer's car show that 78 are college graduates. Test the manufacturer's claim at the 0.01 level of significance.
5. A local university reports that less than 10% of its students enrolled in the remedial mathematics course fail the course. During the last semester a random sampling of 300 students enrolled in this remedial course revealed that 25 failed the course. Test the university's claim at the 0.01 level of significance.

7

Additional Statistical Techniques and Related Prepackaged Programs

7.1 THE CHI-SQUARE (χ^2) DISTRIBUTION

Chi square (χ^2) is a statistic that allows us to test for differences between observed frequencies in an actual distribution and frequencies in an expected or hypothetical distribution. The chi square statistic is computed as follows:

$$\chi^2 = \sum_{i=1}^{n} \frac{(f_i - e_i)^2}{e_i}$$

where $f_1, f_2 \cdots f_n$ are the frequencies of the actual (sample) distribution, $e_1, e_2, \ldots, e_n$ are the frequencies of the expected distribution, and n is the number of outcomes.

After computing the chi square statistic we may (1) test a chance, motivated, or equal frequency hypothesis, or (2) test some predetermined or *a priori* hypothesis based on previous research or derived from some theory. Another use of chi square is (3) testing for the significance of differences between two or more groups with respect to certain characteristics when the groups have been classified into categories. We may also use chi square (4) to discover whether an observed set of frequencies and those to be expected on the basis of a normal distribution differ significantly. We shall consider here only the uses of chi square for analysis of frequencies in (1) and (2) above. We will compare our observed χ^2–value with the critical χ^2–value given in Table F.6 (Appendix) for the desired level of significance where $df = n - 1$ (df = degrees of freedom).

Sample Problems

1. The voters of New Jersey must vote Yes or No on a referendum for a public transportation bond issue. *The Jersey Journal* conducts a random sampling of New Jersey voters and finds 210 of 500 Hudson County voters, 200 of 460 Essex County voters, and 300 of 520 Bergen County voters favor the bond issue. Test the motivated hypothesis at the 0.01 level of significance that the proportions of voters favoring the bond issue differs among Hudson, Essex, and Bergen counties.

Solution

(a) H_m: All proportions not equal.
H_o: $p_1 = p_2 = p_3$

(b) The expected number of frequencies for an event $E = np$, where n = the number of trials and p = the probability of success on any one trial.
The observed probability of getting a Yes vote is

$$\bar{p} = \frac{210 + 200 + 300}{500 + 460 + 520} = \frac{710}{1480} \approx 0.48,$$

where $\bar{p}$ = the assumed common population proportion.

Therefore, the expected number of Yes votes are:

$$E(\text{Hudson County}) = 500(0.48) = 240$$
$$E(\text{Essex County}) = 460(0.48) \approx 221$$
$$E(\text{Bergen County}) = 520(0.48) \approx 250$$

(c) Now we have the following information:

	Hudson	Essex	Bergen
Observed frequencies	210	200	300
Expected frequencies	240	221	250

(d) Compute the observed χ^2-value.

Observed (f)	Expected (e)	$f - e$	$(f - e)^2$	$(f - e)^2/e$
210	240	−30	900	3.75
200	221	−21	441	2.00
300	250	+50	2500	10.00

$$\text{Observed } \chi^2 = \sum \frac{(f - e)^2}{e} = 3.75 + 2.00 + 10.00$$

$$= 15.75$$

(e) From Table F.6 (Appendix) we can see that the critical χ^2-value at the 0.01 level of significance, where df $= n - 1 = 3 - 1 = 2$, is $\chi^2_{.01} = 9.210$.

(f) Conclusion:

$$\begin{array}{ccc} \text{Observed } \chi^2_{.01} & > & \text{critical } \chi^2_{0.01} \\ 15.75 & > & 9.210 \end{array}$$

Therefore, we can reject H_o at the 0.01 level of significance. Thus, the proportions of voters favoring the bond issue are not the same in the three counties.

Note: We reject H_o *only if* the observed χ^2-value *exceeds* the critical χ^2-value. See Fig. 7.1. Therefore, we accept H_o only if the observed χ^2-value is less than or equal to the critical χ^2-value. The value of

$$\text{Observed } \chi^2 = \sum_{i=1}^{n} \frac{(f_i - e_i)^2}{e_i}$$

will be *large* when the differences between the observed frequencies and expected frequencies are large; and the value of the observed χ^2 will be *small* when the differences between the observed frequencies and the expected frequencies are small.

2. The probability that machine A will produce a defective radio tube on any one trial is 0.001. A random sampling of 5000 radio tubes produced by machine A each day for five consecutive days shows the following number of observed defective radio tubes:

	MON.	TUES.	WED.	THURS.	FRI.
Defective tubes	2	3	8	5	8

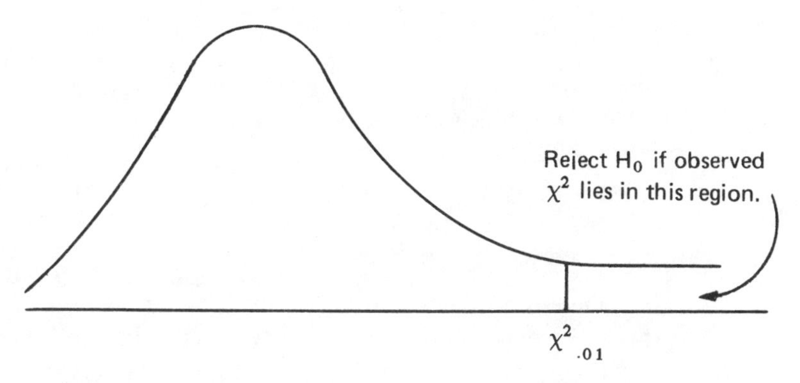

Fig. 7.1. Chi-square distribution for the 0.01 level of significance

Test the motivated hypothesis at the .05 level of significance that the five-day sample of radio tubes differs from the expected ratio for defective tubes.

Solution

(a) $H_m: p \neq 0.001$
$H_o: p = 0.001$

(b) The expected number of defective tubes for each day is $E = (5000)(0.001) = 5.000$

(c) Now we have the following information:

	Mon.	Tues.	Wed.	Thurs.	Fri.
Observed frequencies	2	3	8	5	8
Expected frequencies	5	5	5	5	5

(d) Compute the observed χ^2-value.

Observed (f)	Expected (e)	$f - e$	$(f - e)^2$	$(f - e)^2/e$
2	5	−3	9	1.8
3	5	−2	4	0.8
8	5	3	9	1.8
5	5	0	0	0.0
8	5	3	9	1.8

$$\text{Observed } \chi^2 = \sum \frac{(f - e)^2}{e}$$

$$= 1.8 + 0.8 + 1.8 + 0 + 1.8 = 6.2$$

(e) From Table F.6 we can see that the critical $\chi^2_{.05}$ where df $= 5 - 1 = 4$ is $\chi^2_{.05} = 9.488$.

(f) Conclusion:

Observed $\chi^2_{.05} <$ critical $\chi^2_{.05}$
$6.2 < 9.488$

Therefore, we cannot reject H_o at the 0.05 level of significance. Thus, there is no significant difference between observed and expected frequencies, or no difference in number of defective tubes produced in sample and number produced in previous groups.

3. The State Health Department reports that four of every five school children between the ages of 10 and 12 receive the proper nutrition. A testing company randomly selects pupils from six state schools and has doctors examine them. The doctors report the following information:

	School Number					
	1	2	3	4	5	6
Total children in sample	200	300	400	200	300	400
Number of children with proper nutrition	150	200	350	180	250	290

Test the motivated hypothesis at the 0.05 level of significance that the sample of pupils differs from the Health Department's expected ratio.

Solution

(a) H_m: $p \neq 0.80$
H_o: $p = 0.80$

(b) The expected number of students with proper nutrition for each school is as follows:

$$E(\text{school 1}) = 4/5(200) = 160$$
$$E(\text{school 2}) = 4/5(300) = 240$$
$$E(\text{school 3}) = 4/5(400) = 320$$
$$E(\text{school 4}) = 4/5(200) = 160$$
$$E(\text{school 5}) = 4/5(300) = 240$$
$$E(\text{school 6}) = 4/5(400) = 320$$

(c) Therefore, we have this information:

	School Number					
	1	2	3	4	5	6
Observed frequencies	150	200	350	180	250	290
Expected frequencies	160	240	320	160	240	320

(d) Compute the observed χ^2-value.

Observed (f)	Expected (e)	$(f - e)$	$(f - e)^2$	$(f - e)^2/e$
150	160	−10	100	0.625
200	240	−40	1600	6.667
350	320	+30	900	2.813
180	160	+20	400	2.500
250	240	+10	100	0.417
290	320	−30	900	2.813

$$\text{Observed } \chi^2 = \sum \frac{(f - e)^2}{e}$$

$$= 0.625 + 6.667 + 2.813 + 2.500 + 0.417 + 2.813$$

$$= 15.835$$

(e) From Table F.6 we can see that the critical $\chi^2_{.05}$ where df = 6 − 1 = 5 is $\chi^2_{.05} = 11.070$.

(f) Conclusion:

Observed $\chi^2_{.05}$ > critical $\chi^2_{.05}$
15.835 > 11.070

Therefore, we can reject H_o at the 0.05 level of significance. Thus, there is a significant difference between children receiving proper nutrition in observed sample and children in previous groups.

Problem Set 7.1

1. A survey of residents in New York City found that 150 of 200 Queens County residents, 300 of 600 Manhattan residents, and 200 of 300 Bronx residents attended church services every week. Test the motivated hypothesis that the residents of Queens, Manhattan, and Bronx counties do not have the same church attendance habits at the 0.01 level of significance.

2. The State Health Department reports that six of every ten high school students smoke cigarettes. A random sampling of five state high schools shows the following results:

	School Number				
	1	2	3	4	5
Number in sample	100	100	50	200	300
Number in sample who smoke	50	70	25	130	185

Test the motivated hypothesis at the 0.05 level of significance that the sample of students in five high schools differs from the Health Department's expected ratio.

3. The Fresh Smelling Soap Company reports that one of every three housewives uses its fabric softener. A random sampling of housewives shows the following results:

	Supermarket Number			
	1	2	3	4
Number of housewives in sample	50	30	100	50
Number in sample who use Fresh Smelling softener	20	8	20	10

Test the motivated hypothesis at the 0.05 level of significance that the random sample of housewives differs from the company's claimed ratio.

4. Given the following information:

	Math Course			
	MA10	MA20	MA21	MA22
Number of students in sample	100	100	100	100
Number of students in sample who passed course last year	35	45	70	80
Number of students expected to pass course	50	50	60	70

5. In a sample of 800 women, 500 preferred tea brand A and 300 preferred tea brand B. Test at the 0.05 level of significance whether there is a difference among the women in their preference for brand A. The expected frequencies are 400 for each brand.
   (a) Test the motivated hypothesis at the 0.05 level of significance that the number of students in the sample who passed the course last year differs from the number of students expected to pass the course.
   (b) Test the hypothesis in (a) at the 0.01 level of significance.
6. A recent survey disclosed the following results:

	Oil Company			
	A	B	C	D
Number of supervisors in sample	100	100	100	100
Number of supervisors in sample with a college degree	70	60	85	90

Test the motivated hypothesis that there is a difference between the proportions of supervisors with college degrees at the 0.05 level of significance.

7.2 A PREPACKAGED PROGRAM FOR COMPUTING THE OBSERVED CHI SQUARE (χ^2) VALUE

Name: STAT10

Description: This program computes the observed χ^2- (chi square) value for a given set of observed frequencies.

Instructions: Enter the data beginning in line 150 as follows:

150 DATA N,P,M1,F1,M2,F2, . . ., MN,FN

where

N = number of distinct samples

$M1, M2, \ldots, MN$ = number of observations in each distinct sample

$F1, F2, \ldots, FN$ = number of observed frequencies in each distinct sample

$$P = \begin{cases} \text{value of probability of success on one trial, if known, or} \\ \text{0 if probability of success on one trial is unknown} \end{cases}$$

Note: Line 200 is the END statement.

Sample Problems

User's Input

```
GET-STAT10

150 DATA 3, 0, 500, 210, 460, 200, 520, 300
151 DATA 5, .001, 5000, 2, 5000, 3, 5000, 8, 5000, 5, 5000, 8
RUN
```

Computer Output

```
OBSERVED CHI-SQUARE VALUE = 15.8951
OBSERVED CHI-SQUARE VALUE = 6.2

LINE     5:   END OF DATA

TIME 0.0 SECS.
```

Let us look at the flowcharting of STAT10 (Fig. 7.2).
The coding of STAT10 is as follows:

```
2 DIM M(30),F(30),E(30)
5 READ N,P
10 FOR I=1 TO N
15    READ M(I),F(I)
20 NEXT I
25 IF P=0 THEN 75
30 FOR I=1 TO N
35    LET E(I)=M(I)*P
40 NEXT I
45 LET S3=0
50 FOR I=1 TO N
55    LET S3=S3+(F(I)-E(I))↑2/E(I)
60 NEXT I
```

```
65 PRINT "OBSERVED CHI-SQUARE VALUE ="S3
70 GO TO 5
75 LET S1=0
80 LET S2=0
85 FOR I=1 TO N
90     LET S1=S1+F(I)
95     LET S2=S2+M(I)
100 NEXT I
105 LET P=S1/S2
110 GO TO 30
150 DATA
200 END
```

Problem Set 7.2

1. Check your manual computations of critical chi-square values for each of the six exercises in Problem Set 7.1.
2. Given the following data:

	1	2	3	4	5	6	7
Observed frequencies	50	125	90	215	201	203	190
Expected frequencies	100	100	100	200	200	200	200

(a) Use STAT10 to test the motivated hypothesis that the observed sample does not differ from the expected sample at the 0.05 level of significance.
(b) Repeat part (a) at the 0.01 level of significance.

7.3 INTRODUCTION TO LINEAR REGRESSION

Regression is a statistical technique that allows for predicting the value of one variable, given the value of another variable, if a relationship between the two variables is known.

Examples:

(a) Given a high school student's cumulative average, we might want to predict his/her college grade-point average (GPA).
(b) Given a person's pulse rate in beats per minute, we might want to predict his/her height.
(c) Given the number of years a city employee attended college, we might want to predict his/her salary.

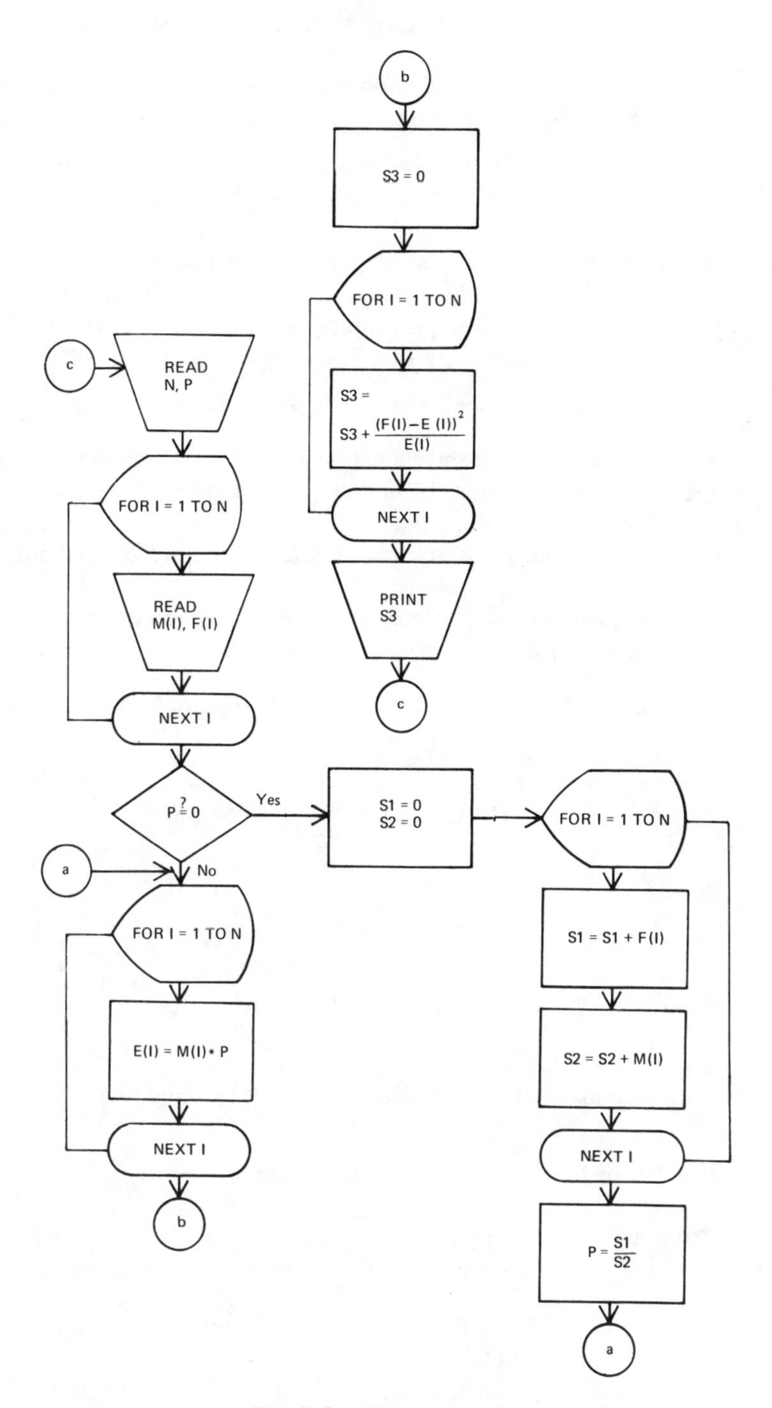

Fig. 7.2. Flowchart 19

(d) Given the age of a 4000-pound automobile, we might want to predict the average number of miles per gallon (mpg) it gets.

The easiest way to predict the value of one variable, given the value of another variable, is to use the general form of *a linear equation in two variables:* $y = a + bx$, where a and b are any real numbers, x is the given (independent) variable, and y is the predicted (dependent) variable.

The equation $y = a + bx$, where a and b are any real numbers, has an infinite number of ordered pair solutions (x,y) which lie on a straight line in the Cartesian plane (see Fig. 7.3). For example:

$$y = 0 + 2x = 2x, \quad y = -2 + 3x$$

As you may recall from elementary algebra, given the equation $y = a + bx$, a is the y-intercept and b is the slope of the straight line represented by $y = a + bx$. That is, $(0,a)$ is the point where the line crosses the y-axis and, given any two points on the line, $b =$ (change in y)/(change in x).

Now suppose we were given the information related to the four prediction examples at the beginning of this article:

Example (a)

x High School Average	y College GPA
70	2.1
75	2.6
80	2.9
85	3.3
90	3.8

Example (b)

x Pulse Rate, Beats per Minute	y Height, cm
59	175
65	176
73	168
72	170
68	172
60	170

Example (c)

x Number of Years Attended College	y Annual Salary
1	$12,000
2	14,000
3	13,000
4	17,000

Example (d)

x Age of 4000-pound Auto, Years	y Average Mpg
1	14.2
2	13.0
3	10.0
4	9.6

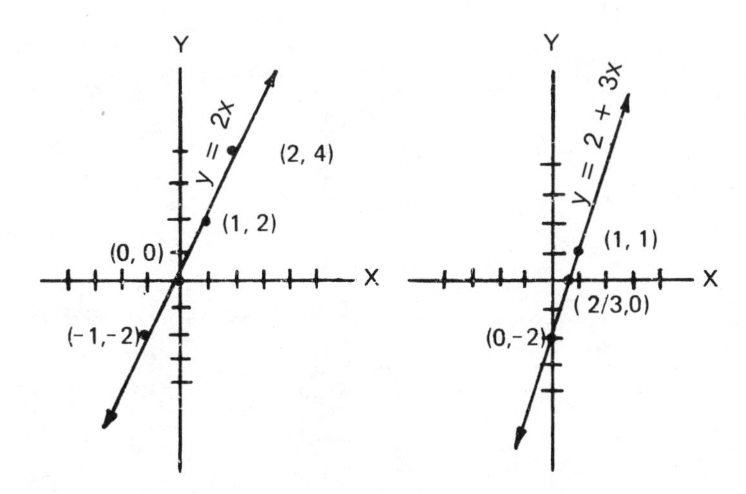

Fig. 7.3

We graph this given information as in Fig. 7.4. The dotted lines $y_p = a_1 + b_1x$, $y_p = a_2 + b_2x$, $y_p = a_3 + b_3x$, $y_p = a_4 + b_4x$ represent the *estimated regression line*. That is, we would like to use the given information to *estimate* a linear equation, $y_p = a + bx$, that will allow us to predict the value of y (dependent variable), given the value of x (independent variable). In order to find the equation of the estimated regression line, $y_p = a + bx$, we must find the y-intercept (value of a) and the slope (value of b) for the straight line represented by the equation $y_p = a + bx$.

7.4 REGRESSION BY THE METHOD OF LEAST SQUARES

In order to find the value of a and b for the estimating regression line, we must minimize the sum of the squares of the differences between the observed values of y and the value for $y_p = a + bx$ on the estimated regression line. That is, we must minimize $\Sigma[y - (a + bx)]^2$. We need methods from the calculus to complete this minimization. Using these methods from the calculus, we find that the estimating linear equation is

$$y_p = a + bx$$

where $a = \dfrac{(\Sigma y)(\Sigma x^2) - (\Sigma x)(\Sigma xy)}{n(\Sigma x^2) - (\Sigma x)^2}$

$$b = \frac{n(\Sigma xy) - (\Sigma x)(\Sigma y)}{n(\Sigma x^2) - (\Sigma x)^2}$$

n = the number of given ordered pairs

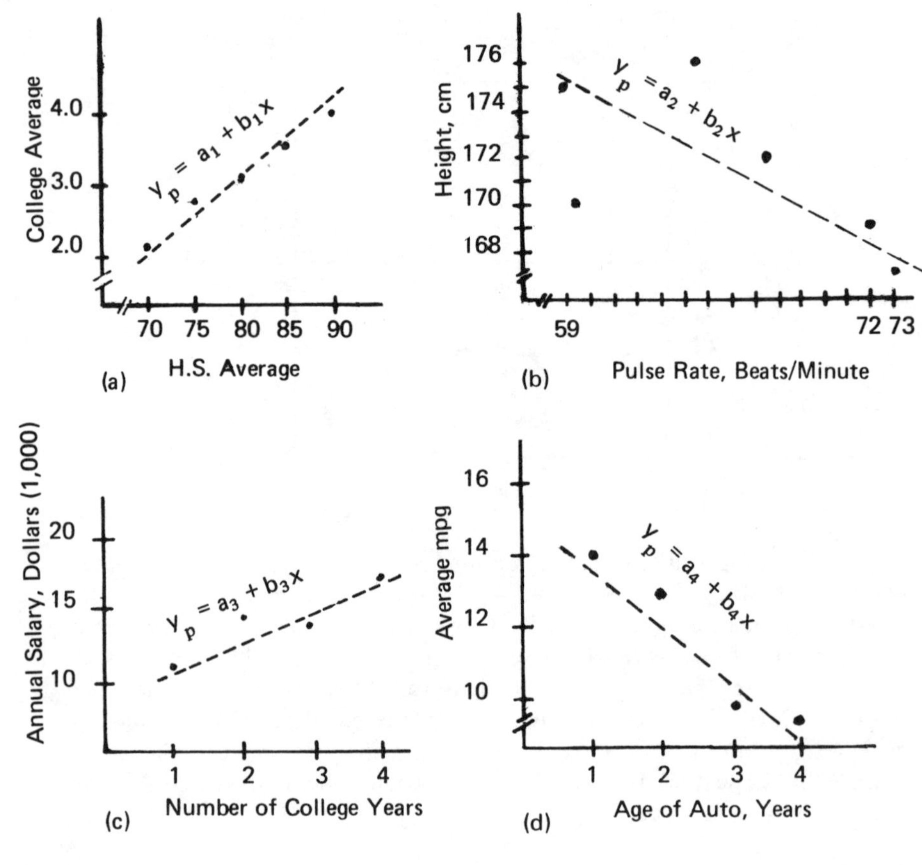

Fig. 7.4

Sample Problems

Let us consider each of the four examples mentioned in Article 7.3.

1. Example (a).

x H.S. Average	y GPA	x^2	xy
70	2.1	4900	147.0
75	2.6	5625	195.0
80	2.9	6400	232.0
85	3.3	7225	280.5
90	3.8	8100	342.0
$\Sigma x = 400$	$\Sigma y = 14.7$	$\Sigma x^2 = 32{,}250$	$\Sigma xy = 1196.5$

For the estimating linear regression line $y_p = a + bx$:

$$a = \frac{(\Sigma y)(\Sigma x^2) - (\Sigma x)(\Sigma xy)}{n(\Sigma x^2) - (\Sigma x)^2}$$

$$= \frac{(14.7)(32{,}250) - (400)(1196.5)}{5(32{,}250) - (400)^2}$$

$$= \frac{474{,}075 - 478{,}600}{161{,}250 - 160{,}000}$$

$$= \frac{-4{,}525}{1{,}250}$$

$$= -3.62$$

and

$$b = \frac{n(\Sigma xy) - (\Sigma x)(\Sigma y)}{n(\Sigma x^2) - (\Sigma x)^2}$$

$$= \frac{5(1196.5) - (400)(14.7)}{5(32{,}250) - (400)^2}$$

$$= \frac{5982.5 - 5880.0}{1250}$$

$$= \frac{102.5}{1250}$$

$$= 0.082$$

Therefore, the estimating linear regression equation is

$$y_p = a + bx \qquad \text{or} \qquad y_p = -3.62 + 0.082x$$

The predicted college grade-point averages for high school students having cumulative averages of 65 and 82, respectively, are

$$y_p = -3.62 + 0.082x \qquad \text{(estimating linear regression equation)}$$

If $x = 65$, then

$$\begin{aligned} y_p &= -3.62 + 0.082(65) \\ &= -3.62 + 5.33 \\ &= 1.71 \end{aligned}$$

If $x = 82$, then

$$\begin{aligned} y_p &= -3.62 + 0.082(82) \\ &= -3.62 + 6.724 \\ &= 3.104 \end{aligned}$$

2. Example (b).

x Pulse Rate, Beats per Minute	y Height, cm	x^2	xy
59	175	3481	10325
65	176	4225	11440
73	168	5329	12264
72	170	5184	12240
68	172	4624	11696
60	170	3600	10200
$\Sigma x = 397$	$\Sigma y = 1031$	$\Sigma x^2 = 26{,}443$	$\Sigma xy = 68{,}165$

For the estimating linear regression equation $y_p = a + bx$;

$$a = \frac{(\Sigma y)(\Sigma x^2) - (\Sigma x)(\Sigma xy)}{n(\Sigma x^2) - (\Sigma x)^2}$$

$$= \frac{(1031)(26{,}443) - (397)(68{,}165)}{6(26{,}443) - (397)^2}$$

$$= \frac{27{,}262{,}733 - 27{,}061{,}505}{158{,}658 - 157{,}609}$$

$$= \frac{201{,}228}{1049}$$

$$= 191.8284$$

and

$$b = \frac{n(\Sigma xy) - (\Sigma x)(\Sigma y)}{n(\Sigma x^2) - (\Sigma x)^2}$$

$$= \frac{6(68{,}165) - (397)(1031)}{1049}$$

$$= \frac{408{,}990 - 409{,}307}{1049}$$

$$= \frac{-317}{1049}$$

$$= -0.3021925$$

or $b \approx -0.302$

Therefore, the estimating linear regression equation is $y_p = a + bx$ or $y_p = 191.8285 - 0.302x$.

The predicted height in centimeters (cm) of people with 65 or 82 pulse beats per minute, respectively, is

$$y_p = 191.8285 - 0.302x \qquad \text{(estimating linear regression equation)}$$

If $x = 65$, then

$$\begin{aligned} y_p &= 191.8285 - 0.302(65) \\ &= 191.8285 - 19.63 \\ &= 172.1985 \text{ cm} \end{aligned}$$

If $x = 82$, then

$$\begin{aligned} y_p &= 191.8285 - 0.302(82) \\ &= 191.8285 - 24.764 \\ &= 167.0645 \text{ cm} \end{aligned}$$

3. Example (c)

x Yrs. of College	y Annual Salary	x^2	xy
1	\$12,000	1	12,000
2	14,000	4	28,000
3	13,000	9	39,000
4	17,000	16	68,000
$\Sigma x = 10$	$\Sigma y = 56{,}000$	$\Sigma x^2 = 30$	$\Sigma xy = 147{,}000$

For the estimating linear regression equation $y_p = a + bx$,

$$\begin{aligned} a &= \frac{(\Sigma y)(\Sigma x^2) - (\Sigma x)(\Sigma xy)}{n(\Sigma x^2) - (\Sigma x)^2} \\ &= \frac{(56{,}000)(30) - (10)(147{,}000)}{4(30) - (10)^2} \\ &= \frac{1{,}680{,}000 - 1{,}470{,}000}{120 - 100} \\ &= \frac{210{,}000}{20} \\ &= 10{,}500 \end{aligned}$$

and
$$b = \frac{n(\Sigma xy) - (\Sigma x)(\Sigma y)}{n(\Sigma x^2) - (\Sigma x)^2}$$
$$= \frac{4(147{,}000) - 10(56{,}000)}{20}$$
$$= \frac{588{,}000 - 560{,}000}{20}$$
$$= \frac{28{,}000}{20}$$
$$= 1400$$

Therefore, the estimating linear regression equation is

$$y_p = a + bx \qquad \text{or} \qquad y_p = 10{,}500 + 1{,}400x$$

The predicted annual salary for employees with 1½ years or 7 years of college, respectively, is

$$y_p = 10{,}500 + 1{,}400x \qquad \text{(estimating linear regression equation)}$$

If $x = 1.5$, then

$$\begin{aligned} y_p &= 10{,}500 + 1400(1.5) \\ &= 10{,}500 + 2100 \\ &= \$12{,}600 \end{aligned}$$

If $x = 7$, then

$$\begin{aligned} y_p &= 10{,}500 + 1400(7) \\ &= 10{,}500 + 9{,}800 \\ &= \$20{,}300 \end{aligned}$$

4. Example (d)

x Age of Auto, Years	y Average mpg	x^2	xy
1	14.2	1	14.2
2	13.0	4	26.0
3	10.0	9	30.0
4	9.6	16	38.4
$\Sigma x = 10$	$\Sigma y = 46.8$	$\Sigma x^2 = 30$	$\Sigma xy = 108.6$

For the estimating linear regression equation $y_p = a + bx$,

$$a = \frac{(\Sigma y)(\Sigma x^2) - (\Sigma x)(\Sigma xy)}{n(\Sigma x^2) - (\Sigma x)^2}$$

$$= \frac{(46.8)(30) - (10)(108.6)}{4(30) - (10)^2}$$

$$= \frac{1404 - 1086}{120 - 100}$$

$$= \frac{318}{20}$$

$$= 15.9$$

and

$$b = \frac{n(\Sigma xy) - (\Sigma x)(\Sigma y)}{n(\Sigma x^2) - (\Sigma x)^2}$$

$$= \frac{4(108.6) - (10)(46.8)}{4(30) - (10)^2}$$

$$= \frac{434.4 - 468}{20}$$

$$= \frac{-33.6}{20}$$

$$= -1.68$$

Therefore, the estimating linear regression equation is $y_p = a + bx$ or $y_p = 15.9 - 1.68x$.

The predicted average miles per gallon attained by a 4000-pound automobile that is 1.5 or 3.5 years old, respectively, is

$y_p = 15.9 - 1.68x$ (estimating linear regression equation)

If $x = 1.5$, then

$$\begin{aligned} y_p &= 15.9 - 1.68(1.5) \\ &= 15.9 - 2.52 \\ &= 13.38 \text{ mpg} \end{aligned}$$

If $x = 3.5$, then

$$\begin{aligned} y_p &= 15.9 - 1.68(3.5) \\ &= 15.9 - 5.88 \\ &= 10.02 \text{ mpg} \end{aligned}$$

Problem Set 7.3

1. Seven people are enrolled in an experimental speed-reading program. The following table shows the increase in the number of words per minute they could read and the number of weeks they were enrolled in the program:

x Number of Weeks	y Increase in Words per Minute
1	10
2	25
3	22
4	35
5	60
6	55
9	80

(a) Find the estimating linear regression equation to predict the increase in words read per minute, given the number of weeks of participation in the program.
(b) Use the estimating linear regression equation obtained in (a) to predict the increase in the number of words read per minute for a person enrolled in the experimental program for $3\frac{1}{2}$ weeks and also for a person enrolled in the program for 8 weeks.

2. From the freshman students enrolled at Jersey City State College who attended Hoboken High School, eight were randomly selected. The following table indicates their freshman grade-point averages at Jersey City State and their class rank upon graduation from Hoboken High School.

x Class Rank	y Freshman GPA
20	3.5
100	2.5
10	3.4
50	3.2
65	3.0
210	1.8
70	2.8
150	2.6

(a) Find the estimating linear regression equation to predict a student's freshman GPA at Jersey City State College, given his/her class rank at Hoboken High School.

(b) Use the estimating linear regression equation in (a) to predict the freshman GPA at Jersey City State of students whose class ranks at Hoboken High School were 200, 300, 5.

3. The following table illustrates the average number of cigarettes smoked daily by a person and his/her corresponding age at death as determined by a recent survey.

x Average Number of Cigarettes Daily	y Age at Death
10	65
20	58
30	50
40	51
50	45
5	70
8	65

(a) Find the estimating linear regression equation to predict a person's age at death, given the average number of cigarettes he/she smokes daily.

(b) Use the estimating linear regression equation in (a) to predict the age at death of persons smoking an average of 15, 25, and 0 cigarettes daily.

4. The Strongman Tool and Die Company lists the following information, which shows the age of a machine and the corresponding number of 4-inch bolts it produces in 8 hours.

x Age in Years	y Number of 4-in. Bolts in 8 Hours
½	2500
1	2400
2	2000
3	1800
4	1500

(a) Find the estimating linear regression equation to predict the number of 4-inch bolts produced by a machine in 8 hours, given the age in years of the machine.

(b) Use the estimating linear regression equation in (a) to predict the number of 4-inch bolts produced in 8 hours by machines that are 1½ years old; 2⅓ years old; 2 months old.

5. The following table illustrates the amount of money a producer invested in a Broadway play and the corresponding net return on his investment.

x Investment	y Net Return
50,000	100,000
200,000	100,000
70,000	30,000
500,000	−100,000
250,000	1,000,000

(a) Find the estimating linear equation to predict the net return on a Broadway producer's investment, given the amount of the investment.

(b) Use your estimating equation determined in (a) to predict the net returns on a producer's investments of $800,000; $1,000,000; $450,000.

6. The following table lists the ages and second-hand prices charged for the Hurricane sedan.

Age, years	Price, dollars
1	$2500
3	2175
8	1375
9	1125
5	1525
5	1600
1	2375

(a) Determine the estimating linear equation to predict the second-hand price charged for the Hurricane car, given the car's age in years.

(b) Use the linear equation you determined in (a) to predict the second-hand prices charged for Hurricane cars that are 10 years old; 2 years old; 7 years old.

7.5 A PREPACKAGED PROGRAM FOR FINDING THE ESTIMATING LINEAR REGRESSION EQUATION

Name: STAT11

Description: This program computes the slope and y-intercept for an estimating linear regression equation.

Instructions: Enter the data beginning in line 150 as follows:

```
150 DATA N,X1,Y1,X2,Y2,...,XN,YN
```

where

N = number of given ordered pairs

$X1,X2,...,XN$ = values of the independent variable

$Y1,Y2,...,YN$ = values of the dependent variable (variable to be predicted)

Note: Line 200 is the END statement.

Sample Problems

User's Input

```
GET-STAT11

150 DATA 5,70,2.1,75,2.6,80,2.9,85,3.3,90,3.8
151 DATA 6,'59,175,65,176,73,168,72,170,68,172,60,170
152 DATA 4,1,12000,2,14000,3,13000,4,17000
153 DATA 4,1,14.2,2,13,3,10,4,9.6
RUN
```

Computer Output

```
Y=-3.6199    + 8.19969E-02    X

Y= 191.817     -.302193    X

Y= 10500    + 1400    X

Y= 15.9    -1.67999    X


LINE      5:    END OF DATA

TIME 0.0 SECS.
```

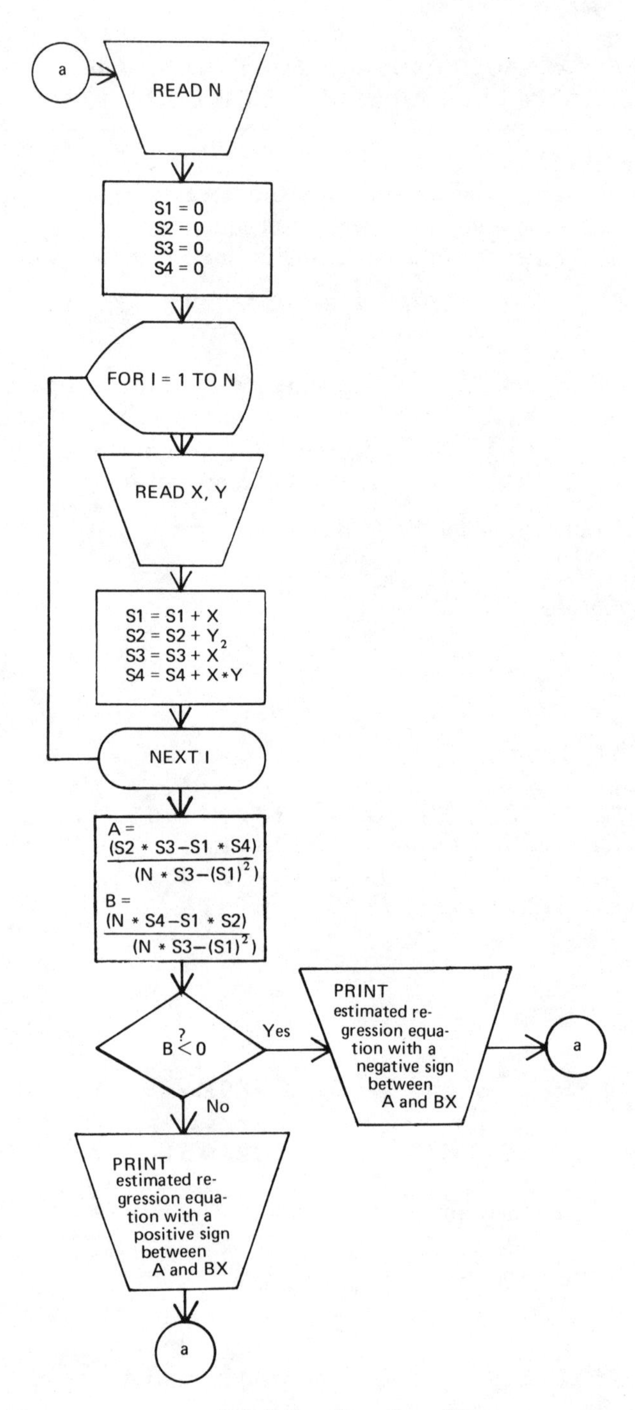

Fig. 7.5. Flowchart 20

Let us look at the flowcharting of STAT11 (Fig. 7.5).
The coding of STAT11 is as follows:

```
5 READ N
10 LETS1=0
15 LET S2=0
20 LETS3=0
25 LET S4=0
30 FOR I=1 TO N
35     READ X,Y
40     LET S1=S1+X
45     LET S2=S2+Y
50     LET S3=S3+X↑2
55     LET S4=S4+X*Y
60 NEXT I
65 LET A=(S2*S3-S1*S4)/(N*S3-(S1)↑2)
70 LET B=(N*S4-S1*S2)/(N*S3-(S1)↑2)
75 IF B<0 THEN 90
80 PRINT "Y="A;"+"B;"X"
85 GO TO 95
90 PRINT "Y="A;B;"X"
95 PRINT
100 GO TO 5
200 END
```

Problem Set 7.4

1. Use STAT11 to compute the estimating linear regression equation for each of the problems in Problem Set 7.3 (that is, just do part (a) of each problem)

2. While you have the results to Problem 1 and STAT11 on your terminal, add the following statements to STAT11:

```
96  READ X
97  LET Y=A+B*X
98  PRINT "THE GIVEN VALUE OF X="X,"THE PREDICTED
VALUE OF Y=",Y
100  GO TO 96
```

Now change each of the DATA statements used in Problem 1 by adding the given values of *X* for part (b) of each problem in Problem Set 7.3.

For example, your first DATA statement for part (a) of Problem 1 in Problem Set 7.3 was

```
150  DATA 7,1,10,2,25,3,22,4,35,5,60,6,55,9,80
```

Change it to

```
150  DATA 7,1,10,2,25,3,22,4,35,5,60,6,55,9,80,3.5,8
```

After changing your DATA statement RUN the new STAT11.

7.6 THE COEFFICIENT OF CORRELATION

Once we have estimated a regression equation, we should be able to decide how well it fits the given data; that is, we can determine a numerical measure of the *linear relationship* between the two variables. Let us consider Example (a) in Article 7.3, in which we attempted to predict a student's freshman grade-point average, given his/her high school cumulative average. We were given the following data:

x H.S. Average	y GPA
70	2.1
75	2.6
80	2.9
85	3.3
90	3.8

Using the least squares method discussed in Article 7.4, we estimated the linear regression line

$$y_p = a + bx = -3.62 + 0.082x$$

The value

$$d = \frac{\Sigma[y - (a + bx)]^2}{\Sigma(y - \bar{y})^2}$$

represents the proportion of variation between the estimated values of y and the actual values of y in our estimated regression equation. Then the value $r = \pm\sqrt{1 - d}$, which is called the *coefficient of correlation*, represents the proportion of total y variation within the x relationship. Table 7.1 lists the data for Example (a) discussed above.

Table 7.1

x, H.S. Avg.	y, GPA	$(a + bx)$	$y - (a + bx)$	$(y - (a + bx))^2$	$y - \bar{y}$	$(y - \bar{y})^2$
70	2.1	2.12	−0.02	0.0004	−0.84	0.7056
75	2.6	2.53	+0.07	0.0049	−0.34	0.1156
80	2.9	2.94	−0.04	0.0016	−0.04	0.0016
85	3.3	3.35	−0.05	0.0025	+0.36	0.1296
90	3.8	3.76	+0.04	0.0016	+0.86	0.7396
$\Sigma x = 400$	$\Sigma y = 14.7$			$\Sigma(y - (a + bx))^2 = 0.0110$		$\Sigma(y - \bar{y})^2 = 1.6920$

$$y_p = a + bx \qquad \bar{y} = \frac{14.7}{5} = 2.94$$

$$y_p = -3.62 + 0.082x$$

Therefore, the coefficient of correlation is

$$r = +\sqrt{1 - \frac{0.011}{1.692}} = +\sqrt{1 - 0.0065011}$$

$$= +\sqrt{0.9934989} = 0.9967441$$

The coefficient of correlation is given the sign of b in $y_p = a + bx$, which represents the slope of the estimated regression line. If the coefficient of correlation is

(a) $r \approx \pm 1$, the variation of the y-values is strongly correlated to the x-values.

(b) $r \approx 0$, little of the variation of the y-values is correlated to the x-values.

Thus, the sign of the correlation shows us the direction of the relationship. When our coefficient is *positive*, the values of one variable increase and the corresponding values of the other variable also tend to increase. When our coefficient is *negative*, the values of one variable increase and the corresponding values of the other variable tend to decrease. If the correlation is zero, then changes in the values of one variable do not produce any discernible corresponding changes in the values of the other variable.

Mathematicians have developed an alternative formula for computing the *coefficient of correlation*, which automatically gives r the sign of the slope in our estimated regression equation. The formula most widely used for the coefficient of correlation is

$$r = \frac{n(\Sigma xy) - (\Sigma x)(\Sigma y)}{(\sqrt{n(\Sigma x^2) - (\Sigma x)^2})(\sqrt{n(\Sigma y^2) - (\Sigma y)^2})}$$

Let us illustrate the use of formula (1) with Example (a) of Article 7.3, which predicted GPA from high school average. Data are listed in Table 7.2.

Table 7.2

x H.S. Average	y GPA	xy	x^2	y^2
70	2.1	147.0	4900	4.41
75	2.6	195.0	5625	6.76
80	2.9	232.0	6400	8.41
85	3.3	280.5	7225	10.89
90	3.8	342.0	8100	14.44
$\Sigma x = 400$	$\Sigma y = 14.7$	$\Sigma xy = 1196.5$	$\Sigma x^2 = 32{,}250$	$\Sigma y^2 = 44.91$

$$\begin{aligned} r &= \frac{n(\Sigma xy) - (\Sigma x)(\Sigma y)}{(\sqrt{n(\Sigma x^2) - (\Sigma x)^2})(\sqrt{n(\Sigma y^2) - (\Sigma y)^2})} \\ &= \frac{5(1196.5) - 400(14.7)}{(\sqrt{5(32{,}250) - (400)^2})(\sqrt{5(44.91) - (14.7)^2})} \\ &= \frac{5982.5 - 5880.0}{(\sqrt{161{,}250 - 160{,}000})(\sqrt{224.55 - 216.09})} \\ &= \frac{102.5}{(\sqrt{1250})(\sqrt{8.46})} \\ &= \frac{102.5}{(35.355339)(2.9086079)} \\ &= \frac{102.5}{102.83481} \\ &= 0.9967441 \approx 0.997 \end{aligned}$$

which is the same value we computed using $r = \pm\sqrt{1 - d}$. As you can see, the coefficient of correlation is $r \approx 1$; therefore, there is a strong *positive correlation* between the variations of y and their relationship to the x-values (that is, small values of x seem to indicate small values of y and large values of x seem to indicate large values of y).

Now let us look at Example (d) in Article 7.3, in which we determined a linear regression equation to estimate the average miles per gallon (mpg) of a 4000-pound automobile, given its age in years. Data are listed in Table 7.3.

Table 7.3

x Age in yrs.	y Avg. mpg	xy	x^2	y^2
1	14.2	14.2	1	201.64
2	13.0	26.0	4	169.00
3	10.0	30.0	9	100.00
4	9.6	38.4	16	92.16
$\Sigma x = 10$	$\Sigma y = 46.8$	$\Sigma xy = 108.6$	$\Sigma x^2 = 30$	$\Sigma y^2 = 562.8$

The coefficient of correlation is

$$r = \frac{n(\Sigma xy) - (\Sigma x)(\Sigma y)}{(\sqrt{n(\Sigma x^2) - (\Sigma x)^2})(\sqrt{n(\Sigma y^2) - (\Sigma y)^2})}$$

$$= \frac{4(108.6) - (10)(46.8)}{(\sqrt{4(30) - (10)^2})(\sqrt{4(562.8) - (46.8)^2})}$$

$$= \frac{434.4 - 468}{(\sqrt{120 - 100})(\sqrt{2251.2 - 2190.24})}$$

$$= \frac{-33.6}{(\sqrt{20})(\sqrt{60.96})}$$

$$= \frac{-33.6}{(4.4721359)(7.8076885)}$$

$$= \frac{-33.6}{34.917044}$$

$$= -0.9622807 \approx -0.962$$

As you can see, the coefficient of correlation r is very close to -1; therefore, there is a strong *negative correlation* or *inverse relationship* between the variations of y and their relationships to the x values (that is, small values of x seem to indicate large values of y, and large values of x seem to indicate small values of y).

Suppose we were given the following information:

x Age of car in yrs.	y Average mpg
1	15.4
2	15.0
3	15.1
4	14.8

Table 7.4

x	y	xy	x^2	y^2
1	15.4	15.4	1	237.16
2	15.0	30.0	4	225.00
3	15.4	46.2	9	237.16
4	14.8	59.2	16	219.04
$\Sigma x = 10$	$\Sigma y = 60.6$	$\Sigma xy = 150.8$	$\Sigma x^2 = 30$	$\Sigma y^2 = 918.36$

Given the data in Table 7.4, let us compute the coefficient of correlation:

$$
\begin{aligned}
r &= \frac{n(\Sigma xy) - (\Sigma x)(\Sigma y)}{(\sqrt{n(\Sigma x^2) - (\Sigma x)^2})\ (\sqrt{n(\Sigma y^2) - (\Sigma y)^2})} \\
&= \frac{4(150.8) - (10)(60.6)}{(\sqrt{4(30) - (10)^2})(\sqrt{4(918.36) - (60.6)^2})} \\
&= \frac{603.2 - 606}{(\sqrt{20})(\sqrt{1.08})} \\
&= \frac{-2.8}{(4.4721359)(1.0392304)} \\
&= \frac{-2.8}{4.6475795} \\
&= -0.6024641 \approx -0.602
\end{aligned}
$$

Notice that the coefficient r is not close to ± 1, nor is it close to 0. Therefore, we do not have a strong positive correlation or a strong negative correlation, and we do not have a zero correlation between the x and y values. Table F.7 (Appendix) gives us critical values for $r_{.025}$ to test for significant correlations at the 0.05 level of significance [Fig.

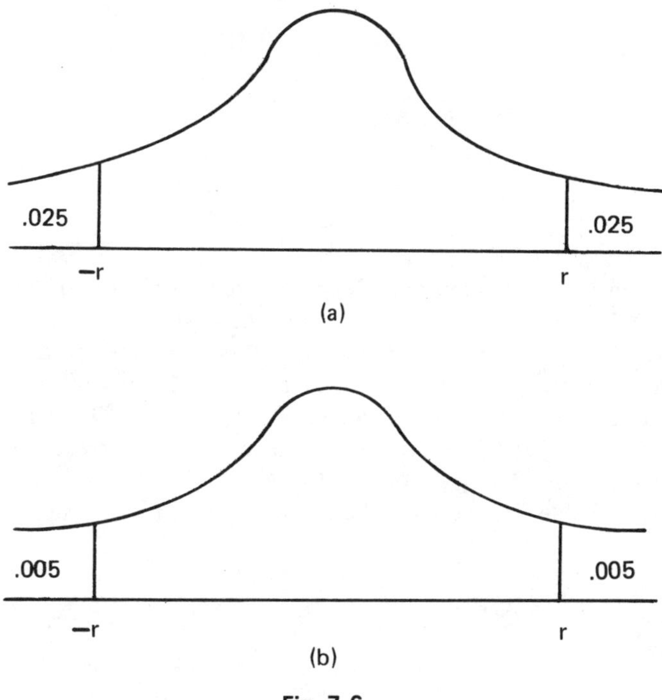

Fig. 7.6

7.6(a)] and $r_{.005}$ to test for significant 0.01 level of significance [see Fig. 7.6(b)]. Therefore, if we computed the coefficient of correlation $r = -0.6024641 \approx -0.602$ for four given ordered pairs, and wanted a correlation significant at the 0.05 level of significance, we would move vertically in Table F.7 to $n = 4$ and horizontally to $r_{.025}$ to find our critical $r_{.05} = \pm 0.950$. If our observed $r < -0.950$, we would have a *negative* correlation significant at the 0.05 level of significance. If our observed $r > +0.950$, we would have a positive correlation significant at the 0.05 level of significance. Finally, if our observed $-0.950 < r < +0.950$, we would have a correlation that was not significant at the 0.05 level of significance.

In Example (d) presented above, our observed $r \approx -0.602$ and $-0.950 < -0.602 < +0.950$. Therefore, we have a negative correlation that is not significant at the 0.05 level of significance.

Now, let us consider Examples (a) and (d) in Article 7.3 at the 0.01 level of significance. From Table 7.1 we have the following data for Example (a).

x H.S. Avg.	y GPA
70	2.1
75	2.6
80	2.9
85	3.3
90	3.8

Given $n = 5$, we computed our observed coefficient of correlation $r = 0.9967441 \approx 0.997$. Table F.7 shows us that for $n = 5$, the critical $r_{.05} = \pm 0.878$. Our observed $r \approx 0.997 > 0.878$; therefore, we have *positive* correlation that is significant at the 0.05 level of significance. Also we can see that for $n = 5$, the critical $r_{.01} = \pm 0.959$. Our observed $r \approx 0.997 > 0.959$; therefore, we have a *positive* correlation that is significant at the 0.01 level of significance.

From Table 7.3, we have the following data for Example (d):

x Age of Auto, Yrs.	y Avg. mpg
1	14.2
2	13.0
3	10.0
4	9.6

Given $n = 4$, we computed our observed coefficient of correlation $r = -0.9622807 \approx -0.962$. From Table F.7 we see that for $n = 4$, the critical $r_{.05} = \pm 0.950$. Our observed $r \approx -0.962 < -0.950$; therefore, we have a *negative* correlation that is significant at the 0.05 level of significance. Also we can see that for $n = 4$, the critical $r_{.01} = \pm 0.999$. Our observed $r \approx -0.962 \nless -0.999$; therefore, we have a *negative* correlation that is *not* significant at the 0.01 level of significance.

In general, a correlation coefficient of size 0.40 to 0.60 is considered useful for group type predictions. A correlation coefficient r of size 0.60 to 0.80 may be substantially utilized for both group type and individual predictions. A correlation coefficient greater than 0.80 is said to have a high, or strong, relationship that has definite value in individual predictions.

7.7 A PREPACKAGED PROGRAM FOR COMPUTING THE COEFFICIENT OF CORRELATION

Name: STAT12

Description: This program computes the coefficient of correlation for a set of ordered pairs (x,y), where x is the independent variable and y is the dependent (predicted) variable.

Instructions: Enter the data beginning in line 150 as follows:

150 DATA N,X1,Y1,X2,Y2,...,XN,YN

where

N = number of given ordered pairs

$X1,X2,...,XN$ = values of the independent variable

$Y1,Y2,...,YN$ = values of the dependent (predicted) variable.

Note: Line 200 is the END statement.

Sample Problems

User's Input

```
GET-STAT12

150 DATA 5,70,2.1,75,2.6,80,2.9,85,3.3,90,3.8
151 DATA 4,1,14.2,2,13,3,10,4,9.6
RUN
```

Computer Output

```
COEFF. OF CORRELATION R= .996715
COEFF. OF CORRELATION R=-.962273

LINE      5:   END OF DATA

TIME 0.0 SECS.
```

Let us look at the flowcharting of STAT12 (Fig. 7.7).

The coding of STAT12 is as follows:

```
5 READ N
10 LET S1=0
15 LET S2=0
20 LET S3=0
25 LET S4=0
30 LET S5=0
35 FOR I=1 TO N
40 READ X,Y
45     LET S1=S1+X
50     LET S2=S2+Y
55     LET S3=S3+X*Y
60     LET S4=S4+X↑2
65     LET S5=S5+Y↑2
70 NEXT I
75 LET R=(N*S3-S1*S2)/(SQR(N*S4-S1↑2)
         *SQR(N*S5-S2↑2))
80 PRINT "COEFF. OF CORRELATION R="R
85 GO TO 5
150 DATA
200 END
```

Problem Set 7.5

1. For each of the problems in Problem Set 7.3, proceed as follows:
   (a) Use STAT12 to compute the coefficient of correlation for the given sets of data.
   (b) Use Table F.7 (Appendix) to determine whether or not the coefficient of correlation is significant at the 0.05 level.
   (c) Determine whether or not the coefficient of correlation is significant at the 0.01 level.

7.8 ANALYSIS OF VARIANCE: ONE-WAY CLASSIFICATION

In Chapter 6 we tested for significant differences between two sample means. The *analysis of variance procedure* will allow us to test for significant differences between three or more sample means. Experiments that use *one* independent variable are said to involve one basis of, or *one-way, classification.* Although we will not consider the topic here, the analysis of variance may also be used in analyzing data obtained from experiments that involve more than one basis of classification.

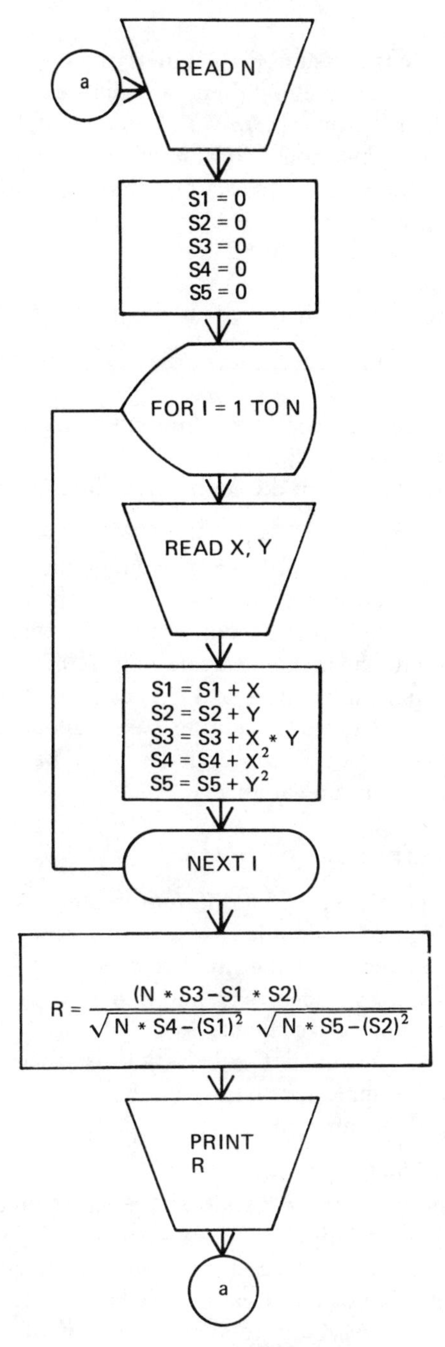

Fig. 7.7. Flowchart 21

There are *two* basic assumptions underlying the method of analysis of variance. First, this method assumes that each of our samples considered is obtained from a *normally distributed* population. It is also assumed that all populations considered have the same standard deviation, that is, $SD = \sigma_1 = \sigma_2 = \sigma_3$, and that the scores in the samples have been randomly selected.

Examples

We might want to see if there is a significant difference in the

1. Results of four different doctors' dietary plans for losing excess weight.
2. Achievement of seniors from three different high schools on a national college entrance examination.
3. Quality of radio tubes produced by five different machines in the same factory.

Testing for significant differences among three or more sample means requires more numerical calculations than does testing for significant differences between two means. We will use a four-step algorithm to facilitate the manual calculations in the analysis of variance procedure. Note that the analysis of variance consists of obtaining two independent estimates of variance, one based upon variation between groups (between-group variance), and the other based upon variation within groups (within-group variance).

FOUR-STEP ALGORITHM

Say we are given $x_1, x_2, \ldots, x_k$ different treatments (different dietary plans, different high school student scores, different machine outputs), where $n_1, n_2, \ldots, n_k$ represent the number of subjects or items in these treatments $x_1, x_2, \ldots, x_k$, respectively, and $n = n_1 + n_2 + \cdots + n_k$ is the total number of subjects or items. We want to test the motivated hypothesis H_m: all means not equal (that is, there is a significant difference in the k-sample means), where H_o: $\mu_1 = \mu_2 = \cdots = \mu_k$. We use the following algorithm:

1. Compute the following statistics:
   (a) *Within sum of squares* = variation 1 + variation 2 + $\cdots$ + variation k, where variation $i = \Sigma x_i^2 - (\Sigma x_i)^2/n_i$ (that is, the sum of the squared deviations of each score in a treatment from the arithmetic mean score of that treatment, where $i = 1, 2, 3, \ldots k$).
   (b) *Total sum of squares* = total variation of all scores = $\Sigma x^2 - (\Sigma x)^2/n$ (that is, the sum of the squared deviations of each score in a treatment from the grand mean of all treatments).
   (c) *Between sum of squares* = (total sum of squares) — (within sum of squares).

(d) *Between mean square (larger variance) = (between sum of squares)/(k − 1)*. (This statistic represents the larger estimate of the population variance.)

(e) *Within mean square (smaller variance) = (within sum of squares)/(n − k)*. (This statistic represents the smaller population variance estimate.)

2. Compute the observed F (variance) RATIO:

$$F_{k-1,n-k} = \frac{\text{BETWEEN MEAN SQUARE}}{\text{WITHIN MEAN SQUARE}}$$

where $k - 1$ = degrees of freedom for the *between mean square*, and $n - k$ = degrees of freedom for the *within mean square*.

3. Look up the critical $F_{k-1,n-k}$ score in Table F.8 (Appendix) for the desired level of significance.

4. Conclusion:

(a) Either we can reject H_o if our calculated F falls in this area, that is, if our calculated F ratio > critical F ratio in the table of F values (Table F.8, Appendix), or

(b) We cannot reject H_o.

Sample Problem

Seven patients from each of four different dietary plans were selected at random. At the end of three weeks, the doctors recorded the number of pounds the seven patients had lost (see Table 7.5).

Table 7.5

Patient	Plan 1 (x_1)	Plan 2 (x_2)	Plan 3 (x_3)	Plan 4 (x_4)
1	10	22	5	14
2	15	13	22	12
3	8	9	19	11
4	20	11	18	10
5	7	20	10	8
6	9	20	10	9
7	11	15	10	8

Test the motivated hypothesis at the 0.01 level of significance that the dietary plans do not have differential effects on losing weight (H_0: $\mu_1 = \mu_2 = \mu_3 = \mu_4$; H_m = All means not equal).

Solution

1. Compute the necessary statistics, using the data in Table 7.6, and

$$\text{Variation } i = \Sigma x_i^2 - \frac{(\Sigma x_i)^2}{n_i}$$

Table 7.6

x_1	x_1^2	x_2	x_2^2	x_3	x_3^2	x_4	x_4^2
10	100	22	484	5	25	14	196
15	225	13	169	22	484	12	144
8	64	9	81	19	361	11	121
20	400	11	121	18	324	10	100
7	49	20	400	10	100	8	64
9	81	20	400	10	100	9	81
11	121	15	225	10	100	8	64
$\Sigma x_1=$ 80	$\Sigma x_1^2=$ 1040	$\Sigma x_2=$ 110	$\Sigma x_2^2=$ 1880	$\Sigma x_3=$ 94	$\Sigma x_3^2=$ 1494	$\Sigma x_4=$ 72	$\Sigma x_4^2=$ 770

Then

Variation 1 = 1040 − (80)²/7 = 1040 − 914.29 = 125.71
Variation 2 = 1880 − (110)²/7 = 1880 − 1728.57 = 151.43
Variation 3 = 1494 − (94)²/7 = 1494 − 1262.29 = 231.71
Variation 4 = 770 − (72)²/7 = 770 − 740.57 = 29.43

(Note: $k = 4$, $n_1 = n_2 = n_3 = n_4 = 7$, $n = 28$.)

(a) *Within sum of squares* = $\sum_{i=1}^{k}$ variation i = 125.71 + 151.43 + 231.71 + 29.43 = 538.28.

(b) *Total sum of squares* = total variation = $x^2 - (\Sigma x)^2/n$ = (1040 + 1880 + 1494 + 770) − (80 + 110 + 94 + 72)²/28 = 5184 − (356)²/28 = 5184 − 4526.29 = 657.71.

(c) *Between sum of squares* = (total sum of squares) − (within sum of squares) = 657.71 − 538.28 = 119.43.

(d) *Between mean square* = (between sum of squares)/$(k-1)$ = 119.43/(4 − 1) = 39.81.

(e) *Within mean square* = (within sum of squares)/$(n-k)$ = 538.28/(28 − 4) ≈ 22.43.

2. Compute observed F ratio:

Observed $F_{k-1,n-k}$ = (between mean square)/(within mean square)
Observed $F_{3,24} = 39.81/22.43 \approx 1.77$

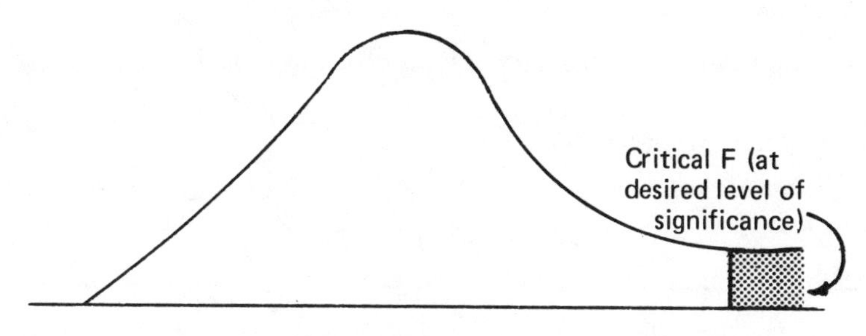

Fig. 7.8

3. From Table F.8 we see that the critical $F_{3,24} = 3.01$ at the 0.05 level of significance. (Move horizontally to df = 3 and vertically to df = 24 to find $F_{3,24} = 3.01$.)

4. Conclusion:

$$\text{Calculated } F_{3,24} \not> \text{critical } F_{3,24}$$
$$1.77 \not> 3.01$$

Therefore we *cannot* reject H_o.

Thus, there is no significant difference in the weight loss means among the four dietary plans.

Note: Table F.8 shows us that the critical $F_{3,24} = 4.72$ at the 0.01 level of significance, and $1.77 \not> 4.72$. Therefore, we could *not* reject H_o at the 0.01 level of significance.

Sample Problem

Now suppose five students from each of three different high schools are randomly selected after completing a national college entrance exam and their scores are recorded as follows:

Student	H.S. 1	H.S. 2	H.S. 3
1	500	400	600
2	450	395	625
3	525	420	490
4	600	450	520
5	650	595	550

Test the motivated hypothesis that the mean scores of students differ among the three different high schools at the 0.01 level of significance.

Solution

1. Compute the necessary statistics, using data in Table 7.7, and

$$\text{Variation } i = \Sigma x_i^2 - \frac{(\Sigma x_i)^2}{n_i}$$

Table 7.7

x_1	x_1^2	x_2	x_2^2	x_3	x_3^2
500	250,000	400	160,000	600	360,000
450	202,500	395	156,025	625	390,625
525	275,625	420	176,400	490	240,100
600	360,000	450	202,500	520	270,400
650	422,500	595	354,025	550	302,500
$\Sigma x_1 =$ 2,725	$\Sigma x_1^2 =$ 1,510,625	$\Sigma x_2 =$ 2,260	$\Sigma x_2^2 =$ 1,048,950	$\Sigma x_3 =$ 2,785	$\Sigma x_3^2 =$ 1,563,625

Then

Variation 1 = 1,510,625 − (2,725)²/5 = 25,500
Variation 2 = 1,048,950 − (2,260)²/5 = 27,430
Variation 3 = 1,563,625 − (2,785)²/5 = 12,380

(Note: $k = 3$, $n_1 = n_2 = n_3 = 5$, $n = 15$.)

(a) *Within sum of squares* = $\sum_{i=1}^{k}$ variation i = 25,500 + 27,430 + 12,380 = 65,310.

(b) *Total sum of squares* = total variation = $\Sigma x^2 - (\Sigma x)^2/n$ = (1,510,625 + 1,048,950 + 1,563,625) − (2725 + 2260 + 2785)²/15 = 4123200 − (7770)²/15 = 4,123,200 − 4,024,860 = 98,340.

(c) *Between sum of squares* = (total sum of squares) − (within sum of squares) = 98,340 − 65,310 = 33,030.

(d) *Between mean square* = (between sum of squares)/(k − 1) = 33,030/(3 − 1) = 16,515.

(e) *Within mean square* = (within sum of squares)/($n - k$) = 65,310/(15 − 3) = 5,442.50.

2. Compute observed F ratio:

$$\text{Observed } F_{k-1,n-k} = \frac{\text{between mean square}}{\text{within mean square}}$$

$$\text{Observed } F_{2,12} = \frac{16{,}515}{5{,}442.50} \approx 3.03$$

3. From Table F.8 (Appendix) we see that the critical $F_{2,12} = 6.93$ at the 0.01 level of significance (move horizontally to df = 2 and vertically to df = 12 to find $F_{2,12} = 6.93$).

4. Conclusion:

$$\begin{array}{ccc} \text{Calculated } F_{2,12} & \not> & \text{critical } F_{2,12} \\ 3.03 & \not> & 6.93 \end{array}$$

Therefore, we *cannot* reject H_o at the 0.01 level of significance. Thus, there are no significant differences in the mean scores of students among the three high schools.

Problem Set 7.6

1. Three different machines (A, B, and C) produce radio tubes for the Now Sound Radio Corporation. For each of five consecutive days, 100 radio tubes are selected from each of the three machines and are tested for defects. The following table indicates the number of defective radio tubes for each machine for five consecutive days.

Day	Machine A	Machine B	Machine C
1	2	2	6
2	7	1	3
3	8	5	2
4	3	3	1
5	5	3	10

(a) Test the motivated hypothesis that the mean numbers of defective radio tubes differ among the three machines at the 0.05 level of significance.

(b) Now test the motivated hypothesis at the 0.01 level of significance.

2. Four different bus companies service the route from Quietown, U.S.A., to Swingtown, U.S.A. Three trips from Quietown to Swingtown are randomly selected for each of the four bus companies, and the length of time in minutes for each trip is recorded as follows:

Trip	Bus Co. 1	Bus Co. 2	Bus Co. 3	Bus Co. 4
1	50	50	58	50
2	52	57	50	50
3	54	45	49	50

Test the motivated hypothesis that there are differences in the mean times among the four bus companies at the 0.01 level of significance.

3. Six students are randomly selected from each of three different speed-reading schools. Their increase in number of words read per minute is recorded below after three weeks in each school:

Student	School 1	School 2	School 3
1	22	30	25
2	20	30	25
3	30	21	30
4	35	35	30
5	20	22	32
6	25	30	32

Test the motivated hypothesis that there are significant differences among the mean increases of words per minute for students enrolled in the three different schools at the 0.05 level of significance.

4. The following final examination information is given for four different sections of STAT1 at a local university.

Section 1	Section 2	Section 3	Section 4
80	70	80	80
60	76	50	75
55	90	75	60
90	60	70	90
85	52	62	75

Test the motivated hypothesis that there exist differences in the mean final exam scores among the four different sections of STAT1:

(a) At the 0.05 level of significance.

(b) At the 0.01 level of significance.

5. The table below lists the weekly salaries of four salesmen for the Atlas Aluminum Siding Company for the past three weeks. Calculate F and (a) test at the 0.05 level of significance the motivated hypothesis that differences exist among the average weekly salaries of Atlas salesmen for this given period of three weeks; (b) also test the motivated hypothesis at the 0.01 level of significance.

Axelrod	Inge	Miller	Williams
$226	$251	$208	$247
218	237	191	218
252	237	222	209

6. Twenty-four homogeneous fifth-grade pupils were divided into three equal subgroups for the purpose of comparing the effectiveness of three distinct methods of teaching reading. Each group was taught by one of these methods, and at the end of the school year the pupils were given the same reading test. The test scores were reported as follows:

Method 1	Method 2	Method 3
75	90	77
91	75	84
80	80	83
82	83	90
79	78	85
81	76	84
83	84	87
80	88	81

Test at the 0.05 level of significance the motivated hypothesis that differences exist in the mean test scores of pupil groups taught reading by the three different methods.

7.9 A PREPACKAGED PROGRAM FOR COMPUTING THE OBSERVED F (VARIANCE) RATIO

Name: STAT13

Description: This program computes the calculated F ratio in an analysis of variance procedure.

Instructions: Enter the data beginning in line 180 as follows:

180 DATA K,T1,T2,...,TK,X1,X2,...,XN

where

K = number of different treatments

$T1$ = number of scores from the first treatment

$T2$ = number of scores from the second treatment, etc., . . .
TK = number of scores from the last treatment.
$X1,X2,...,XN$ = listing of all the scores.
First list the scores from the first treatment; then list the scores from the second treatment; finally list the scores from the last treatment.

Note: Line 200 is the END statement.

Sample Problem

User's Input

```
GET-STAT13

180 DATA 4,7,7,7,7,10,15,8,20,7,9,11
181 DATA 22,13,9,11,20,20,15
182 DATA 5,22,19,18,10,10,10
183 DATA 14,12,11,10,8,9,8
RUN
```

Computer Output

```
OBSERVED F= 1.77495
DF. FOR NUMERATOR = 3
DF. FOR DENOMINATOR= 24

LINE      5:    END OF DATA

TIME 0.0 SECS.
```

Now let us look at the flowcharting of STAT13 (Fig. 7.9). The coding of STAT13 is as follows:

```
2 DIM A(10), B(10), T(10), V(10)
5 READ K
10 LET N= 0
15 LET S4=0
```

```
20 LET S5 =0
25 LET S6=0
30 FOR M= 1 TO K
35 LET A(M)=0
40 LET B(M)=0
45 READ T(M)
50 LET N=N+T(M)
55 NEXT M
60 FOR J= 1 TO K
70 FOR I= 1 TO T(J)
75 READ X
85   LET A(J)=A(J)+X
90   LET B(J)=B(J)+X↑2
95   LET S4=S4+X
100  LET S5=S5+X↑2
120 NEXT I
122 NEXT J
125 FOR L=1 TO K
130   LET V(L)=B(L)-A(L)↑2/T(L)
135   LET S6=S6+V(L)
140 NEXT L
145 LET S7=S5-S4↑2/N
150 LET B=S7-S6
155 LET F=(B/(K-1))/(S6/(N-K))
160 PRINT "OBSERVED F="F
165 PRINT "DF. FOR NUMERATOR ="K-1
170 PRINT "DF. FOR DENOMINATOR="N-K
175 GO TO 5
200 END
```

Problem Set 7.7

Use STAT13 to check your manual calculations for each of the problems in Problem Set 7.6.

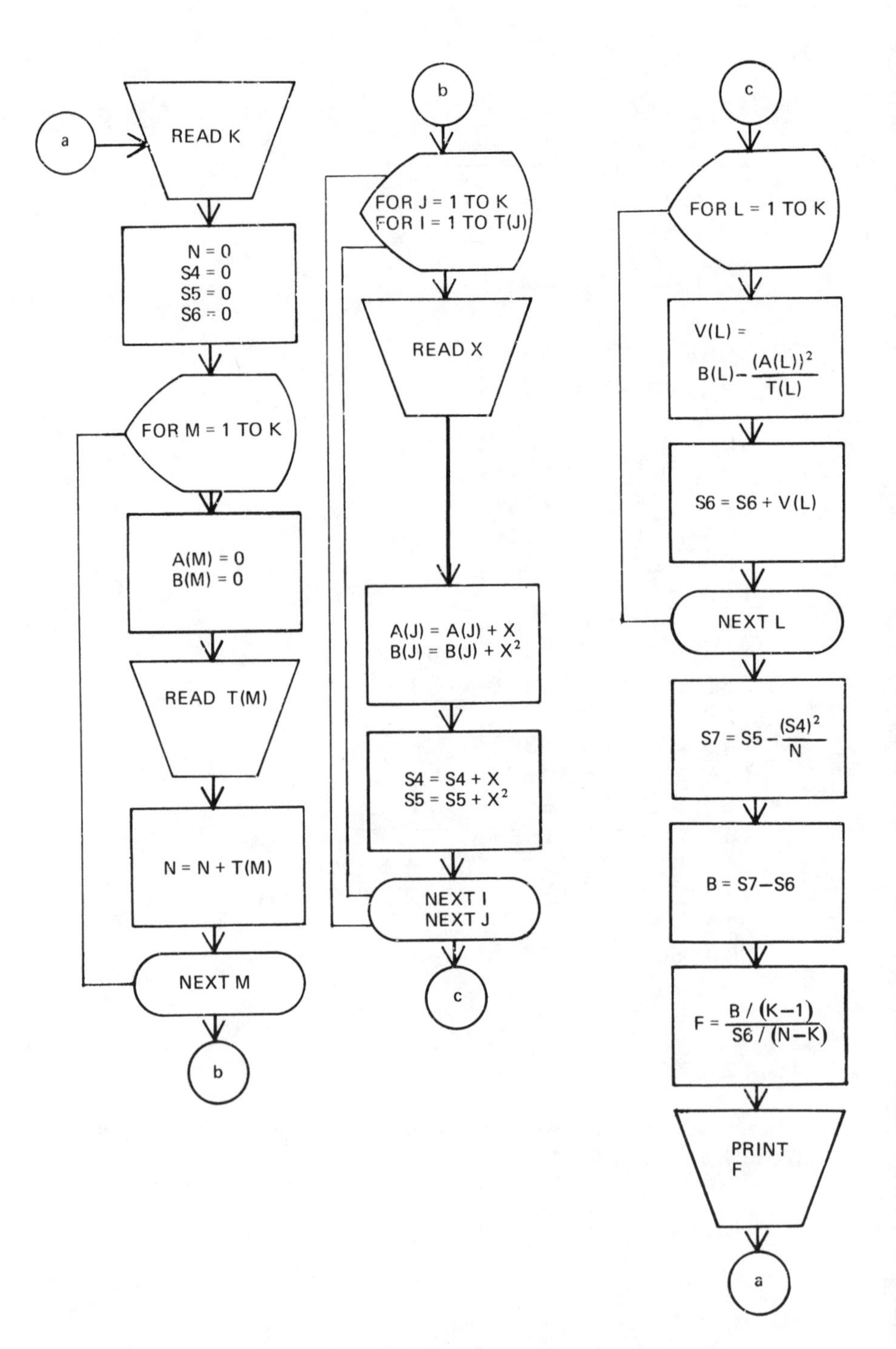

Fig. 7.9. Flowchart 22

Appendix A
Additional Basic Statements and Functions

The STOP Statment

Most Basic compilers will not accept more than one END statement. If you want the computer to stop executing a program before it reaches the END or last statement in the program, you can use the STOP statement.

General Form:

Line number STOP

For example,

```
75  STOP
```

The REM (Remark) Statement

The REM statement is used to place important remarks in a Basic program. Any number of remarks may be inserted in a Basic program to explain the nature of the program; give directions for general usage of the program; identify important parts of the program; and explain the user's input and/or the computer output of data. This statement is a nonexecutable statement.

General Form:

Line number REM remarks

For example:

```
10    REM THIS PROGRAM PRINTS A TABLE OF
15    REM SQUARES AND SQUARE ROOTS FOR THE
20    REM COUNTING NUMBERS FROM 1 TO 100.
25    PRINT "N↑2", "SQR(N)"
30    FOR I = 1 TO 100
40        PRINT I↑2, SQR(I)
50    NEXT I
60    END
```

The RESTORE Statement

General Form:

n RESTORE

where n is the statement number.

For example:

```
220 RESTORE
```

The RESTORE statement commands the computer to restore the data pointer to the beginning of the DATA list. That is, after a set of DATA has been used, it can be used again in the same Basic program.

The DEF (Definition) Statement

General Form:

line number DEF $FNm(n)$ = rule for a function of n,
where m and n are any letters of the alphabet.

The DEF statement tells the computer how to evaluate a function, given a particular argument for n. Computation takes place only when a statement containing DEF $FNm(n)$ is assigned a particular value for n and is subsequently executed.

For example, suppose we want to use the DEF statement to print out a table of the first ten counting numbers and their squares.

Solution:

```
10 DEF FNA(X)=X↑2
20 FOR I=1 TO 10
30     PRINT I, FNA(I)
40 NEXT I
50 END


RUN

                  15:48    06/24/75

 1                         1
 2                         4
 3                         9
 4                         16
 5                         25
 6                         36
 7                         49
 8                         64
 9                         81
 10                        100

TIME 0.0 SECS.
```

The GO SUB (Subroutine) and RETURN Statements

In many computer programs a certain sequence of statements must be executed more than once. A programmer can use a subroutine to avoid retyping this sequence of statements.

General Forms and Sample Diagram:

line number GO SUB line number

line number RETURN

```
10
20
30    GO SUB 150
40
50
60    GO SUB 150
70
80
```

```
                 90
                 100
                 110   GO SUB 150
                 120
                 130
                 140   STOP
                ┌150
Subroutine      │160
150             │170
                └180   RETURN
                 190
                 200   END
```

If this sample were an actual program, statement 30 would command the computer to execute statements 150, 160, 170, and then return to execute statement 40; statement 60 would command the computer to execute statements 150, 160, 170, and then return to execute statement 70; statement 110 would command the computer to execute statements 150, 160, 170, and then return to execute statement 120.

Samples Basic Program

```
10 LET R=50
15 LET D=324
20 GO SUB 100
30 LET R=73
35 LET D=2134
40 GO SUB 100
50 LET R=55
55 LET D=1074
60 GO SUB 100
70 STOP
80 REM    SUBROUTINE 100 COMPUTES AND PRINTS TRAVEL TIME
90 REM             GIVEN DISTANCE AND TIME
100 LET T=D/R
110 PRINT "DISTANCE="D,"RATE="R,"TIME="T
120 RETURN
130 END


RUN

               15:27    06/24/75

DISTANCE= 324             RATE= 50             TIME= 6.48
DISTANCE= 2134            RATE= 73             TIME= 29.2329
DISTANCE= 1074            RATE= 55             TIME= 19.5273

TIME 0.0 SECS.
```

THE GREATEST INTEGER FUNCTION

General Form:

INT (X)

where X is any real number, commands the computer to compute the greatest integer less than or equal to X.

Sample Basic Program

```
10  PRINT INT(0);INT(8)                          -0    8
20  PRINT INT(0.9);INT(8.7);INT(-1);INT(-45)     0    8      -1    -45
30  PRINT INT(87.88);INT(3/2)                   -87  -1
40  END
```

THE RANDOM NUMBER FUNCTION

General Form:

RND (X)

where X is any real number. (The X is a meaningless but necessary part of the function.) This instruction commands the computer to take a random number between 0 and 1.

Sample Basic Program	Computer Output	
5 FOR I=1 TO 4	.21132	.14464
10 PRINT RND(8),RND(27)	.852625	.927054
15 NEXT I	.162866	.433095
20 END	.563933	.20965

Many problems associated with probability and statistics require random-number selections, but not necessarily random numbers between 0 and 1. Suppose we needed a random number between 0 and 9 inclusive. RND(X) gives us numbers of the form:

$.0xxx\cdots$
$.1xxx\cdots$
$.2xxx\cdots$ (where $x = 0, 1, 2, \ldots, 9$)

.

.

.

$.9xxx\cdots$

If we combine INT(X) and RND(X) in the following fashion:

INT (10*RND(X))

where X is any real number, the computer would select a random number between 0 and 9, inclusive. Can you convince yourself this is true? Look at the form illustration above.

If we wanted a random number between 1 and 10, inclusive, we would use the following statement:

INT(10*RND(X)+1)

Convince yourself that this is true.

In general, if we want a random number between B and A + B − 1 inclusive, we can use the following formula:

INT (A*RND(X)+B)

where X is a meaningless but necessary real number.

Example

Wanted: A Basic program to print out two random numbers between 27 and 54 inclusive.

Solution:

```
B=27      A+B-1=54
         A+27-1=54
            A+26=54
               A=28

2  PRINT INT (28*RND(2)+27)
4  PRINT INT (28*RND(3)+27)
6  END
```

Exercise

Run the program above on a computer terminal. Run it several times.

COMPUTER APPLICATION

A state lottery number is selected each month. The lottery number consists of six digits, each digit a number from 0 to 9, inclusive.

Wanted: A flowchart and a Basic program to select the lottery digits from right to left (that is, select the one's digit first, the ten's digit second, ...) and then print out the lottery number in its proper order (that is, the hundred-thousand's digit first, the ten-thousand's digit second, ...).

Solution

Flowchart

Let N1 = a random number between 0 and 9 inclusive

Let N2 = a random number between 0 and 9 inclusive

.

.

.

Let N6 = a random number between 0 and 9 inclusive

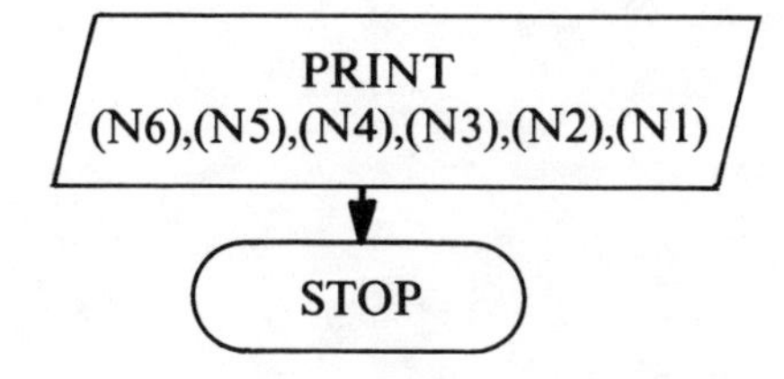

Coding

```
10  LET N1 = INT (10*RND (2))
12  LET N2 = INT (10*RND (4))
14  LET N3 = INT (10*RND (5))
16  LET N4 = INT (10*RND (6))
18  LET N5 = INT (10*RND (7))
20  LET N6 = INT (10*RND (8))
22  PRINT N6;N5;N4;N3;N2;N1
24  END
```

Exercise

Run this program.

Practice Problems

1. Give the computer output for the following program:

```
 5  PRINT ABS(-7.34); INT(-7.34)
10  PRINT INT(-15.32*10 + .6)/10
15  PRINT INT(SQR(4))
20  PRINT INT(-.5*SQR(4 + 21))/(-5)
25  END
```

2. Punch the following program on paper tape and then run it four times on a computer terminal:

```
10  PRINT RND (2), RND(3), RND(27)
15  END
```

3. Write and run a Basic program to print out a random number between

(a) 0 and 99 inclusive.
(b) 1 and 99 inclusive.
(c) 1 and 6 inclusive.
(d) 55 and 65 inclusive.

Appendix B
Common Basic Compiler Error Messages

UNDEFINED VARIABLE
UNACCEPTABLE VARIABLE NAME
SYNTAX ERROR
MISSING LEFT (RIGHT) PARENTHESIS
MISSING OR ILLEGAL TO
MISSING LEFT (RIGHT) QUOTE MARKS
ILLEGAL OR MISSING THEN
ILLEGAL OR MISSING STEP
FOR WITHOUT MATCHING NEXT
NEXT WITHOUT MATCHING FOR
SQUARE ROOT OF A NEGATIVE ARGUMENT
UNDEFINED STATEMENT REFERRED TO
GO SUB WITHOUT MATCHING RETURN
RETURN WITHOUT MATCHING GO SUB
SUBSCRIPTED VARIABLE TOO LARGE
SUBSCRIPTED VARIABLE NOT DIMENSIONED
SUBSCRIPTED VARIABLE EXCEEDS ITS DIMENSION
UNACCEPTABLE OR MISSING SUBSCRIPT
SIN OR TAN ARGUMENT IS TOO LARGE
BAD FORMAT
BAD INPUT
LAST STATEMENT NOT END
BAD DATA
EXPONENTIAL OVERFLOW - WARNING
FUNCTION DEFINED MORE THAN ONCE
LOG OF A NEGATIVE ARGUMENT

MATRIX ON BOTH SIDES
MATRIX NOT SQUARE
ILLEGAL OR MISSING FUNCTION NAME
ILLEGAL OR MISSING RELATIONAL OPERATOR
ALMOST A SINGULAR MATRIX
ILLEGAL OR MISSING BINARY OPERATOR
NO MORE DATA
NO MORE STORAGE
OVERFLOW - WARNING
UNDERFLOW - WARNING
MATRIX HAS A ZERO DIMENSION
DIMENSIONS DO NOT ALLOW MULTIPLICATION
ILLEGAL OR MISSING STRING DIMENSION

Appendix C
A List of Basic Program Statements

Type of Statement	Sample	Reference
READ	5 READ X,Y 13 READ T(1),A,Z(4)	Art. 1.5
INPUT	11 INPUT X,T,R 29 INPUT T(5),S(9)	Art. 1.5
DATA	112 DATA 2,3.2,97.533,8	Art. 1.5
LET	27 LET X=2+3↑2 94 LET T=A+3*B(4)-7	Art. 1.5
PRINT	35 PRINT 2+3↑5-27 40 PRINT T(3),X 143 PRINT "SUM1=", S	Art. 1.5
GO TO	22 GO TO 5	Art. 1.5
IF-THEN	25 IF N>27 THEN 9 37 IF A-B=C+7 THEN 10	Art. 1.5
FOR	12 FOR X=1 TO 23 33 FOR Y=5 TO 7 STEP .1 95 FOR T=R TO S STEP2	Art. 1.5

Type of Statement	Sample	Reference
NEXT	55 NEXT Y 66 NEXT S2	Art. 1.5
END	200 END	Art. 1.5
DIM	2 DIM A(5),B(34),T(2,3)	Art. 1.7
STOP	123 STOP	Appendix A
REM	5 REM THIS IS QUADRATIC PROGRAM	Appendix A
DEF	57 DEF FNA(T)=T↑2-3*T+2	Appendix A
GO SUB	10 GO SUB 222	Appendix A
RETURN	226 RETURN	Appendix A
RESTORE	99 RESTORE	Appendix A

Appendix D Statistical and Probability Formulas and Algorithms for Computations

CHAPTER 2

Measures of Central Tendency for Ungrouped Data

ARITHMETIC MEAN

$$\overline{X} = \frac{\sum_{i=1}^{n} X_i}{n}$$

where X_i = value of individual measurement
n = total number of individual measurements

MEDIAN

$$Md = \begin{cases} X_{n+1}/2 & \text{if } n \text{ is odd} \\ (X_{n/2} + X_{(n+2)/2})/2 & \text{if } n \text{ is even} \end{cases}$$

where X_n = value of individual measurement
n = total number of individual measurements

Measures for Deviation for Ungrouped Data

VARIANCE

$$s^2 = \frac{\sum_{i=1}^{n} (X_i - \bar{X})^2}{n}$$

where X_i = value of individual measurement
$\overline{X}$ = arithmetic mean of total number of measurements
n = total number of measurements

VARIANCE *(Alternate Formula)*

$$s^2 = \frac{\sum_{i=1}^{n} (X_i)^2 - \left[\left(\sum_{i=1}^{n} X_i\right)^2 / n\right]}{n}$$

STANDARD DEVIATION

$$s = \sqrt{\text{variance}}$$

Variance s^2 is computed by either of the preceding formulas.

Measures of Central Tendency for Grouped Data

ARITHMETIC MEAN

$$X = \frac{\sum_{i=1}^{n} X_i f_i}{n}$$

where x_i = midpoint of each class interval
f_i = corresponding frequencies in each class interval
n = total frequencies

MEDIAN OR 50TH PERCENTILE

$$P_x = l + \frac{w(p - cf)}{f}$$

where = lower class boundary of the class containing the position of the desired percentile
w = width of class interval
p = position of desired percentile, where $p = n(x/100)$
cf = number of cumulative frequencies in the class one step below the class containing the position of the desired percentile
f = number of frequencies in the class containing the position of the desired percentile
n = total number of frequencies

Measures of Deviation for Grouped Data

VARIANCE

$$s^2 = \frac{\sum_{i=1}^{n}(x_i^2 f_i) - \sum_{i=1}^{n}(x_i f_i)^2/n}{n}$$

where x_i = midpoint of each class interval
f_i = number of frequencies in each corresponding class interval
n = total number of frequencies in the distribution

STANDARD DEVIATION

$$s = \sqrt{\text{variance}}$$

Variance s^2 for grouped data is computed by the preceding formula.

CHAPTER 3

FUNDAMENTAL PRINCIPLE OF COUNTING

$$n_1 \cdot n_2 \cdot n_3 \cdots n_k$$

where n_1 = number of ways a task can be performed
n_2 = number of ways a second task can be performed *after* first task has been completed
n_3 = number of ways a third task can be performed *after* completion of first and second tasks
n_k = number of ways the kth task can be performed

FACTORIAL NOTATION

$$x! = x(x - 1)(x - 2) \cdots (1)$$
$$n! = n(n - 1)!$$
$$0! = 1$$
$$1! = 1$$

PERMUTATIONS (*Order* is *important.*)

$$_nP_r = n(n-1)(n-2)\cdots(n-r+1)$$

$$_nP_r = \frac{n!}{(n-r)!}$$

The number of permutations of n objects, where a_1 are alike, a_2 are alike, . . ., a_m are alike is

$$\frac{_nP_n}{a_1!a_2!\cdots a_m!} = \frac{n!}{a_1!a_2!\cdots a_m!}$$

COMBINATIONS (*Order is* not *important.*)

$$_nC_r = \frac{_nP_r}{r!} \quad \text{or} \quad _nC_r = \frac{n!}{r!(n-r)!}$$

THE BINOMIAL THEOREM

In general, the binomial theorem states

$$(g+d)^n = g^n + \frac{n}{1}g^{n-1}d + \frac{n(n-1)}{1(2)}g^{n-2}d^2 + \cdots + d^n$$

where $n \varepsilon \{1,2,3\cdots\}$; or

$$(g+d)^n = {_nC_0}g^n + {_nC_1}g^{n-1}d + {_nC_2}g^{n-2}d + \cdots + {_nC_n}d^n$$

where $n \varepsilon \{1,2,3\cdots\}$.

The term containing g^r in the expansion of $(g+d)^r$ is equal to $_nC_rg^rd^{n-r}$.

CHAPTER 4

PROBABILITY OF AN EVENT E

Given a sample space $S = \{e_1, e_2, e_3, \ldots, e_n\}$ associated with some experiment and some event $E = \{o_1, o_2, o_3, \ldots, o_m\}$ such that $E \subset S$, then the *probability* of event E is denoted by $P(E) = p(e_1) + p(e_2) + p(e_3) + \cdots + p(o_m)$.

Note: If each of the sample points in S has the same probability [that is, $p(o_1) = p(o_2) = \cdots = p(o_m)$], then $P(E) = m/n$, where $m =$ the number of sample points in event E and $n =$ the number of sample points in the sample space S.

Probability Formulas

Given E and F are events in some finite sample space, then

$$\begin{aligned}
P(S) &= 1 \\
P(E') &= 1 - P(E) \\
P(E) &= 1 - P(E') \\
P(E \cup F) &= P(E) + P(F) - P(E \cap F) \\
P(E|F) &= \frac{P(E \cap F)}{P(F)} \\
P(F|E) &= \frac{P(E \cap F)}{P(E)}
\end{aligned}$$

Bayes' Theorem

Given a sample space S and events $E_1, E_2, \ldots, E_n, D$ in S such that $E_1 \cup E_2 \cup \cdots \cup E_n = S$ and $E_1 \cap E_2 \cap \cdots E_n = \phi$ if $i \neq j$, then

$$P(E_i|D) = \frac{P(E_i)P(D|E_i)}{P(E_1)P(D|E_1) + P(E_2)P(D|E_2) + \cdots + P(E_n)P(D|E_n)}$$

for any $i = 1,2,3, \ldots, n$.

Probability of r Successes in n Trials of a Binomial Experiment

The probability of r successes in n trials of a binomial experiment, where $p =$ probability of success on any one trial and $q =$ probability of failure on any one trial, is

$$_nC_n p^r q^{n-r} \qquad p + q = 1$$
$$q = 1 - p$$

CHAPTER 5

Given a random variable X defined on a sample space S associated with an experiment, with a probability distribution $\{(X_i, p(X_i))\}$, where

n = the number of trials, p = the probability of success, q = probability of failure, we have determined:

Measure	In General	If the Experiment is Binomi
The MEAN of X	$\mu = \sum_{i=1}^{n} X_i p(X_i)$	$\mu = np$
The VARIANCE of X	$\sigma^2 = \sum_{i=1}^{n} (X_i - \mu)^2 p(X_i)$	$\sigma^2 = npq$
The STANDARD DEVIATION of X	$\sigma = \sqrt{\sigma^2}$	$\sigma = \sqrt{npq}$

CONVERSION OF NORMAL (X) DISTRIBUTION SCORE TO STANDARD NORMAL (Z) DISTRIBUTION SCORE

$$Z = \frac{X - \mu}{\sigma}$$

where X = value assumed by random variable
μ = population mean
σ = population standard deviation

NORMALIZING RAW TEST SCORES

To normalize a raw test score X, one must compute a normal score $N = \mu + Z\sigma$, where
$Z = (X - \overline{X})/s$
μ = population mean (or an estimate of the population mean)
σ = population standard deviation (or an estimate of the population standard deviation)

CHAPTER 6

OBSERVED Z-SCORE

The observed Z-score for the difference between the sample mean and the population mean, where $n \geq 30$, is

$$Z = \frac{\overline{X} - \mu}{\sigma / \sqrt{n}}$$

UNBIASED STANDARD DEVIATION FORMULA TO ESTIMATE THE POPULATION STANDARD DEVIATION

$$s' = \sqrt{\frac{\sum_{i=1}^{n}(X_i - \bar{X})^2}{n - 1}} \quad \text{or} \quad s' = \sqrt{\frac{n}{n-1}}(s)$$

OBSERVED t-SCORE

In testing for differences between a sample mean and a population mean, where the size of the sample $n < 30$, the observed t = score is computed as follows:

$$t = \frac{\bar{X} - \mu}{s'/\sqrt{n}}$$

where $s' = \sqrt{\frac{n}{n-1}}(s)$

$n < 30$
$\text{df} = n - 1$

ALGORITHM FOR HYPOTHESIS TESTING

1. State the motivated hypothesis (H_m) and the null hypothesis (H_o).
2. Decide whether you are using a one-tail test or a two-tail test.
3. Decide whether to use a critical Z-score ($n \geq 30$) or a critical t-score ($n < 30$).
4. Find the critical Z-score (Table 5 in Appendix F) or the critical t-score (Table 5) for the desired level of significance.
5. Compute the observed Z_i score or the observed t_i score.
6. State your conclusion (that is, we *can* reject H_o or we *cannot* reject H_o).

OBSERVED Z-, t-SCORES FOR DIFFERENCES BETWEEN A SAMPLE MEAN ($\bar{X}$) AND A POPULATION MEAN (μ)

$$Z_{\text{obs}} = \frac{\bar{X} - \mu}{\sigma/\sqrt{n}}$$

where $n \geq 30$, and

$$t_{\text{obs}} = \frac{\bar{X} - \mu}{s'/\sqrt{n}}$$

where $n < 30$

$$s' = \sqrt{\frac{n}{n-1}}\,(s)$$

$$\text{df} = n - 1$$

OBSERVED Z-, *t*-SCORES FOR DIFFERENCES BETWEEN TWO SAMPLE MEANS ($\bar{Y}_1$ AND $\bar{Y}_2$)

$$Z_{\text{obs}} = \frac{\bar{Y}_1 - \bar{Y}_2}{\sqrt{(s_1^2/n_1) + (s_2^2/n_2)}}$$

where $n_1 \geq 30$ and $n_2 \geq 30$.

$$t_{\text{obs}} = \frac{\bar{Y}_1 - \bar{Y}_2}{\sqrt{((n_1)s_1^2 + (n_2)s^2_2)/(n_1 + n_2 - 2)}\,(\sqrt{(1/n_1) + (1/n_2)})}$$

where $n_1 < 30$ or $n_2 < 30$

$$\text{df} = n_1 + n_2 - 2$$

$$s^2 = \frac{\Sigma(Y - \bar{Y})^2}{n}$$

CHAPTER 7

CHI SQUARE

$$\chi^2 = \sum_{i=1}^{n} \frac{(f_i - e_i)^2}{e_i}$$

where f_i = frequencies of observed (sample) distribution
e_i = frequencies of expected distribution
n = number of outcomes

ESTIMATED LINEAR REGRESSION EQUATION

$$y_p = a + bx$$

where y_p = value for dependent variable
x = value for independent variable

$$\text{(Least squares estimator) } a = \frac{(\Sigma y)(\Sigma x^2) - (\Sigma x)(\Sigma xy)}{n(\Sigma x^2) - (\Sigma x)^2}$$

$$\text{(Least squares estimator) } b = \frac{n(\Sigma xy) - (\Sigma x)(\Sigma y)}{n(\Sigma x^2) - (\Sigma x)^2}$$

where n = number of given ordered pairs

THE COEFFICIENT OF (LINEAR) CORRELATION. (Restriction: $-1 \leq r \leq 1$).

$$r_{xy} = \frac{n(\Sigma xy) - (\Sigma x)(\Sigma y)}{\sqrt{[n(\Sigma x^2) - (\Sigma x)^2]}\sqrt{[n(\Sigma y^2) - (\Sigma y)^2]}}$$

ANALYSIS OF VARIANCE

One-Way Classification Algorithm for Computing the F-Ratio Statistic for Testing Equality of Means

1. Compute the following statistics:
   (a) *Within sum of squares* = variation 1 + variation 2 + $\cdots$ + variation k, where variation $i = \Sigma x_i^2 - ((\Sigma x_i)^2/n_i)$ (that is, the sum of the variations for each treatment).
   (b) *Total sum of squares* = total variation of all the subjects = $\Sigma x^2 - (\Sigma x)^2/n$.
   (c) *Between sum of squares* = (total sum of squares) − (within sum of squares).
   (d) *Between mean square (larger variance)* = (between sum of squares)/$(k - 1)$. (This statistic represents the variance among the means.)
   (e) *Within mean square (smaller variance)* = (within sum of squares)/$(n - k)$. (This statistic represents the variance within the samples.)
2. Compute the observed F (variance) RATIO:

$$F_{k-1,n-k} = \frac{\text{BETWEEN MEAN SQUARE}}{\text{WITHIN MEAN SQUARE}}$$

where $k - 1$ = degrees of freedom for the *between mean square* $n - k$ = degrees of freedom for the *within mean square.*

3. Look up the critical $F_{k-1,n-k}$ score in Table F.8 for the desired level of significance.

4. Conclusion:
   (a) We can reject H_o if the calculated F falls in this area; that is, if the calculated F ratio $>$ critical F ratio in the table of F-values, or
   (b) We *cannot* reject H_o.

Appendix E
Programming Projects

Instructions

Construct a flowchart for each of the following programming projects, code your flowchart in the Basic programming language, and run the program. Each project requires a considerable amount of thought and effort. Any one of the problems can be done by an individual, but it is suggested that all be done by groups of two or more people. It is very unlikely that your first attempt to run the program will be perfect.

An account should be kept of each run, the encountered errors, and your corrections.

Problems

1. Given: The following information for *each student* in a certain two-year college for a given semester: $N, A, B, C, D, F, W, I,$ where

$$N = \begin{cases} 1 & \text{if the student is in his first year} \\ 2 & \text{if the student is in his second year} \\ 3 & \text{if the student is nonmatriculated; and} \\ 4 & \text{signifies that there are no more data to be processed} \end{cases}$$

A = number of A's obtained by the student this semester
B = number of B's obtained by the student this semester
C = number of C's for this semester
D = number of D's for this semester

F = number of F's for this semester
W = number of withdrawals this semester
I = number of incompletes this semester

Wanted: A Basic program to compute and print out:

TOTAL FIRST-YEAR STUDENTS =
TOTAL SECOND-YEAR STUDENTS =
TOTAL NONMATRICULATED STUDENTS =

Output:

	Average Number of						
	A	B	C	D	F	W	I
(a) For first-year students							
(b) For second-year students							
(c) For nonmatriculated students							

Sample Data: 2,2,3,1,1,0,0,1 3,3,2,1,0,0,0,0
1,0,0,1,3,1,1,1 1,0,0,2,2,0,1,0
2,1,2,1,0,1,1,0 4,0,0,0,0,0,0,0
3,0,2,2,0,0,0,1

2. Given: DATA R(1), R(2), R(3), R(4), R(5) (rates of pay) and DATA I, N, G, H, C, D. S for *each employee* in a certain factory, where

I = identification number
N = employee's name
G = gross yearly pay previous to this week
H = number of hours worked this week
C = 1,2,3,4,5 to indicate employee's rate of pay
D = employee's number of dependents
S = 0,1, where 1 implies a savings bond deduction of \$18.75 and 0 implies no savings bond deduction.

Wanted: A Basic program to read and process this information. The following will apply to *each* employee:

(a) A 4½% Social Security deduction (based on gross weekly pay) is made for each employee until his gross yearly pay totals \$7600; beyond this figure there is no deduction.
(b) Income tax deduction:

(1) 33% if 0 dependents (3) 20% if 2 dependents
(2) 25% if 1 dependent (4) 12% if 3 dependents

(c) Regular time is 35 hours and each overtime hour is paid at regular time and one-half.

Output:

ID NAME WEEKS GROSS PAY WEEKS NET PAY YEARLY GROSS PAY

Sample Data: 2.75, 3.50, 3.75, 4.25, 4.75
882, F. JONES, 4711.11, 42, 4, 2, 0
961, H. CAPPS, 6800, 48, 5, 3, 1

3. Given: DATA N, A(1), A(2), . . ., A(N)

Wanted: A Basic program to sort and arrange N numbers: moreover (a) from smallest to largest and (b) from largest to smallest.

Output:

$X_1, X_2, \ldots$ (smallest to largest)
$Y_1, Y_2, \ldots$ (largest to smallest)

Sample Data: 5, 7, 8, 2, 10, 2
12, 2, 7, 1, 99, 72, 84, 6, 8, 16, 11, 10, 101

4. Given: The Taylor Series for approximating SINE(X) and COSINE(X):

$$\sin(X) = X - \frac{X^3}{3!} + \frac{X^5}{5!} - \frac{X^7}{7!} + \cdots$$

$$\cos(X) = 1 - \frac{X^2}{2!} + \frac{X^4}{4!} - \frac{X^6}{6!} + \cdots$$

where X is in radian measure.

Wanted: A Basic program that will approximate SINE(X) and COSINE(X) in the following manner: Have the computer approximate the SINE(X) and the COSINE(X) by adding or subtracting terms until $|\text{term}| < 0.000001$.

Output:

Degrees	Radians	SINE(X)	COSINE(X)
0			
1			
2			
.			
.			
.			
45			

JOB COMPLETED BY ________.
(Name)

5. Given: DATA X, Y, where X = the total of your purchases at a department store, and Y = what you offer to pay the cashier. Assume $Y \geq X$.

Wanted: A Basic program to compute and print out the following information:

TOTAL OF PURCHASES = OFFER TO PAY =
NUMBER OF BILLS AND/OR COINS (change):

Twenties =	Quarters =
Tens =	Dimes =
Fives =	Nickels =
Ones =	Pennies =
Halves =	

Sample Data: 26.50, 50
15.99, 50
15.99, 100

Appendix F
Tables of Statistics

Table 1. Squares and Square Roots for Numbers 1 Through 1,000

n	n^2	$\sqrt{n}$	$\sqrt{10n}$
1	1	1.000 000	3.162 278
2	4	1.414 214	4.472 136
3	9	1.732 051	5.477 226
4	16	2.000 000	6.324 555
5	25	2.236 068	7.071 068
6	36	2.449 490	7.745 967
7	49	2.645 751	8.366 600
8	64	2.828 427	8.944 272
9	81	3.000 000	9.486 833
10	100	3.162 278	10.00000
11	121	3.316 625	10.48809
12	144	3.464 102	10.95445
13	169	3.605 551	11.40175
14	196	3.741 657	11.83216
15	225	3.872 983	12.24745
16	256	4.000 000	12.64911
17	289	4.123 106	13.03840
18	324	4.242 641	13.41641
19	361	4.358 899	13.78405
20	400	4.472 136	14.14214
21	441	4.582 576	14.49138
22	484	4.690 416	14.83240
23	529	4.795 832	15.16575
24	576	4.898 979	15.49193
25	625	5.000 000	15.81139
26	676	5.099 020	16.12452
27	729	5.196 152	16.43168
28	784	5.291 503	16.73320
29	841	5.385 165	17.02939
30	900	5.477 226	17.32051
31	961	5.567 764	17.60682
32	1 024	5.656 854	17.88854
33	1 089	5.744 563	18.16590
34	1 156	5.830 952	18.43909
35	1 225	5.916 080	18.70829
36	1 296	6.000 000	18.97367
37	1 369	6.082 763	19.23538
38	1 444	6.164 414	19.49359
39	1 521	6.244 998	19.74842
40	1 600	6.324 555	20.00000
41	1 681	6.403 124	20.24846
42	1 764	6.480 741	20.49390
43	1 849	6.557 439	20.73644
44	1 936	6.633 250	20.97618
45	2 025	6.708 204	21.21320
46	2 116	6.782 330	21.44761
47	2 209	6.855 655	21.67948
48	2 304	6.928 203	21.90890
49	2 401	7.000 000	22.13594
50	2 500	7.071 068	22.36068
51	2 601	7.141 428	22.58318
52	2 704	7.211 103	22.80351
53	2 809	7.280 110	23.02173
54	2 916	7.348 469	23.23790
55	3 025	7.416 198	23.45208
56	3 136	7.483 315	23.66432
57	3 249	7.549 834	23.87467
58	3 364	7.615 773	24.08319
59	3 481	7.681 146	24.28992
60	3 600	7.745 967	24.49490
61	3 721	7.810 250	24.69818
62	3 844	7.874 008	24.89980
63	3 969	7.937 254	25.09980
64	4 096	8.000 000	25.29822
65	4 225	8.062 258	25.49510
66	4 356	8.124 038	25.69047
67	4 489	8.185 353	25.88436
68	4 624	8.246 211	26.07681
69	4 761	8.306 624	26.26785
70	4 900	8.366 600	26.45751
71	5 041	8.426 150	26.64583
72	5 184	8.485 281	26.83282
73	5 329	8.544 004	27.01851
74	5 476	8.602 325	27.20294
75	5 625	8.660 254	27.38613
76	5 776	8.717 798	27.56810
77	5 929	8.774 964	27.74887
78	6 084	8.831 761	27.92848
79	6 241	8.888 194	28.10694
80	6 400	8.944 272	28.28427
81	6 561	9.000 000	28.46050
82	6 724	9.055 385	28.63564
83	6 889	9.110 434	28.80972
84	7 056	9.165 151	28.98275
85	7 225	9.219 544	29.15476
86	7 396	9.273 618	29.32576
87	7 569	9.327 379	29.49576
88	7 744	9.380 832	29.66479
89	7 921	9.433 981	29.83287
90	8 100	9.486 833	30.00000
91	8 281	9.539 392	30.16621
92	8 464	9.591 663	30.33150
93	8 649	9.643 651	30.49590
94	8 836	9.695 360	30.65942
95	9 025	9.746 794	30.82207
96	9 216	9.797 959	30.98387
97	9 409	9.848 858	31.14482
98	9 604	9.899 495	31.30495
99	9 801	9.949 874	31.46427
100	10 000	10.00000	31.62278
101	10 201	10.04988	31.78050
102	10 404	10.09950	31.93744
103	10 609	10.14889	32.09361
104	10 816	10.19804	32.24903
105	11 025	10.24695	32.40370
106	11 236	10.29563	32.55764
107	11 449	10.34408	32.71085
108	11 664	10.39230	32.86335
109	11 881	10.44031	33.01515

Squares and Square Roots (Continued)

n	n^2	$\sqrt{n}$	$\sqrt{10n}$
110	12 100	10.48809	33.16625
111	12 321	10.53565	33.31666
112	12 544	10.58301	33.46640
113	12 769	10.63015	33.61547
114	12 996	10.67708	33.76389
115	13 225	10.72381	33.91165
116	13 456	10.77033	34.05877
117	13 689	10.81665	34.20526
118	13 924	10.86278	34.35113
119	14 161	10.90871	34.49638
120	14 400	10.95445	34.64102
121	14 641	11.00000	34.78505
122	14 884	11.04536	34.92850
123	15 129	11.09054	35.07136
124	15 376	11.13553	35.21363
125	15 625	11.18034	35.35534
126	15 876	11.22497	35.49648
127	16 129	11.26943	35.63706
128	16 384	11.31371	35.77709
129	16 641	11.35782	35.91657
130	16 900	11.40175	36.05551
131	17 161	11.44552	36.19392
132	17 424	11.48913	36.33180
133	17 689	11.53256	36.46917
134	17 956	11.57584	36.60601
135	18 225	11.61895	36.74235
136	18 496	11.66190	36.87818
137	18 769	11.70470	37.01351
138	19 044	11.74734	37.14835
139	19 321	11.78983	37.28270
140	19 600	11.83216	37.41657
141	19 881	11.87434	37.54997
142	20 164	11.91638	37.68289
143	20 449	11.95826	37.81534
144	20 736	12.00000	37.94733
145	21 025	12.04159	38.07887
146	21 316	12.08305	38.20995
147	21 609	12.12436	38.34058
148	21 904	12.16553	38.47077
149	22 201	12.20656	38.60052
150	22 500	12.24745	38.72983
151	22 801	12.28821	38.85872
152	23 104	12.32883	38.98718
153	23 409	12.36932	39.11521
154	23 716	12.40967	39.24283
155	24 025	12.44990	39.37004
156	24 336	12.49000	39.49684
157	24 649	12.52996	39.62323
158	24 964	12.56981	39.74921
159	25 281	12.60952	39.87480
160	25 600	12.64911	40.00000
161	25 921	12.68858	40.12481
162	26 244	12.72792	40.24922
163	26 569	12.76715	40.37326
164	26 806	12.80625	40.49691
165	27 225	12.84523	40.62019
166	27 556	12.88410	40.74310
167	27 889	12.92285	40.86563
168	28 224	12.96148	40.98780
169	28 561	13.00000	41.10961
170	28 900	13.03840	41.23106
171	29 241	13.07670	41.35215
172	29 584	13.11488	41.47288
173	29 929	13.15295	41.59327
174	30 276	13.19091	41.71331
175	30 625	13.22876	41.83300
176	30 976	13.26650	41.95235
177	31 329	13.30413	42.07137
178	31 684	13.34166	42.19005
179	32 041	13.37909	42.30839
180	32 400	13.41641	42.42641
181	32 761	13.45362	42.54409
182	33 124	13.49074	42.66146
183	33 489	13.52775	42.77850
184	33 856	13.56466	42.89522
185	34 225	13.60147	43.01163
186	34 596	13.63818	43.12772
187	34 969	13.67479	43.24350
188	35 344	13.71131	43.35897
189	35 721	13.74773	43.47413
190	36 100	13.78405	43.58899
191	36 481	13.82027	43.70355
192	36 864	13.85641	43.81780
193	37 249	13.89244	43.93177
194	37 636	13.92839	44.04543
195	38 025	13.96424	44.15880
196	38 416	14.00000	44.27189
197	38 809	14.03567	44.38468
198	39 204	14.07125	44.49719
199	39 601	14.10674	44.60942
200	40 000	14.14214	44.72136
201	40 401	14.17745	44.83302
202	40 804	14.21267	44.94441
203	41 209	14.24781	45.05552
204	41 616	14.28286	45.16636
205	42 025	14.31782	45.27693
206	42 436	14.35270	45.38722
207	42 849	14.38749	45.49725
208	43 264	14.42221	45.60702
209	43 681	14.45683	45.71652
210	44 100	14.49138	45.82576
211	44 521	14.52584	45.93474
212	44 944	14.56022	46.04346
213	45 369	14.59452	46.15192
214	45 796	14.62874	46.26013
215	46 225	14.66288	46.36809
216	46 656	14.69694	46.47580
217	47 089	14.73092	46.58326
218	47 524	14.76482	46.69047
219	47 961	14.79865	46.79744
220	48 400	14.83240	46.90416
221	48 841	14.86607	47.01064
222	49 284	14.89966	47.11688
223	49 729	14.93318	47.22288
224	50 176	14.96663	47.32864
225	50 625	15.00000	47.43416
226	51 076	15.03330	47.53946
227	51 529	15.06652	47.64452
228	51 984	15.09967	47.74935
229	52 441	15.13275	47.85394

Squares and Square Roots (Continued)

n	n^2	$\sqrt{n}$	$\sqrt{10n}$
230	52 900	15.16575	47.95832
231	53 361	15.19868	48.06246
232	53 824	15.23155	48.16638
233	54 289	15.26434	48.27007
234	54 756	15.29706	48.37355
235	55 225	15.32971	48.47680
236	55 696	15.36229	48.57983
237	56 169	15.39480	48.68265
238	56 644	15.42725	48.78524
239	57 121	15.45962	48.88763
240	57 600	15.49193	48.98979
241	58 081	15.52417	49.09175
242	58 564	15.55635	49.19350
243	59 049	15.58846	49.29503
244	59 536	15.62050	49.39636
245	60 025	15.65248	49.49747
246	60 516	15.68439	49.59839
247	61 009	15.71623	49.69909
248	61 504	15.74802	49.79960
249	62 001	15.77973	49.89990
250	62 500	15.81139	50.00000
251	63 001	15.84298	50.09990
252	63 504	15.87451	50.19960
253	64 009	15.90597	50.29911
254	64 516	15.93738	50.39841
255	65 025	15.96872	50.49752
256	65 536	16.00000	50.59644
257	66 049	16.03122	50.69517
258	66 564	16.06238	50.79370
259	67 081	16.09348	50.89204
260	67 600	16.12452	50.99020
261	68 121	16.15549	51.08816
262	68 644	16.18641	51.18594
263	69 169	16.21727	51.28353
264	69 696	16.24808	51.38093
265	70 225	16.27882	51.47815
266	70 756	16.30951	51.57519
267	71 289	16.34013	51.67204
268	71 824	16.37071	51.76872
269	72 361	16.40122	51.86521
270	72 900	16.43168	51.96152
271	73 441	16.46208	52.05766
272	73 984	16.49242	52.15362
273	74 529	16.52271	52.24940
274	75 076	16.55295	52.34501
275	75 625	16.58312	52.44044
276	76 176	16.61325	52.53570
277	76 729	16.64332	52.63079
278	77 284	16.67333	52.72571
279	77 841	16.70329	52.82045
280	78 400	16.73320	52.91503
281	78 961	16.76305	53.00943
282	79 524	16.79286	53.10367
283	80 089	16.82260	53.19774
284	80 656	16.85230	53.29165
285	81 225	16.88194	53.38539
286	81 796	16.91153	53.47897
287	82 369	16.94107	53.57238
288	82 944	16.97056	53.66563
289	83 521	17.00000	53.75872
290	84 100	17.02939	53.85165
291	84 681	17.05872	53.94442
292	85 264	17.08801	54.03702
293	85 849	17.11724	54.12947
294	86 436	17.14643	54.22177
295	87 025	17.17556	54.31390
296	87 616	17.20465	54.40588
297	88 209	17.23369	54.49771
298	88 804	17.26268	54.58938
299	89 401	17.29162	54.68089
300	90 000	17.32051	54.77226
301	90 601	17.34935	54.86347
302	91 204	17.37815	54.95453
303	91 809	17.40690	55.04544
304	92 416	17.43560	55.13620
305	93 025	17.46425	55.22681
306	93 636	17.49286	55.31727
307	94 249	17.52142	55.40758
308	94 864	17.54993	55.49775
309	95 481	17.57840	55.58777
310	96 100	17.60682	55.67764
311	96 721	17.63519	55.76737
312	97 344	17.66352	55.85696
313	97 969	17.69181	55.94640
314	98 596	17.72005	56.03570
315	99 225	17.74824	56.12486
316	99 856	17.77639	56.21388
317	100 489	17.80449	56.30275
318	101 124	17.83255	56.39149
319	101 761	17.86057	56.48008
320	102 400	17.88854	56.56854
321	103 041	17.91647	56.65686
322	103 684	17.94436	56.74504
323	104 329	17.97220	56.83309
324	104 976	18.00000	56.92100
325	105 625	18.02776	57.00877
326	106 276	18.05547	57.09641
327	106 929	18.08314	57.18391
328	107 584	18.11077	57.27128
329	108 241	18.13836	57.35852
330	108 900	18.16590	57.44563
331	109 561	18.19341	57.53260
332	110 224	18.22087	57.61944
333	110 889	18.24829	57.70615
334	111 556	18.27567	57.79273
335	112 225	18.30301	57.87918
336	112 896	18.33030	57.96551
337	113 569	18.35756	58.05170
338	114 244	18.38478	58.13777
339	114 921	18.41195	58.22371
340	115 600	18.43909	58.30952
341	116 281	18.46619	58.39521
342	116 964	18.49324	58.48077
343	117 649	18.52026	58.56620
344	118 336	18.54724	58.65151
345	119 025	18.57418	58.73670
346	119 716	18.60108	58.82176
347	120 409	18.62794	58.90671
348	121 104	18.65476	58.99152
349	121 801	18.68154	59.07622

Squares and Square Roots (Continued)

n	n^2	$\sqrt{n}$	$\sqrt{10n}$
350	122 500	18.70829	59.16080
351	123 201	18.73499	59.24525
352	123 904	18.76166	59.32959
353	124 609	18.78829	59.41380
354	125 316	18.81489	59.49790
355	126 025	18.84144	59.58188
356	126 736	18.86796	59.66574
357	127 449	18.89444	59.74948
358	128 164	18.92089	59.83310
359	128 881	18.94730	59.91661
360	129 600	18.97367	60.00000
361	130 321	19.00000	60.08328
362	131 044	19.02630	60.16644
363	131 769	19.05256	60.24948
364	132 496	19.07878	60.33241
365	133 225	19.10497	60.41523
366	133 956	19.13113	60.49793
367	134 689	19.15724	60.58052
368	135 424	19.18333	60.66300
369	136 161	19.20937	60.74537
370	136 900	19.23538	60.82763
371	137 641	19.26136	60.90977
372	138 384	19.28730	60.99180
373	139 129	19.31321	61.07373
374	139 876	19.33908	61.15554
375	140 625	19.36492	61.23724
376	141 376	19.39072	61.31884
377	142 129	19.41649	61.40033
378	142 884	19.44222	61.48170
379	143 641	19.46792	61.56298
380	144 400	19.49359	61.64414
381	145 161	19.51922	61.72520
382	145 924	19.54482	61.80615
383	146 689	19.57039	61.88699
384	147 456	19.59592	61.96773
385	148 225	19.62142	62.04837
386	148 996	19.64688	62.12890
387	149 769	19.67232	62.20932
388	150 544	19.69772	62.28965
389	151 321	19.72308	62.36986
390	152 100	19.74842	62.44998
391	152 881	19.77372	62.52999
392	153 664	19.79899	62.60990
393	154 449	19.82423	62.68971
394	155 236	19.84943	62.76942
395	156 025	19.87461	62.84903
396	156 816	19.89975	62.92853
397	157 609	19.92486	63.00794
398	158 404	19.94994	63.08724
399	159 201	19.97498	63.16645
400	160 000	20.00000	63.24555
401	160 801	20.02498	63.32456
402	161 604	20.04994	63.40347
403	162 409	20.07486	63.48228
404	163 216	20.09975	63.56099
405	164 025	20.12461	63.63961
406	164 836	20.14944	63.71813
407	165 649	20.17424	63.79655
408	166 464	20.19901	63.87488
409	167 281	20.22375	63.95311
410	168 100	20.24846	64.03124
411	168 921	20.27313	64.10928
412	169 744	20.29778	64.18723
413	170 569	20.32240	64.26508
414	171 396	20.34699	64.34283
415	172 225	20.37155	64.42049
416	173 056	20.39608	64.49806
417	173 889	20.42058	64.57554
418	174 724	20.44505	64.65292
419	175 561	20.46949	64.73021
420	176 400	20.49390	64.80741
421	177 241	20.51828	64.88451
422	178 084	20.54264	64.96153
423	178 929	20.56696	65.03845
424	179 776	20.59126	65.11528
425	180 625	20.61553	65.19202
426	181 476	20.63977	65.26868
427	182 329	20.66398	65.34524
428	183 184	20.68816	65.42171
429	184 041	20.71232	65.49809
430	184 900	20.73644	65.57439
431	185 761	20.76054	65.65059
432	186 624	20.78461	65.72671
433	187 489	20.80865	65.80274
434	188 356	20.83267	65.87868
435	189 225	20.85665	65.95453
436	190 096	20.88061	66.03030
437	190 969	20.90454	66.10598
438	191 844	20.92845	66.18157
439	192 721	20.95233	66.25708
440	193 600	20.97618	66.33250
441	194 481	21.00000	66.40783
442	195 364	21.02380	66.48308
443	196 249	21.04757	66.55825
444	197 136	21.07131	66.63332
445	198 025	21.09502	66.70832
446	198 916	21.11871	66.78323
447	199 809	21.14237	66.85806
448	200 704	21.16601	66.93280
449	201 601	21.18962	67.00746
450	202 500	21.21320	67.08204
451	203 401	21.23676	67.15653
452	204 304	21.26029	67.23095
453	205 209	21.28380	67.30527
454	206 116	21.30728	67.37952
455	207 025	21.33073	67.45369
456	207 936	21.35416	67.52777
457	208 849	21.37756	67.60178
458	209 764	21.40093	67.67570
459	210 681	21.42429	67.74954
460	211 600	21.44761	67.82330
461	212 521	21.47091	67.89698
462	213 444	21.49419	67.97058
463	214 369	21.51743	68.04410
464	215 296	21.54066	68.11755
465	216 225	21.56386	68.19091
466	217 156	21.58703	68.26419
467	218 089	21.61018	68.33740
468	219 024	21.63331	68.41053
469	219 961	21.65641	68.48357

Squares and Square Roots (Continued)

n	n^2	$\sqrt{n}$	$\sqrt{10n}$
470	220 900	21.67948	68.55655
471	221 841	21.70253	68.62944
472	222 784	21.72556	68.70226
473	223 729	21.74856	68.77500
474	224 676	21.77154	68.84766
475	225 625	21.79449	68.92024
476	226 576	21.81742	68.99275
477	227 529	21.84033	69.06519
478	228 484	21.86321	69.13754
479	229 441	21.88607	69.20983
480	230 400	21.90890	69.28203
481	231 361	21.93171	69.35416
482	232 324	21.95450	69.42622
483	233 289	21.97726	69.49820
484	234 256	22.00000	69.57011
485	235 225	22.02272	69.64194
486	236 196	22.04541	69.71370
487	237 169	22.06808	69.78539
488	238 144	22.09072	69.85700
489	239 121	22.11334	69.92853
490	240 100	22.13594	70.00000
491	241 081	22.15852	70.07139
492	242 064	22.18107	70.14271
493	243 049	22.20360	70.21396
494	244 036	22.22611	70.28513
495	245 025	22.24860	70.35624
496	246 016	22.27106	70.42727
497	247 009	22.29350	70.49823
498	248 004	22.31591	70.56912
499	249 001	22.33831	70.63993
500	250 000	22.36068	70.71068
501	251 001	22.38303	70.78135
502	252 004	22.40536	70.85196
503	253 009	22.42766	70.92249
504	254 016	22.44994	70.99296
505	255 025	22.47221	71.06335
506	256 036	22.49444	71.13368
507	257 049	22.51666	71.20393
508	258 064	22.53886	71.27412
509	259 081	22.56103	71.34424
510	260 100	22.58318	71.41428
511	261 121	22.60531	71.48426
512	262 144	22.62742	71.55418
513	263 169	22.64950	71.62402
514	264 196	22.67157	71.69379
515	265 225	22.69361	71.76350
516	266 256	22.71563	71.83314
517	267 289	22.73763	71.90271
518	268 324	22.75961	71.97222
519	269 361	22.78157	72.04165
520	270 400	22.80351	72.11103
521	271 441	22.82542	72.18033
522	272 484	22.84732	72.24957
523	273 529	22.86919	72.31874
524	274 576	22.89105	72.38784
525	275 625	22.91288	72.45688
526	276 676	22.93469	72.52586
527	277 729	22.95648	72.59477
528	278 784	22.97825	72.66361
529	279 841	23.00000	72.73239

n	n^2	$\sqrt{n}$	$\sqrt{10n}$
530	280 900	23.02173	72.80110
531	281 961	23.04344	72.86975
532	283 024	23.06513	72.93833
533	284 089	23.08679	73.00685
534	285 156	23.10844	73.07530
535	286 225	23.13007	73.14369
536	287 296	23.15167	73.21202
537	288 369	23.17326	73.28028
538	289 444	23.19483	73.34848
539	290 521	23.21637	73.41662
540	291 600	23.23790	73.48469
541	292 681	23.25941	73.55270
542	293 764	23.28089	73.62065
543	294 849	23.30236	73.68853
544	295 936	23.32381	73.75636
545	297 025	23.34524	73.82412
546	298 116	23.36664	73.89181
547	299 209	23.38803	73.95945
548	300 304	23.40940	74.02702
549	301 401	23.43075	74.09453
550	302 500	23.45208	74.16198
551	303 601	23.47339	74.22937
552	304 704	23.49468	74.29670
553	305 809	23.51595	74.36397
554	306 916	23.53720	74.43118
555	308 025	23.55844	74.49832
556	309 136	23.57965	74.56541
557	310 249	23.60085	74.63243
558	311 364	23.62202	74.69940
559	312 481	23.64318	74.76630
560	313 600	23.66432	74.83315
561	314 721	23.68544	74.89993
562	315 844	23.70654	74.96666
563	316 969	23.72762	75.03333
564	318 096	23.74868	75.09993
565	319 225	23.76973	75.16648
566	320 356	23.79075	75.23297
567	321 489	23.81176	75.29940
568	322 624	23.83275	75.36577
569	323 761	23.85372	75.43209
570	324 900	23.87467	75.49834
571	326 041	23.89561	75.56454
572	327 184	23.91652	75.63068
573	328 329	23.93742	75.69676
574	329 476	23.95830	75.76279
575	330 625	23.97916	75.82875
576	331 776	24.00000	75.89466
577	332 929	24.02082	75.96052
578	334 084	24.04163	76.02631
579	335 241	24.06242	76.09205
580	336 400	24.08319	76.15773
581	337 561	24.10394	76.22336
582	338 724	24.12468	76.28892
583	339 889	24.14539	76.35444
584	341 056	24.16609	76.41989
585	342 225	24.18677	76.48529
586	343 396	24.20744	76.55064
587	344 569	24.22808	76.61593
588	345 744	24.24871	76.68116
589	346 921	24.26932	76.74634

Squares and Square Roots (Continued)

n	n^2	$\sqrt{n}$	$\sqrt{10n}$
590	348 100	24.28992	76.81146
591	349 281	24.31049	76.87652
592	350 464	24.33105	76.94154
593	351 649	24.35159	77.00649
594	352 836	24.37212	77.07140
595	354 025	24.39262	77.13624
596	355 216	24.41311	77.20104
597	356 409	24.43358	77.26578
598	357 604	24.45404	77.33046
599	358 801	24.47448	77.39509
600	360 000	24.49490	77.45967
601	361 201	24.51530	77.52419
602	362 404	24.53569	77.58866
603	363 609	24.55606	77.65307
604	364 816	24.57641	77.71744
605	366 025	24.59675	77.78175
606	367 236	24.61707	77.84600
607	368 449	24.63737	77.91020
608	369 664	24.65766	77.97435
609	370 881	24.67793	78.03845
610	372 100	24.69818	78.10250
611	373 321	24.71841	78.16649
612	374 544	24.73863	78.23043
613	375 769	24.75884	78.29432
614	376 996	24.77902	78.35815
615	378 225	24.79919	78.42194
616	379 456	24.81935	78.48567
617	380 689	24.83948	78.54935
618	381 924	24.85961	78.61298
619	383 161	24.87971	78.67655
620	384 400	24.89980	78.74008
621	385 641	24.91987	78.80355
622	386 884	24.93993	78.86698
623	388 129	24.95997	78.93035
624	389 376	24.97999	78.99367
625	390 625	25.00000	79.05694
626	391 876	25.01999	79.12016
627	393 129	25.03997	79.18333
628	394 384	25.05993	79.24645
629	395 641	25.07987	79.30952
630	396 900	25.09980	79.37254
631	398 161	25.11971	79.43551
632	399 424	25.13961	79.49843
633	400 689	25.15949	79.56130
634	401 956	25.17936	79.62412
635	403 225	25.19921	79.68689
636	404 496	25.21904	79.74961
637	405 769	25.23886	79.81228
638	407 044	25.25866	79.87490
639	408 321	25.27845	79.93748
640	409 600	25.29822	80.00000
641	410 881	25.31798	80.06248
642	412 164	25.33772	80.12490
643	413 449	25.35744	80.18728
644	414 736	25.37716	80.24961
645	416 025	25.39685	80.31189
646	417 316	25.41653	80.37413
647	418 609	25.43619	80.43631
648	419 904	25.45584	80.49845
649	421 201	25.47548	80.56054
650	422 500	25.49510	80.62258
651	423 801	25.51470	80.68457
652	425 104	25.53429	80.74652
653	426 409	25.55386	80.80842
654	427 716	25.57342	80.87027
655	429 025	25.59297	80.93207
656	430 336	25.61250	80.99383
657	431 649	25.63201	81.05554
658	432 964	25.65151	81.11720
659	434 281	25.67100	81.17881
660	435 600	25.69047	81.24038
661	436 921	25.70992	81.30191
662	438 244	25.72936	81.36338
663	439 569	25.74879	81.42481
664	440 896	25.76820	81.48620
665	442 225	25.78759	81.54753
666	443 556	25.80698	81.60882
667	444 889	25.82634	81.67007
668	446 224	25.84570	81.73127
669	447 561	25.86503	81.79242
670	448 900	25.88436	81.85353
671	450 241	25.90367	81.91459
672	451 584	25.92296	81.97561
673	452 929	25.94224	82.03658
674	454 276	25.96151	82.09750
675	455 625	25.98076	82.15838
676	456 976	26.00000	82.21922
677	458 329	26.01922	82.28001
678	459 684	26.03843	82.34076
679	461 041	26.05763	82.40146
680	462 400	26.07681	82.46211
681	463 761	26.09598	82.52272
682	465 124	26.11513	82.58329
683	466 489	26.13427	82.64381
684	467 856	26.15339	82.70429
685	469 225	26.17250	82.76473
686	470 596	26.19160	82.82512
687	471 969	26.21068	82.88546
688	473 344	26.22975	82.94577
689	474 721	26.24881	83.00602
690	476 100	26.26785	83.06624
691	477 481	26.28688	83.12641
692	478 864	26.30589	83.18654
693	480 249	26.32489	83.24662
694	481 636	26.34388	83.30666
695	483 025	26.36285	83.36666
696	484 416	26.38181	83.42661
697	485 809	26.40076	83.48653
698	487 204	26.41969	83.54639
699	488 601	26.43861	83.60622
700	490 000	26.45751	83.66600
701	491 401	26.47640	83.72574
702	492 804	26.49528	83.78544
703	494 209	26.51415	83.84510
704	495 616	26.53300	83.90471
705	497 025	26.55184	83.96428
706	498 436	26.57066	84.02381
707	499 849	26.58947	84.08329
708	501 264	26.60827	84.14274
709	502 681	26.62705	84.20214

Squares and Square Roots (Continued)

n	n^2	$\sqrt{n}$	$\sqrt{10n}$
710	504 100	26.64583	84.26150
711	505 521	26.66458	84.32082
712	506 944	26.68333	84.38009
713	508 369	26.70206	84.43933
714	509 796	26.72078	84.49852
715	511 225	26.73948	84.55767
716	512 656	26.75818	84.61678
717	514 089	26.77686	84.67585
718	515 524	26.79552	84.73488
719	516 961	26.81418	84.79387
720	518 400	26.83282	84.85281
721	519 841	26.85144	84.91172
722	521 284	26.87006	84.97058
723	522 729	26.88866	85.02941
724	524 176	26.90725	85.08819
725	525 625	26.92582	85.14693
726	527 076	26.94439	85.20563
727	528 529	26.96294	85.26429
728	529 984	26.98148	85.32292
729	531 441	27.00000	85.38150
730	532 900	27.01851	85.44004
731	534 361	27.03701	85.49854
732	535 824	27.05550	85.55700
733	537 289	27.07397	85.61542
734	538 756	27.09243	85.67380
735	540 225	27.11088	85.73214
736	541 696	27.12932	85.79044
737	543 169	27.14774	85.84870
738	544 644	27.16616	85.90693
739	546 121	27.18455	85.96511
740	547 600	27.20294	86.02325
741	549 081	27.22132	86.08136
742	550 564	27.23968	86.13942
743	552 049	27.25803	86.19745
744	553 536	27.27636	86.25543
745	555 025	27.29469	86.31338
746	556 516	27.31300	86.37129
747	558 009	27.33130	86.42916
748	559 504	27.34959	86.48699
749	561 001	27.36786	86.54479
750	562 500	27.38613	86.60254
751	564 001	27.40438	86.66026
752	565 504	27.42262	86.71793
753	567 009	27.44085	86.77557
754	568 516	27.45906	86.83317
755	570 025	27.47726	86.89074
756	571 536	27.49545	86.94826
757	573 049	27.51363	87.00575
758	574 564	27.53180	87.06320
759	576 081	27.54995	87.12061
760	577 600	27.56810	87.17798
761	579 121	27.58623	87.23531
762	580 644	27.60435	87.29261
763	582 169	27.62245	87.34987
764	583 696	27.64055	87.40709
765	585 225	27.65863	87.46428
766	586 756	27.67671	87.52143
767	588 289	27.69476	87.57854
768	589 824	27.71281	87.63561
769	591 361	27.73085	87.69265

n	n^2	$\sqrt{n}$	$\sqrt{10n}$
770	592 900	27.74887	87.74964
771	594 441	27.76689	87.80661
772	595 984	27.78489	87.86353
773	597 529	27.80288	87.92042
774	599 076	27.82086	87.97727
775	600 625	27.83882	88.03408
776	602 176	27.85678	88.09086
777	603 729	27.87472	88.14760
778	605 284	27.89265	88.20431
779	606 841	27.91057	88.26098
780	608 400	27.92848	88.31761
781	609 961	27.94638	88.37420
782	611 524	27.96426	88.43076
783	613 089	27.98214	88.48729
784	614 656	28.00000	88.54377
785	616 225	28.01785	88.60023
786	617 796	28.03569	88.65664
787	619 369	28.05352	88.71302
788	620 944	28.07134	88.76936
789	622 521	28.08914	88.82567
790	624 100	28.10694	88.88194
791	625 681	28.12472	88.93818
792	627 264	28.14249	88.99438
793	628 849	28.16026	89.05055
794	630 436	28.17801	89.10668
795	632 025	28.19574	89.16277
796	633 616	28.21347	89.21883
797	635 209	28.23119	89.27486
798	636 804	28.24889	89.33085
799	638 401	28.26659	89.38680
800	640 000	28.28427	89.44272
801	641 601	28.30194	89.49860
802	643 204	28.31960	89.55445
803	644 809	28.33725	89.61027
804	646 416	28.35489	89.66605
805	648 025	28.37252	89.72179
806	649 636	28.39014	89.77750
807	651 249	28.40775	89.83318
808	652 864	28.42534	89.88882
809	654 481	28.44293	89.94443
810	656 100	28.46050	90.00000
811	657 721	28.47806	90.05554
812	659 344	28.49561	90.11104
813	660 969	28.51315	90.16651
814	662 596	28.53069	90.22195
815	664 225	28.54820	90.27735
816	665 856	28.56571	90.33272
817	667 489	28.58321	90.38805
818	669 124	28.60070	90.44335
819	670 761	28.61818	90.49862
820	672 400	28.63564	90.55385
821	674 041	28.65310	90.60905
822	675 684	28.67054	90.66422
823	677 329	28.68798	90.71935
824	678 976	28.70540	90.77445
825	680 625	28.72281	90.82951
826	682 276	28.74022	90.88454
827	683 929	28.75761	90.93954
828	685 584	28.77499	90.99451
829	687 241	28.79236	91.04944

Squares and Square Roots (Continued)

n	n^2	$\sqrt{n}$	$\sqrt{10n}$
830	688 900	28.80972	91.10434
831	690 561	28.82707	91.15920
832	692 224	28.84441	91.21403
833	693 889	28.86174	91.26883
834	695 556	28.87906	91.32360
835	697 225	28.89637	91.37833
836	698 896	28.91366	91.43304
837	700 569	28.93095	91.48770
838	702 244	28.94823	91.54234
839	703 921	28.96550	91.59694
840	705 600	28.98275	91.65151
841	707 281	29.00000	91.70605
842	708 964	29.01724	91.76056
843	710 649	29.03446	91.81503
844	712 336	29.05168	91.86947
845	714 025	29.06888	91.92388
846	715 716	29.08608	91.97826
847	717 409	29.10326	92.03260
848	719 104	29.12044	92.08692
849	720 801	29.13760	92.14120
850	722 500	29.15476	92.19544
851	724 201	29.17190	92.24966
852	725 904	29.18904	92.30385
853	727 609	29.20616	92.35800
854	729 316	29.22328	92.41212
855	731 025	29.24038	92.46621
856	732 736	29.25748	92.52027
857	734 449	29.27456	92.57429
858	736 164	29.29164	92.62829
859	737 881	29.30870	92.68225
860	739 600	29.32576	92.73618
861	741 321	29.34280	92.79009
862	743 044	29.35984	92.84396
863	744 769	29.37686	92.89779
864	746 496	29.39388	92.95160
865	748 225	29.41088	93.00538
866	749 956	29.42788	93.05912
867	751 689	29.44486	93.11283
868	753 424	29.46184	93.16652
869	755 161	29.47881	93.22017
870	756 900	29.49576	93.27379
871	758 641	29.51271	93.32738
872	760 384	29.52965	93.38094
873	762 129	29.54657	93.43447
874	763 876	29.56349	93.48797
875	765 625	29.58040	93.54143
876	767 376	29.59730	93.59487
877	769 129	29.61419	93.64828
878	770 884	29.63106	93.70165
879	772 641	29.64793	93.75500
880	774 400	29.66479	93.80832
881	776 161	29.68164	93.86160
882	777 924	29.69848	93.91486
883	779 689	29.71532	93.96808
884	781 456	29.73214	94.02127
885	783 225	29.74895	94.07444
886	784 996	29.76575	94.12757
887	786 769	29.78255	94.18068
888	788 544	29.79933	94.23375
889	790 321	29.81610	94.28680
890	792 100	29.83287	94.33981
891	793 881	29.84962	94.39280
892	795 664	29.86637	94.44575
893	797 449	29.88311	94.49868
894	799 236	29.89983	94.55157
895	801 025	29.91655	94.60444
896	802 816	29.93326	94.65728
897	804 609	29.94996	94.71008
898	806 404	29.96665	94.76286
899	808 201	29.98333	94.81561
900	810 000	30.00000	94.86833
901	811 801	30.01666	94.92102
902	813 604	30.03331	94.97368
903	815 409	30.04996	95.02631
904	817 216	30.06659	95.07891
905	819 025	30.08322	95.13149
906	820 836	30.09983	95.18403
907	822 649	30.11644	95.23655
908	824 464	30.13304	95.28903
909	826 281	30.14963	95.34149
910	828 100	30.16621	95.39392
911	829 921	30.18278	95.44632
912	831 744	30.19934	95.49869
913	833 569	30.21589	95.55103
914	835 396	30.23243	95.60335
915	837 225	30.24897	95.65563
916	839 056	30.26549	95.70789
917	840 889	30.28201	95.76012
918	842 724	30.29851	95.81232
919	844 561	30.31501	95.86449
920	846 400	30.33150	95.91663
921	848 241	30.34798	95.96874
922	850 084	30.36445	96.02083
923	851 929	30.38092	96.07289
924	853 776	30.39737	96.12492
925	855 625	30.41381	96.17692
926	857 476	30.43025	96.22889
927	859 329	30.44667	96.28084
928	861 184	30.46309	96.33276
929	863 041	30.47950	96.38465
930	864 900	30.49590	96.43651
931	866 761	30.51229	96.48834
932	868 624	30.52868	96.54015
933	870 489	30.54505	96.59193
934	872 356	30.56141	96.64368
935	874 225	30.57777	96.69540
936	876 096	30.59412	96.74709
937	877 969	30.61046	96.79876
938	879 844	30.62679	96.85040
939	881 721	30.64311	96.90201
940	883 600	30.65942	96.95360
941	885 481	30.67572	97.00515
942	887 364	30.69202	97.05668
943	889 249	30.70831	97.10819
944	891 136	30.72458	97.15966
945	893 025	30.74085	97.21111
946	894 916	30.75711	97.26253
947	896 809	30.77337	97.31393
948	898 704	30.78961	97.36529
949	900 601	30.80584	97.41663

Squares and Square Roots (Continued)

n	n^2	$\sqrt{n}$	$\sqrt{10n}$
950	902 500	30.82207	97.46794
951	904 401	30.83829	97.51923
952	906 304	30.85450	97.57049
953	908 209	30.87070	97.62172
954	910 116	30.88689	97.67292
955	912 025	30.90307	97.72410
956	913 936	30.91925	97.77525
957	915 849	30.93542	97.82638
958	917 764	30.95158	97.87747
959	919 681	30.96773	97.92855
960	921 600	30.98387	97.97959
961	923 521	31.00000	98.03061
962	925 444	31.01612	98.08160
963	927 369	31.03224	98.13256
964	929 296	31.04835	98.18350
965	931 225	31.06445	98.23441
966	933 156	31.08054	98.28530
967	935 089	31.09662	98.33616
968	937 024	31.11270	98.38699
969	938 961	31.12876	98.43780
970	940 900	31.14482	98.48858
971	942 841	31.16087	98.53933
972	944 784	31.17691	98.59006
973	946 729	31.19295	98.64076
974	948 676	31.20897	98.69144
975	950 625	31.22499	98.74209
976	952 576	31.24100	98.79271
977	954 529	31.25700	98.84331
978	956 484	31.27299	98.89388
979	958 441	31.28898	98.94443
980	960 400	31.30495	98.99495
981	962 361	31.32092	99.04544
982	964 324	31.33688	99.09591
983	966 289	31.35283	99.14636
984	968 256	31.36877	99.19677
985	970 225	31.38471	99.24717
986	972 196	31.40064	99.29753
987	974 169	31.41656	99.34787
988	976 144	31.43247	99.39819
989	978 121	31.44837	99.44848
990	980 100	31.46427	99.49874
991	982 081	31.48015	99.54898
992	984 064	31.49603	99.59920
993	986 049	31.51190	99.64939
994	988 036	31.52777	99.69955
995	990 025	31.54362	99.74969
996	992 016	31.55947	99.79980
997	994 009	31.57531	99.84989
998	996 004	31.59114	99.89995
999	998 001	31.60696	99.94999
1000	1 000 000	31.62278	100.00000

Table 2. Factorials

n	$n!$
0	1
1	1
2	2
3	6
4	24
5	120
6	720
7	5,040
8	40,320
9	362,880
10	3,628,800
11	39,916,800
12	479,001,600
13	6,227,020,800
14	87,178,291,200
15	1,307,674,368,000

Table 3. Binomial Coefficients $_nC_r$

	r										
n	**0**	1	2	3	4	5	6	7	8	9	10
0	**1**										
1	**1**	1									
2	**1**	2	1								
3	**1**	3	3	1							
4	**1**	4	6	4	1						
5	**1**	5	10	10	5	1					
6	**1**	6	15	20	15	6	1				
7	**1**	7	21	35	35	21	7	1			
8	**1**	8	28	56	70	56	28	8	1		
9	**1**	9	36	84	126	126	84	36	9	1	
10	**1**	10	45	120	210	252	210	120	45	10	1
11	**1**	11	55	165	330	462	462	330	165	55	11
12	**1**	12	66	220	495	792	924	792	495	220	66
13	**1**	13	78	286	715	1287	1716	1716	1287	715	286
14	**1**	14	91	364	1001	2002	3003	3432	3003	2002	1001
15	**1**	15	105	455	1365	3003	5005	6435	6435	5005	3003

Table 4. Areas Under the Standard Normal (Z) Curve

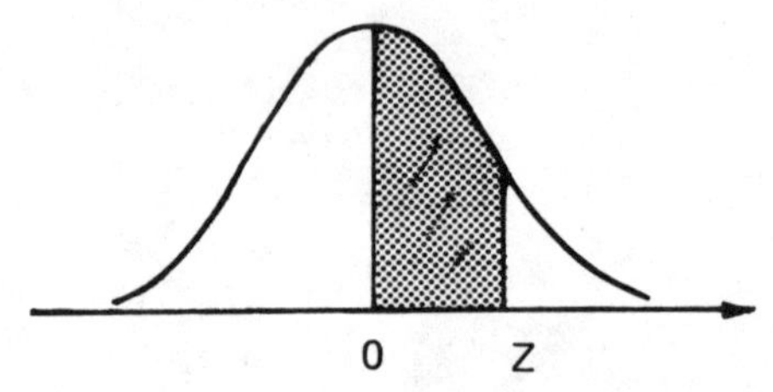

Z	.00	.01	.02	.03	.04	.05	.06	.07	.08	.09
0.0	.0000	.0040	.0080	.0120	.0160	.0199	.0239	.0279	.0319	.0359
0.1	.0398	.0438	.0478	.0517	.0557	.0596	.0636	.0675	.0714	.0753
0.2	.0793	.0832	.0871	.0910	.0948	.0987	.1026	.1064	.1103	.1141
0.3	.1179	.1217	.1255	.1293	.1331	.1368	.1406	.1443	.1480	.1517
0.4	.1554	.1591	.1628	.1664	.1700	.1736	.1772	.1808	.1844	.1879
0.5	.1915	.1950	.1985	.2019	.2054	.2088	.2123	.2157	.2190	.2224
0.6	.2257	.2291	.2324	.2357	.2389	.2422	.2454	.2486	.2518	.2549
0.7	.2580	.2612	.2642	.2673	.2704	.2734	.2764	.2794	.2823	.2852
0.8	.2881	.2910	.2939	.2967	.2995	.3023	.3051	.3078	.3106	.3133
0.9	.3159	.3186	.3212	.3238	.3264	.3289	.3315	.3340	.3365	.3389
1.0	.3413	.3438	.3461	.3485	.3508	.3531	.3554	.3577	.3599	.3621
1.1	.3643	.3665	.3686	.3708	.3729	.3749	.3770	.3790	.3810	.3830
1.2	.3849	.3869	.3888	.3907	.3925	.3944	.3962	.3980	.3997	.4015
1.3	.4032	.4049	.4066	.4082	.4099	.4115	.4131	.4147	.4162	.4177
1.4	.4192	.4207	.4222	.4236	.4251	.4265	.4279	.4292	.4306	.4319
1.5	.4332	.4345	.4357	.4370	.4382	.4394	.4406	.4418	.4429	.4441
1.6	.4452	.4463	.4474	.4484	.4495	.4505	.4515	.4525	.4535	.4545
1.7	.4554	.4564	.4573	.4582	.4591	.4599	.4608	.4616	.4625	.4633
1.8	.4641	.4649	.4656	.4664	.4671	.4678	.4686	.4693	.4699	.4706
1.9	.4713	.4719	.4726	.4732	.4738	.4744	.4750	.4756	.4761	.4767
2.0	.4772	.4778	.4783	.4788	.4793	.4798	.4803	.4808	.4812	.4817
2.1	.4821	.4826	.4830	.4834	.4838	.4842	.4846	.4850	.4854	.4857
2.2	.4861	.4864	.4868	.4871	.4875	.4878	.4881	.4884	.4887	.4890
2.3	.4893	.4896	.4898	.4901	.4904	.4906	.4909	.4911	.4913	.4916
2.4	.4918	.4920	.4922	.4925	.4927	.4929	.4931	.4932	.4934	.4936
2.5	.4938	.4940	.4941	.4943	.4945	.4946	.4948	.4949	.4951	.4952
2.6	.4953	.4955	.4956	.4957	.4959	.4960	.4961	.4962	.4963	.4964
2.7	.4965	.4966	.4967	.4968	.4969	.4970	.4971	.4972	.4973	.4974
2.8	.4974	.4975	.4976	.4977	.4977	.4978	.4979	.4979	.4980	.4981
2.9	.4981	.4982	.4982	.4983	.4984	.4984	.4985	.4985	.4986	.4986
3.0	.49865	.4987	.4987	.4988	.4988	.4989	.4989	.4989	.4990	.4990
3.1	.49903	.4991	.4991	.4991	.4992	.4992	.4992	.4992	.4993	.4993
3.2	.4993129	.4993	.4994	.4994	.4994	.4994	.4994	.4995	.4995	.4995
3.3	.4995166	.4995	.4995	.4996	.4996	.4996	.4996	.4996	.4996	.4997
3.4	.4996631	.4997	.4997	.4497	.4997	.4997	.4997	.4997	.4998	.4998
3.5	.4997674	.4998	.4998	.4998	.4998	.4998	.4998	.4998	.4998	.4998
3.6	.4998409	.4998	.4999	.4999	.4999	.4999	.4999	.4999	.4999	.4999
3.7	.4998922	.4999	.4999	.4999	.4999	.4999	.4999	.4999	.4999	.4999
3.8	.4999277	.4999	.4999	.4999	.4999	.4999	.4999	.5000	.5000	.5000
3.9	.4999519	.5000	.5000	.5000	.5000	.5000	.5000	.5000	.5000	.5000
4.0	.4999683									
4.5	.4999966									
5.0	.4999997133									

Table 5. The t Distribution

d.f.	$t_{.050}$	$t_{.025}$	$t_{.010}$	$t_{.005}$	d.f.
1	6.314	12.706	31.821	63.657	1
2	2.920	4.303	6.965	9.925	2
3	2.353	3.182	4.541	5.841	3
4	2.132	2.776	3.747	4.604	4
5	2.015	2.571	3.365	4.032	5
6	1.943	2.447	3.143	3.707	6
7	1.895	2.365	2.998	3.499	7
8	1.860	2.306	2.896	3.355	8
9	1.833	2.262	2.821	3.250	9
10	1.812	2.228	2.764	3.169	10
11	1.796	2.201	2.718	3.106	11
12	1.782	2.179	2.681	3.055	12
13	1.771	2.160	2.650	3.012	13
14	1.761	2.145	2.624	2.977	14
15	1.753	2.131	2.602	2.947	15
16	1.746	2.120	2.583	2.921	16
17	1.740	2.110	2.567	2.898	17
18	1.734	2.101	2.552	2.878	18
19	1.729	2.093	2.539	2.861	19
20	1.725	2.086	2.528	2.845	20
21	1.721	2.080	2.518	2.831	21
22	1.717	2.074	2.508	2.819	22
23	1.714	2.069	2.500	2.807	23
24	1.711	2.064	2.492	2.797	24
25	1.708	2.060	2.485	2.787	25
26	1.706	2.056	2.479	2.779	26
27	1.703	2.052	2.473	2.771	27
28	1.701	2.048	2.467	2.763	28
29	1.699	2.045	2.462	2.756	29
inf.	1.645	1.960	2.326	2.576	inf.

Table 6. The X^2 Distribution

d.f.	$\chi^2_{.05}$	$\chi^2_{.01}$	*d.f.*
1	3.841	6.635	1
2	5.991	9.210	2
3	7.815	11.345	3
4	9.488	13.277	4
5	11.070	15.086	5
6	12.592	16.812	6
7	14.067	18.475	7
8	15.507	20.090	8
9	16.919	21.666	9
10	18.307	23.209	10
11	19.675	24.725	11
12	21.026	26.217	12
13	22.362	27.688	13
14	23.685	29.141	14
15	24.996	30.578	15
16	26.296	32.000	16
17	27.587	33.409	17
18	28.869	34.805	18
19	30.144	36.191	19
20	31.410	37.566	20
21	32.671	38.932	21
22	33.924	40.289	22
23	35.172	41.638	23
24	36.415	42.980	24
25	37.652	44.314	25
26	38.885	45.642	26
27	40.113	46.963	27
28	41.337	48.278	28
29	42.557	49.588	29
30	43.773	50.892	30

Table 7. Critical Values of r

n	$r_{.025}$	$r_{.005}$	n	$r_{.025}$	$r_{.005}$
3	0.997		**18**	0.468	0.590
4	0.950	0.999	**19**	0.456	0.575
5	0.878	0.959	**20**	0.444	0.561
6	0.811	0.917	**21**	0.433	0.549
7	0.754	0.875	**22**	0.423	0.537
8	0.707	0.834	**27**	0.381	0.487
9	0.666	0.798	**32**	0.349	0.449
10	0.632	0.765	**37**	0.325	0.418
11	0.602	0.735	**42**	0.304	0.393
12	0.576	0.708	**47**	0.288	0.372
13	0.553	0.684	**52**	0.273	0.354
14	0.532	0.661	**62**	0.250	0.325
15	0.514	0.641	**72**	0.232	0.302
16	0.497	0.623	**82**	0.217	0.283
17	0.482	0.606	**92**	0.205	0.267

Table 8. The F Distribution (Values of $F_{.05}$)

Degrees of freedom for denominator	Degrees of freedom for numerator: 1	2	3	4	5	6	7	8	9	10	12	15	20	24	30	40	60	120	∞
1	161	200	216	225	230	234	237	239	241	242	244	246	248	249	250	251	252	253	254
2	18.5	19.0	19.2	19.2	19.3	19.3	19.4	19.4	19.4	19.4	19.4	19.4	19.4	19.5	19.5	19.5	19.5	19.5	19.5
3	10.1	9.55	9.28	9.12	9.01	8.94	8.89	8.85	8.81	8.79	8.74	8.70	8.66	8.64	8.62	8.59	8.57	8.55	8.53
4	7.71	6.94	6.59	6.39	6.26	6.16	6.09	6.04	6.00	5.96	5.91	5.86	5.80	5.77	5.75	5.72	5.69	5.66	5.63
5	6.61	5.79	5.41	5.19	5.05	4.95	4.88	4.82	4.77	4.74	4.68	4.62	4.56	4.53	4.50	4.46	4.43	4.40	4.37
6	5.99	5.14	4.76	4.53	4.39	4.28	4.21	4.15	4.10	4.06	4.00	3.94	3.87	3.84	3.81	3.77	3.74	3.70	3.67
7	5.59	4.74	4.35	4.12	3.97	3.87	3.79	3.73	3.68	3.64	3.57	3.51	3.44	3.41	3.38	3.34	3.30	3.27	3.23
8	5.32	4.46	4.07	3.84	3.69	3.58	3.50	3.44	3.39	3.35	3.28	3.22	3.15	3.12	3.08	3.04	3.01	2.97	2.93
9	5.12	4.26	3.86	3.63	3.48	3.37	3.29	3.23	3.18	3.14	3.07	3.01	2.94	2.90	2.86	2.83	2.79	2.75	2.71
10	4.96	4.10	3.71	3.48	3.33	3.22	3.14	3.07	3.02	2.98	2.91	2.85	2.77	2.74	2.70	2.66	2.62	2.58	2.54
11	4.84	3.98	3.59	3.36	3.20	3.09	3.01	2.95	2.90	2.85	2.79	2.72	2.65	2.61	2.57	2.53	2.49	2.45	2.40
12	4.75	3.89	3.49	3.26	3.11	3.00	2.91	2.85	2.80	2.75	2.69	2.62	2.54	2.51	2.47	2.43	2.38	2.34	2.30
13	4.67	3.81	3.41	3.18	3.03	2.92	2.83	2.77	2.71	2.67	2.60	2.53	2.46	2.42	2.38	2.34	2.30	2.25	2.21
14	4.60	3.74	3.34	3.11	2.96	2.85	2.76	2.70	2.65	2.60	2.53	2.46	2.39	2.35	2.31	2.27	2.22	2.18	2.13
15	4.54	3.68	3.29	3.06	2.90	2.79	2.71	2.64	2.59	2.54	2.48	2.40	2.33	2.29	2.25	2.20	2.16	2.11	2.07
16	4.49	3.63	3.24	3.01	2.85	2.74	2.66	2.59	2.54	2.49	2.42	2.35	2.28	2.24	2.19	2.15	2.11	2.06	2.01
17	4.45	3.59	3.20	2.96	2.81	2.70	2.61	2.55	2.49	2.45	2.38	2.31	2.23	2.29	2.15	2.10	2.06	2.01	1.96
18	4.41	3.55	3.16	2.93	2.77	2.66	2.58	2.51	2.46	2.41	2.34	2.27	2.19	2.15	2.11	2.06	2.02	1.97	1.92
19	4.38	3.52	3.13	2.90	2.74	2.63	2.54	2.48	2.42	2.38	2.31	2.23	2.16	2.11	2.07	2.03	1.98	1.93	1.88
20	4.35	3.49	3.10	2.87	2.71	2.60	2.51	2.45	2.39	2.35	2.28	2.20	2.12	2.08	2.04	1.99	1.95	1.90	1.84
21	4.32	3.47	3.07	2.84	2.68	2.57	2.49	2.42	2.37	2.32	2.25	2.18	2.10	2.05	2.01	1.96	1.92	1.87	1.81
22	4.30	3.44	3.05	2.82	2.66	2.55	2.46	2.40	2.34	2.30	2.23	2.15	2.07	2.03	1.98	1.94	1.89	1.84	1.78
23	4.28	3.42	3.03	2.80	2.64	2.53	2.44	2.37	2.32	2.27	2.20	2.13	2.05	2.01	1.96	1.91	1.86	1.81	1.76
24	4.26	3.40	3.01	2.78	2.62	2.51	2.42	2.36	2.30	2.25	2.18	2.11	2.03	1.98	1.94	1.89	1.84	1.79	1.73
25	4.24	3.39	2.99	2.76	2.60	2.49	2.40	2.34	2.28	2.24	2.16	2.09	2.01	1.96	1.92	1.87	1.82	1.77	1.71
30	4.17	3.32	2.92	2.69	2.53	2.42	2.33	2.27	2.21	2.16	2.09	2.01	1.93	1.89	1.84	1.79	1.74	1.68	1.62
40	4.08	3.23	2.84	2.61	2.45	2.34	2.25	2.18	2.12	2.08	2.00	1.92	1.84	1.79	1.74	1.69	1.64	1.58	1.51
60	4.00	3.15	2.76	2.53	2.37	2.25	2.17	2.10	2.04	1.99	1.92	1.84	1.75	1.70	1.65	1.59	1.53	1.47	1.39
120	3.92	3.07	2.68	2.45	2.29	2.18	2.09	2.02	1.96	1.91	1.83	1.75	1.66	1.61	1.55	1.50	1.43	1.35	1.25
∞	3.84	3.00	2.60	2.37	2.21	2.10	2.01	1.94	1.88	1.83	1.75	1.67	1.57	1.52	1.46	1.39	1.32	1.22	1.00

Table 8. The F Distribution (Values of $F_{.01}$)

Degrees of freedom for numerator (columns); Degrees of freedom for denominator (rows)

	1	2	3	4	5	6	7	8	9	10	12	15	20	24	30	40	60	120	∞
1	4,052	5,000	5,403	5,625	5,764	5,859	5,928	5,982	6,023	6,056	6,106	6,157	6,209	6,235	6,261	6,287	6,313	6,339	6,366
2	98.5	99.0	99.2	99.2	99.3	99.3	99.4	99.4	99.4	99.4	99.4	99.4	99.4	99.5	99.5	99.5	99.5	99.5	99.5
3	34.1	30.8	29.5	28.7	28.2	27.9	27.7	27.5	27.3	27.2	27.1	26.9	26.7	26.6	26.5	26.4	26.3	26.2	26.1
4	21.2	18.0	16.7	16.0	15.5	15.2	15.0	14.8	14.7	14.5	14.4	14.2	14.0	13.9	13.8	13.7	13.7	13.6	13.5
5	16.3	13.3	12.1	11.4	11.0	10.7	10.5	10.3	10.2	10.1	9.89	9.72	9.55	9.47	9.38	9.29	9.20	9.11	9.02
6	13.7	10.9	9.78	9.15	8.75	8.47	8.26	8.10	7.98	7.87	7.72	7.56	7.40	7.31	7.23	7.14	7.06	6.97	6.88
7	12.2	9.55	8.45	7.85	7.46	7.19	6.99	6.84	6.72	6.62	6.47	6.31	6.16	6.07	5.99	5.91	5.82	5.74	5.65
8	11.3	8.65	7.59	7.01	6.63	6.37	6.18	6.03	5.91	5.81	5.67	5.52	5.36	5.28	5.20	5.12	5.03	4.95	4.86
9	10.6	8.02	6.99	6.42	6.06	5.80	5.61	5.47	5.35	5.26	5.11	4.96	4.81	4.73	4.65	4.57	4.48	4.40	4.31
10	10.0	7.56	6.55	5.99	5.64	5.39	5.20	5.06	4.94	4.85	4.71	4.56	4.41	4.33	4.25	4.17	4.08	4.00	3.91
11	9.65	7.21	6.22	5.67	5.32	5.07	4.89	4.74	4.63	4.54	4.40	4.25	4.10	4.02	3.94	3.86	3.78	3.69	3.60
12	9.33	6.93	5.95	5.41	5.06	4.82	4.64	4.50	4.39	4.30	4.16	4.01	3.86	3.78	3.70	3.62	3.54	3.45	3.36
13	9.07	6.70	5.74	5.21	4.86	4.62	4.44	4.30	4.19	4.10	3.96	3.82	3.66	3.59	3.51	3.43	3.34	3.25	3.17
14	8.86	6.51	5.56	5.04	4.70	4.46	4.28	4.14	4.03	3.94	3.80	3.66	3.51	3.43	3.35	3.27	3.18	3.09	3.00
15	8.68	6.36	5.42	4.89	4.56	4.32	4.14	4.00	3.89	3.80	3.67	3.52	3.37	3.29	3.21	3.13	3.05	2.96	2.87
16	8.53	6.23	5.29	4.77	4.44	4.20	4.03	3.89	3.78	3.69	3.55	3.41	3.26	3.18	3.10	3.02	2.93	2.84	2.75
17	8.40	6.11	5.19	4.67	4.34	4.10	3.93	3.79	3.68	3.59	3.46	3.31	3.16	3.08	3.00	2.92	2.83	2.75	2.65
18	8.29	6.01	5.09	4.58	4.25	4.01	3.84	3.71	3.60	3.51	3.37	3.23	3.08	3.00	2.92	2.84	2.75	2.66	2.57
19	8.19	5.93	5.01	4.50	4.17	3.94	3.77	3.63	3.52	3.43	3.30	3.15	3.00	2.92	2.84	2.76	2.67	2.58	2.49
20	8.10	5.85	4.94	4.43	4.10	3.87	3.70	3.56	3.46	3.37	3.23	3.09	2.94	2.86	2.78	2.69	2.61	2.52	2.42
21	8.02	5.78	4.87	4.37	4.04	3.81	3.64	3.51	3.40	3.31	3.17	3.03	2.88	2.80	2.72	2.64	2.55	2.46	2.36
22	7.95	5.72	4.82	4.31	3.99	3.76	3.59	3.45	3.35	3.26	3.12	2.98	2.83	2.75	2.67	2.58	2.50	2.40	2.31
23	7.88	5.66	4.76	4.26	3.94	3.71	3.54	3.41	3.30	3.21	3.07	2.93	2.78	2.70	2.62	2.54	2.45	2.35	2.26
24	7.82	5.61	4.72	4.22	3.90	3.67	3.50	3.36	3.26	3.17	3.03	2.89	2.74	2.66	2.58	2.49	2.40	2.31	2.21
25	7.77	5.57	4.68	4.18	3.86	3.63	3.46	3.32	3.22	3.13	2.99	2.85	2.70	2.62	2.53	2.45	2.36	2.27	2.17
30	7.56	5.39	4.51	4.02	3.70	3.47	3.30	3.17	3.07	2.98	2.84	2.70	2.55	2.47	2.39	2.30	2.21	2.11	2.01
40	7.31	5.18	4.31	3.83	3.51	3.29	3.12	2.99	2.89	2.80	2.66	2.52	2.37	2.29	2.20	2.11	2.02	1.92	1.80
60	7.08	4.98	4.13	3.65	3.34	3.12	2.95	2.82	2.72	2.63	2.50	2.35	2.20	2.12	2.03	1.94	1.84	1.73	1.60
120	6.85	4.79	3.95	3.48	3.17	2.96	2.79	2.66	2.56	2.47	2.34	2.19	2.03	1.95	1.86	1.76	1.66	1.53	1.38
∞	6.63	4.61	3.78	3.32	3.02	2.80	2.64	2.51	2.41	2.32	2.18	2.04	1.88	1.79	1.70	1.59	1.47	1.32	1.00

Answers to Selected Problems

Problem Set 1.1

1.

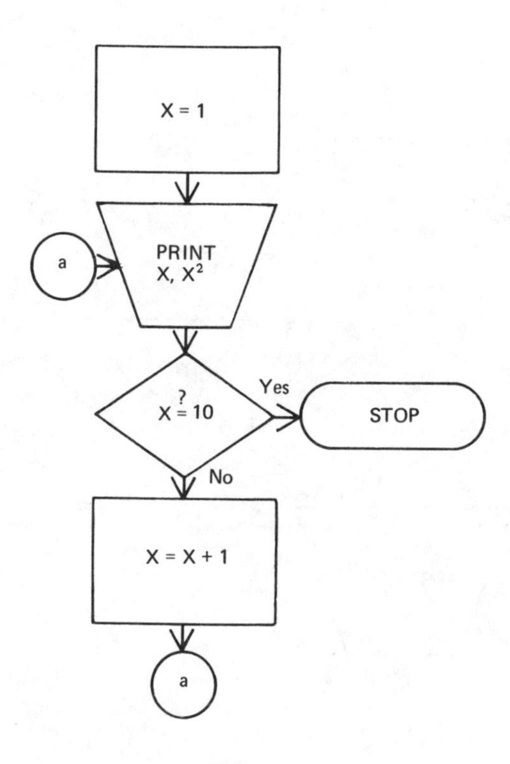

3.

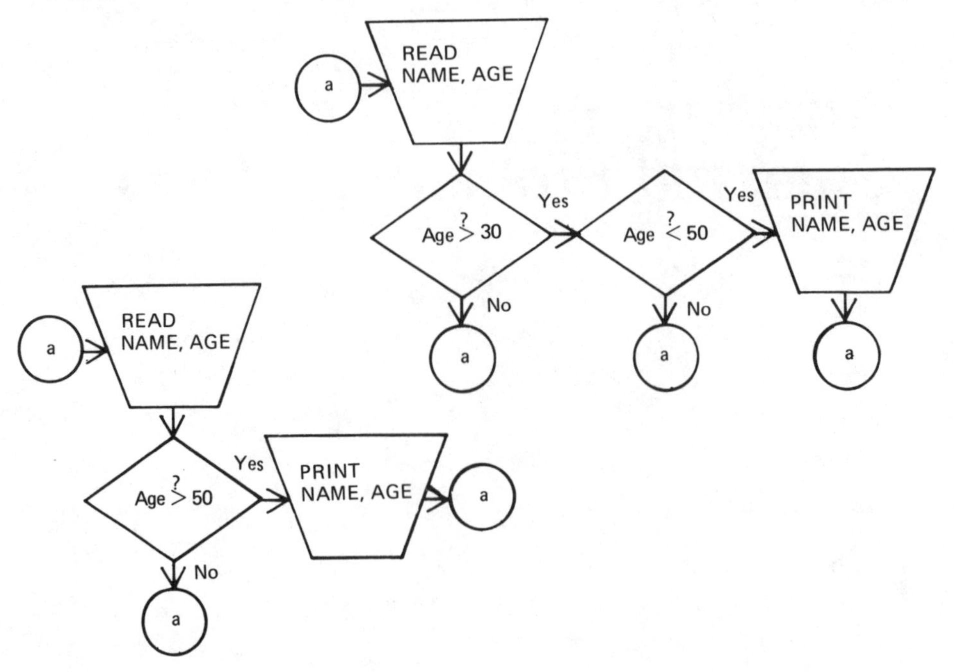
a
READ
NAME, AGE
Age ? > 30
Yes
No
a
Age ? < 50
Yes
No
a
PRINT
NAME, AGE
a
a
READ
NAME, AGE
Age ? > 50
Yes
PRINT
NAME, AGE
a
No
a

5.

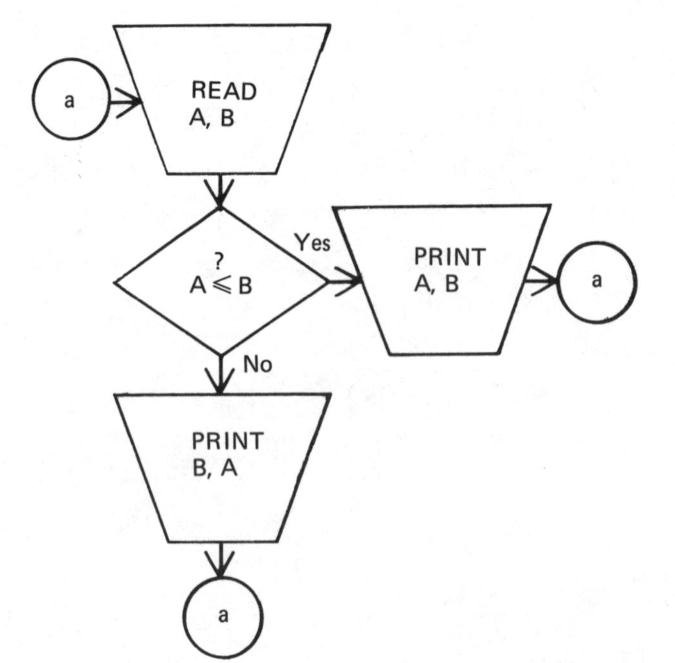
a
READ
A, B
? A ≤ B
Yes
PRINT
A, B
a
No
PRINT
B, A
a

Problem Set 1.2

1.
```
 5 PRINT "FRANK SCALZO - AUGUST 10, 1974"
10 END
```
3.
```
 2 PRINT "FRANK SCALZO"
 5 PRINT "JUNE 6, 1941"
 9 PRINT "33"
12 END
```

Problem Set 1.3

1.
```
(a)  2  LET X=5-2↑3
(b)  5  LET Y=2-17↑3 + 27/101
(c) 10  LET T=(8+7↑2)/21
(d) 15  LET S=A+B-3*C↑2
(e) 27  LET A=(T1+T2+T3+T4)/4
(f) 90  LET Y=SQR(57.321)
(g) 10  LET C=SIN(X) + 3*COS(Y)
(h) 12  LET F=2*SQR(X↑2-1)+2*X-7
(i) 17  LET S=2+7-3↑3 + SQR(94)
(j) 27  LET H=1↑2 + 3↑2 + 5↑2 + 7↑2 + 9↑2
```

Problem Set 1.4

1.
```
 5  READ A,B,C,D
10  IF D=1 THEN 30
15  LET M=(A+B+C)/3
20  PRINT M
25  GO TO 5
30  PRINT "FRANK SCALZO - AUGUST 10, 1974"
35  DATA 1,3,5,0,97,88,21,0,999,275,87,0,0,0,0,1
40  END
```
3.
```
(a)   1     5    16    34
(b)   9  1332     3   148
(c)  -1     2
(d)  -1     2
      2     7
```
5.
```
 2  LET T=0
 5  LET Y=10
 8  LET T=T+Y
11  IF Y=100 THEN 20
14  LET Y=Y+2
```

```
17 GO TO 8
20 PRINT "TOTAL="T
23 END
```

Problem Set 1.5

3. Computer Output:

```
2              4              6
W=22     T=11
12             0              1
W=36     T=7
18            -7             -3
W=103     T=-1
```

5. Computer Output:

```
AVERAGE = 5.666666
AVERAGE = 69.666666
AVERAGE = 62
```

Problem Set 1.6

1.

1	10	99
2	−10	99
1	1	0
2	2	3
3	4	15
4	8	63
5	−9	80

3.

```
100 DATA 4,55,65,72,81
101 DATA 9,61,52,89,94,71,75,88,97,50
RUN

                13:49    06/11/75

N                    X(N)                 X(N)-MEAN
MEAN= 68.25
 1                    55                  -13.25
 2                    65                  -3.25
 3                    72                   3.75
 4                    81                   12.75
N                    X(N)                 X(N)-MEAN
```

```
MEAN= 75.2222
 1                    61                    -14.2222
 2                    52                    -23.2222
 3                    89                     13.7778
 4                    94                     18.7778
 5                    71                    -4.22221
 6                    75                    -.222214
 7                    88                     12.7778
 8                    97                     21.7778
 9                    50                    -25.2222
N                    X(N)                   X(N)-MEAN

LINE    10:   END OF DATA

TIME 0.0 SECS.
```

5.

```
 5 DIM A(7),B(7),C(7)
 10 FOR N=1 TO 7
 20 READ A(N),B(N)
 30 LET C(N)=A(N)-B(N)
 40 NEXT N
 50 FOR N=1 TO 7
 60 PRINT C(N);
 70 NEXT N
 80 DATA 2,9,-7,3,14,22,-81
 90 DATA 19,-8,3,.5,27,-52,8
 100 END
 RUN

                13:58   06/11/75

 -7    -10    -8    -100   -11    -26.5    -60
 TIME 0.0 SECS.
```

Problem Set 2.1

1. (a) 7.6
   (b) 7
   (c) 7
   (d) 7

(e) 4.04
(f) 2.01

3. (a) Section 1: $\overline{X} = 75.4$, Md = 75, Mo = 70
   Section 2: $\overline{X} = 74.8$, Md = 77.5, Mo does not exist.
   (b) Section 1: $s^2 = 29.24$, $s = 5.41$, range = 14
   Section 2: $s^2 = 178.36$, $s = 13.36$, range = 43

5.

Distribution A	Distribution B
$\overline{X} = 15.00$	$\overline{X} = 27.58$
$s = 7.28$	$s = 6.58$
Range = 25.00	Range = 21.00

7. (a) $\overline{X} = 14.91$; Md = 13.00; Mo = 9.00; range = 20.00; $\sigma = 6.72$
   (b) $\overline{X} = 14.09$; Md = 14.00; Mo = 14.00; range = 18.00; $\sigma = 5.35$
   (c) $\overline{X} = 37.68$; Md = 37.00; Mo = 35.00; range = 35.00; $\sigma = 9.47$

Problem Set 2.2

1.

```
150 DATA 10,75,70,82,81,68,70,70,75,81,82
151 DATA 10,75,55,95,80,81,82,70,91,67,52
RUN

                9:28    06/11/75

MEAN= 75.4      VARIANCE= 29.2402     ST. DEV.= 5.40742
MEAN= 74.8      VARIANCE= 178.36  ST. DEV.= 13.3552

LINE      5:    END OF DATA

TIME 0.0 SECS.
```

3.

```
150 DATA 20,3.2,2.5,3.1,2.1,3.3,3.7,1.9,2.2,2.2,2.6,2.1,2.9
151 DATA 2.8,3.6,3.0,2.0,3.5,2.2,2.6,3.1
RUN
```

```
              9:40    06/11/75

MEAN= 2.73      VARIANCE= .308118      ST. DEV.= .555084

LINE      5:    END OF DATA

TIME 0.0 SECS.
```

Problem Set 2.3

1.

(a)

Class	Class Interval	Class Boundaries	f	cf	(x) Midpoint	$\%f$
1	15–17	14.5–17.5	2	2	16	6.66
2	18–20	17.5–20.5	5	7	19	16.66
3	21–23	20.5–23.5	3	10	22	10.00
4	24–26	23.5–26.5	12	22	25	40.00
5	27–29	26.5–29.5	4	26	28	13.33
6	30–32	29.5–32.5	4	30	31	13.33

(b)

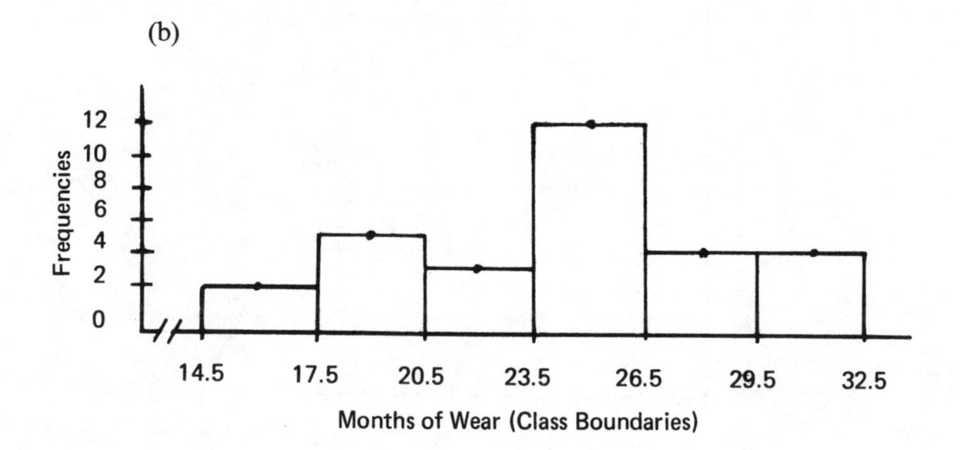

(c)

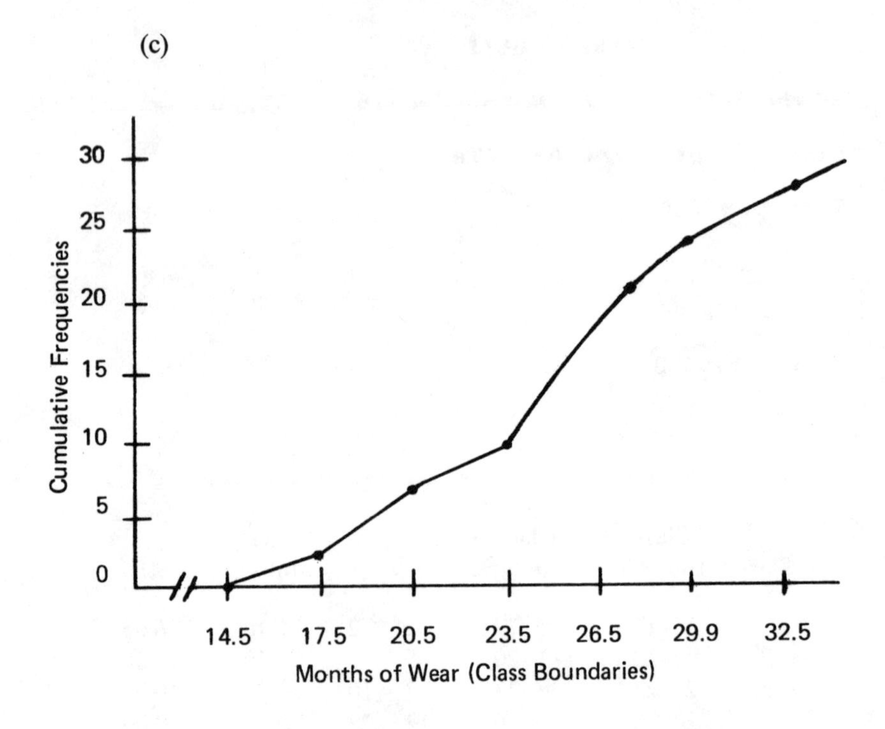
30
25
20
15
10
5
0
Cumulative Frequencies
14.5
17.5
20.5
23.5
26.5
29.9
32.5
Months of Wear (Class Boundaries)

(d)

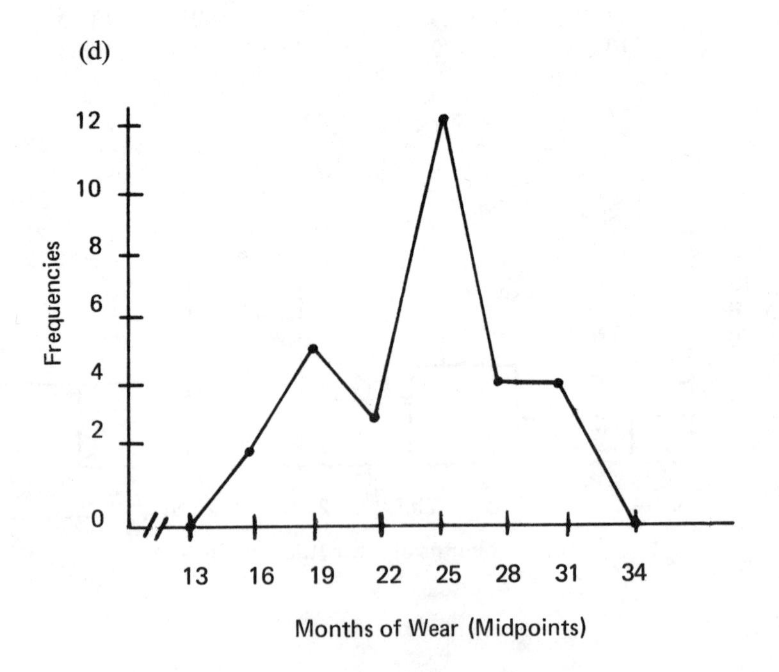
12
10
8
6
4
2
0
Frequencies
13
16
19
22
25
28
31
34
Months of Wear (Midpoints)

(e)

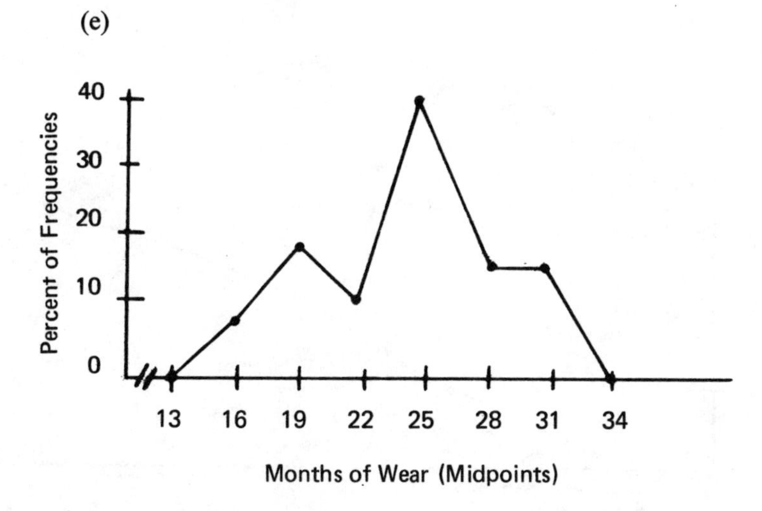

3.

Class	Interval	Boundaries	x	f	cf	$\%f$
1	10–20	9.5–20.5	15	10	10	11.76
2	21–31	20.5–31.5	26	15	25	17.65
3	32–42	31.5–42.5	37	25	50	29.41
4	43–53	42.5–53.5	48	10	60	11.76
5	54–64	53.5–64.5	59	10	70	11.76
6	65–75	64.5–75.5	70	15	85	17.65

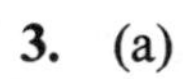
3. (a)

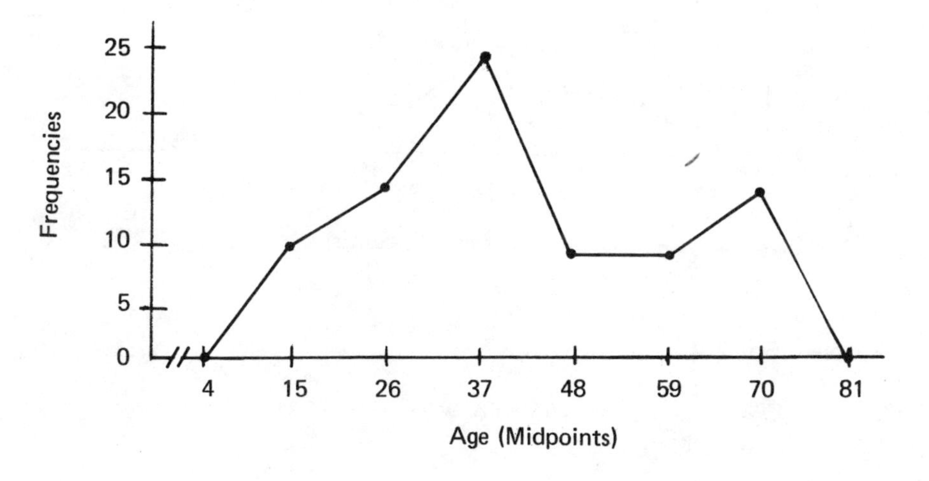

(b)

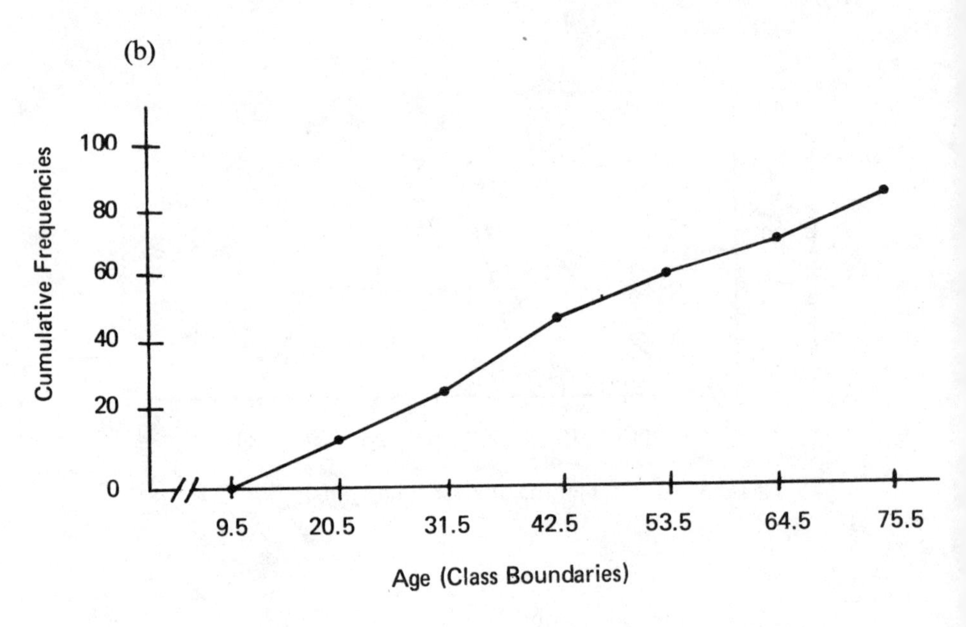

(c)

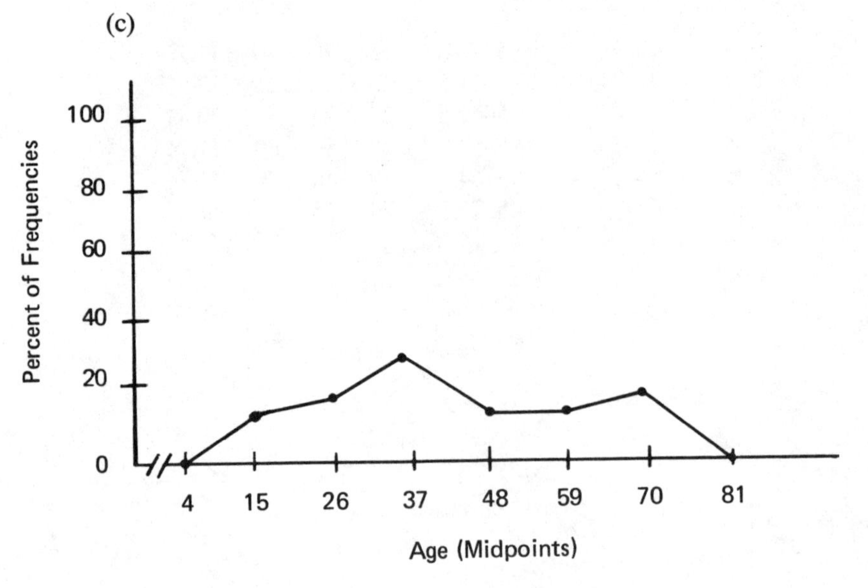

Problem Set 2.4

1. $\overline{X} = 24.3$, Md $= 24.75$, Mo $= 25$, $s^2 = 17.81$,
$s = 4.22$

3. $\overline{X} = 4.8$, Md = 4.0; modal scores are 1 and 4, $s^2 = 14.96$, $s = 3.87$

Problem Set 2.5

1.

```
15Ø DATA 6,2,16,5,19,3,22,12,25,4,28,4,31
RUN

                9:55   Ø6/11/75

MEAN                MEDIAN              VARIANCE            ST.DEV.
 24.3                24.75               17.81               4.22Ø19

LINE    1Ø:   END OF DATA

TIME Ø.Ø SECS.
```

3.

```
15Ø DATA 6,1Ø,1,1Ø,4,5,7,3,1Ø,1,13,1,16
RUN

                9:56   Ø6/11/75

MEAN                MEDIAN              VARIANCE            ST.DEV.
 4.8                 4                   14.96               3.86782

LINE    1Ø:   END OF DATA

TIME Ø.Ø SECS.
```

5.

```
15Ø DATA 9,3,54.5,6,64.5,15,74.5,22,84.5,24,94.5,15,1Ø4.5
151 DATA 1Ø,114.5,4,124.5,1,134.5
RUN

                9:58   Ø6/11/75

MEAN                MEDIAN              VARIANCE            ST.DEV.
 91.4               91.16667             281.39              16.77468

LINE    1Ø:   END OF DATA

TIME Ø.Ø SECS.
```

Problem Set 3.1

1. (a) A = {Shari, Marian, Myself}
(b) B = {1,3,5,7,9}
(c) C = {Johnson, Nixon, Ford}

3. (a) $B \cup C = \{2,3,4,5\}$
(b) $B \cap C = \phi$
(c) $E' = \{1,2,4\}$
(d) $G' = \{2,3,4,5\}$
(e) $U' = \phi$
(f) $H' \cup B = \{2,4\}$
(g) $(F \cup E)' = \{1\}$
(h) $F' \cup E' = \{1,2,3,5\}$
(i) $F' \cap E' = \{1\}$
(j) $(E \cup H) \cup C = \{1,2,3,5\}$
(k) $E \cup (H \cup C) = \{1,2,3,5\}$
(l) $(F \cap H) \cap E = \phi$
(m) $F \cap (H \cap E) = \phi$
(n) $F \cup (H \cap E) = \{2,3,4,5\}$
(o) $(F \cup H) \cap E = \{3,5\}$

5. (a) $A \cap B$
(b) $A \cup B$
(c) A'
(d) $A \cap B'$
(e) $A' \cap B$
(f) $(A \cup B)'$
(g) $(A \cap B') \cup (B \cap A')$

Problem Set 3.2

1. (a) 8
(b) {(A,A,A), (A,A,R), (A,R,A), (A,A,R), (A,R,R), (R,A,R), (R,R,A), (R,R,R)}

3. (a) 380
(b) 19

5. 9•8•7•6 = 3024

Problem Set 3.3

1. (a) 5! = 120
(b) (6!)(3!) = 4320

(c) $6! + 3! = 726$

(d) $\frac{6!}{3!} = 120$

(e) ${}_nP_n = n!$

(f) ${}_nP_0 = 1$

3. $12!$

5. (a) ${}_{25}P_{13}$

(b) 0

(c) $({}_{12}P_5)({}_{20}P_8)$

7. $5! = 120$

Problem Set 3.4

1. (a) ${}_6C_3 = 20$

(b) ${}_6C_3 + {}_6C_4 = 35$

(c) ${}_nC_n = 1$

(d) ${}_nC_0 = 1$

3. ${}_{52}C_5$

5. (a) ${}_{12}C_6$

(b) $({}_5C_4)({}_7C_2) = 105$

(c) $({}_5C_4)({}_7C_2) + ({}_5C_5)({}_7C_1) = 105 + 7 = 112$

(d) $({}_7C_4)({}_5C_2) = 350$

7. (a) $(a + b)^5 = a^5 + 5a^4b + 10a^3b^2 + 10a^2b^3 + 5ab^4 + b^5$

(b) $(2a - b)^4 = 16a^4 - 32a^3b + 24a^2b^2 - 8ab^3 + b^4$

(c) $(x - y)^3 = x^3 - 3x^2y + 3xy^2 - y^3$

(d) $35a^3r^4$

(e) $715g^9d^4$

(f) $20(0.9)^3(0.1)^3 = 0.01458$

(g) $3(0.99)^2(0.01) = 0.029403$

Problem Set 3.5

1.

```
150 DATA 11,3,0
151 DATA 9,5,1
152 DATA 12,4,0
153 DATA 10,7,1
154 DATA 11,7,1
```

```
155 DATA 8,5,0
RUN

                    10:02    06/11/75

P( 11    , 3      )=  990
C( 9     , 5      )=  126
P( 12    , 4      )=  11880
C( 10    , 7      )=  120
C( 11    , 7      )=  330
P( 8     , 5      )=  6720

LINE        5:    END OF DATA

TIME 0.0 SECS.
```

Problem Set 4.1

1. (a)

$$S = \left\{ \begin{matrix} (1,1),(1,2),(1,3),(1,4),(1,5),(1,6), \\ (2,1),(2,2),(2,3),(2,4),(2,5),(2,6), \\ (3,1),(3,2),(3,3),(3,4),(3,5),(3,6), \\ (4,1),(4,2),(4,3),(4,4),(4,5),(4,6), \\ (5,1),(5,2),(5,3),(5,4),(5,5),(5,6), \\ (6,1),(6,2),(6,3),(6,4),(6,5),(6,6) \end{matrix} \right\}$$

The first coordinate gives outcome of first die; second, of second die.

(b) Each sample point has the probability of 1/36.

(c) $E = \{(1,4), (2,3), (3,2), (4,1)\}$

$F = \{(1,1), (1,2), (2,1), (3,1), (2,2), (1,3), (4,1), (3,2), (2,3), (1,4), (5,1), (4,2), (3,3), (2,4), (1,5), (6,1), (5,2), (4,3), (3,4), (2,5), (1,6)\}$

$G = \{(1,6), (2,5), (3,4), (4,3), (5,2), (6,1)\}$

(d) $P(E) = \dfrac{4}{36} = \dfrac{1}{9}$

$P(F) = \dfrac{21}{36} = \dfrac{7}{12}$

$P(G) = \dfrac{6}{36} = \dfrac{1}{6}$

3. (a) ${}_{20}P_5 = 1{,}860{,}480$

(b) $\dfrac{{}_7C_2 \cdot {}_8C_2 \cdot {}_5C_1 \cdot 5!}{{}_{20}P_5} = \dfrac{2{,}940}{15{,}504} = 0.1896$

(c) $\dfrac{{}_5C_5 \cdot 5!}{{}_{20}P_5} = \dfrac{1}{15{,}504} = 0.0000644$

5. (a) $\dfrac{{}_4C_2 \cdot {}_{48}C_1}{{}_{52}C_3} = \dfrac{288}{22{,}100} = 0.013$

(b) $\dfrac{{}_4C_0 \cdot {}_{48}C_3}{{}_{52}C_3} + \dfrac{{}_4C_1 \cdot {}_{48}C_2}{{}_{52}C_3} + \dfrac{{}_4C_2 \cdot {}_{48}C_1}{{}_{52}C_3}$

$= \dfrac{17{,}296 + 4{,}512 + 288}{22{,}100} = \dfrac{22{,}096}{22{,}100} = 0.999$

or $1 - \dfrac{{}_4C_3}{{}_{52}C_3} = \dfrac{22{,}096}{22{,}100} = 0.999$

(c) $\dfrac{{}_{13}C_3}{{}_{52}C_3} = \dfrac{1{,}716}{132{,}600} = 0.013$

(d) $\dfrac{{}_{13}C_1 \cdot {}_{13}C_1 \cdot {}_{13}C_1}{{}_{52}C_3} = \dfrac{2{,}197}{22{,}100} = 0.091$

7. (a) $p(e_5) = 0.1$
(b) $P(E) = 0.7$
(c) $P(F) = 0.5$
(d) $P(E \cup F) = 1$
(e) $P(E \cap F) = 0.2$
(f) $P(E') = 0.3$
(g) $P(F') = 0.5$

Problem Set 4.2

1. (a) $1 - {}_6C_6(9/10)^6(1/10)^0 = 1 - (9/10)^6 = 0.469$
(b) ${}_6C_1(9/10)(1/10)^5 = 54/10^6 = 0.000054$

3. $P(E \cup F) = 0.7$

Problem Set 4.3

1. (a) $P(E/F) = \dfrac{0.1}{0.2} = \dfrac{1}{2}$

(b) $P(F/E) = \frac{0.1}{0.5} = \frac{1}{5}$

(c) Yes (E and F are independent).

3. (a) $P(K/L) = 1/2$
(b) $P(L/K) = 1/2$
(c) No (L and K are not independent).

5. (a)

$$\frac{(0.25)(0.001)}{(0.1)(0.01) + (0.2)(0.01) + (0.15)(0.02) + (0.25)(0.001) + (0.30)(0.02)}$$

$$= \frac{0.00025}{0.01225} = 0.0204081 \approx 0.02$$

(b) $\frac{(0.15)(0.02)}{0.01225} = \frac{0.003}{0.01225} \approx 0.245$

7. (a) $P(E \cap F) = 0.1$
(b) $P(E) = 0.25$

Problem Set 4.4

1. ${}_4C_4(0.8)^4(0.2)^0 = (0.8)^4 = 0.4096$
3. ${}_3C_2(0.99)^2(0.01) + {}_3C_3(0.99)^3(0.01)^0$
$= 0.029403 + 0.970299 = 0.999702$
5. $(0.001)^3 = 0.000000001$ or 10^{-9}

Problem Set 4.5 (using Problem Set 4.4)

1.

```
150 DATA 4,4,.8,0
RUN

          10:31    06/11/75


#TRIALS          #SUCCESSES          PR. OF SUCC.          BINOMIAL PROB.

 4                4                   .8                    .4096
```

```
#TRIALS             #SUCCESSES          PR. OF SUCC.        BINOMIAL PROB.

LINE    35:   END OF DATA

TIME 0.0 SECS.
```

3.

```
150 DATA 3,2,.99,1
151 DATA 3,3,.99,0
RUN

             10:32   06/11/75


#TRIALS             #SUCCESSES          PR. OF SUCC.        BINOMIAL PROB.

 3                   2                   .99  PARTIAL SUM= 2.94030E-02
 3                   3                   .99                  .999702


#TRIALS             #SUCCESSES          PR. OF SUCC.        BINOMIAL PROB.

LINE    35:   END OF DATA

TIME 0.0 SECS.
```

5.

```
150 DATA 3,3,.001,0
RUN

             10:34   06/11/75


#TRIALS             #SUCCESSES          PR. OF SUCC.        BINOMIAL PROB.

 3                   3                   1.00000E-03          9.99999E-10


#TRIALS             #SUCCESSES          PR. OF SUCC.        BINOMIAL PROB.

LINE    35:   END OF DATA

TIME 0.0 SECS.
```

Problem Set 5.1

1. (a) Yes, it is a binomial experiment.

(b) $\{(R,R),(R,N),(N,R),(N,N)\}$
where R = person's back pain which was relieved, n = person's back pain which was not relieved.

(c)

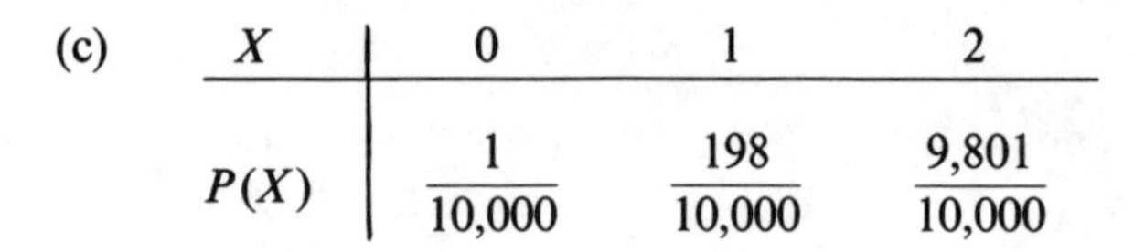

X	0	1	2
$P(X)$	$\frac{1}{10{,}000}$	$\frac{198}{10{,}000}$	$\frac{9{,}801}{10{,}000}$

(d)

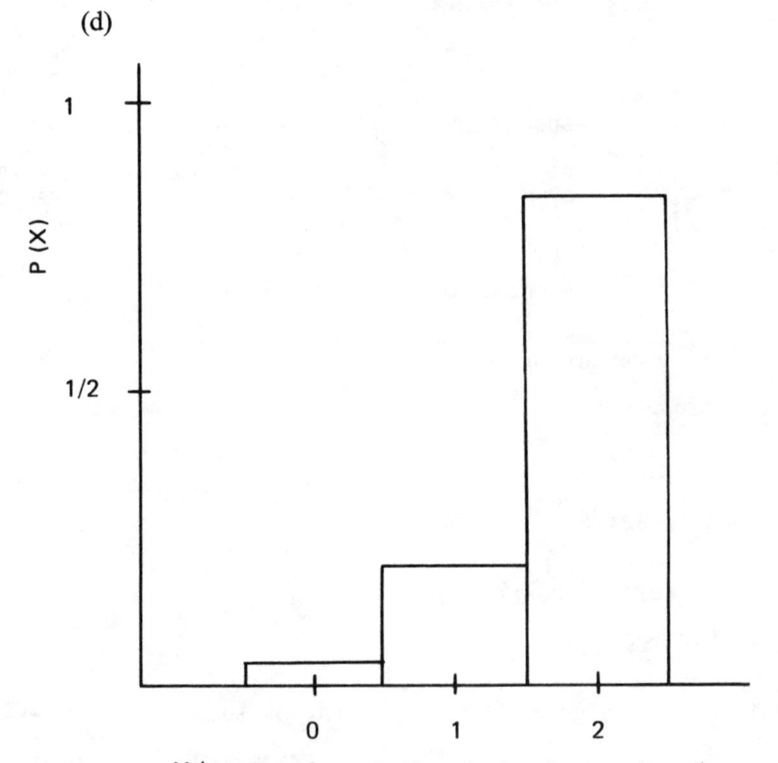

3. (a) Yes, it is a binomial experiment.

(b) {(F,F,F,),(F,F,M),(F,M,F),(M,F,F,),
(M,M,F),(F,M,M),(M,F,M),(M,M,M)}

(c)

X	0	1	2	3
$P(X)$	$\frac{1}{8}$	$\frac{3}{8}$	$\frac{3}{8}$	$\frac{1}{8}$

(d)

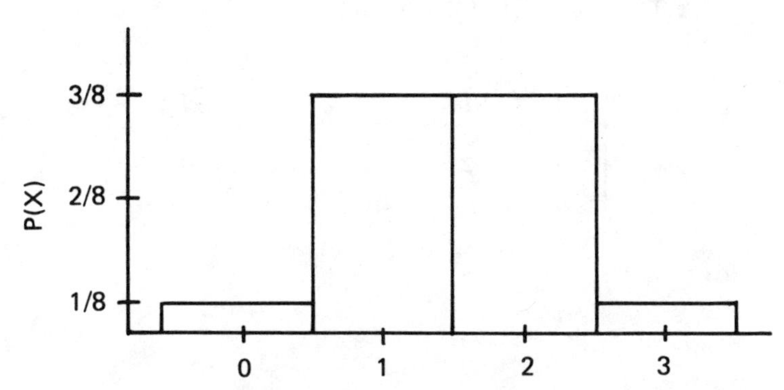

5. (a) Yes, it is a binomial experiment.
(b) }(F,F,F),(F,F,N),(F,N,F),(F,N,N),
(N,F,F),(N,F,N),(N,N,F),(N,N,N)}
where N denotes not obtaining a "five" digit; F denotes obtaining a "five" digit.

(c)

X	0	1	2	3
$P(X)$	$\frac{729}{1000}$	$\frac{243}{1000}$	$\frac{27}{1000}$	$\frac{1}{1000}$

(d)

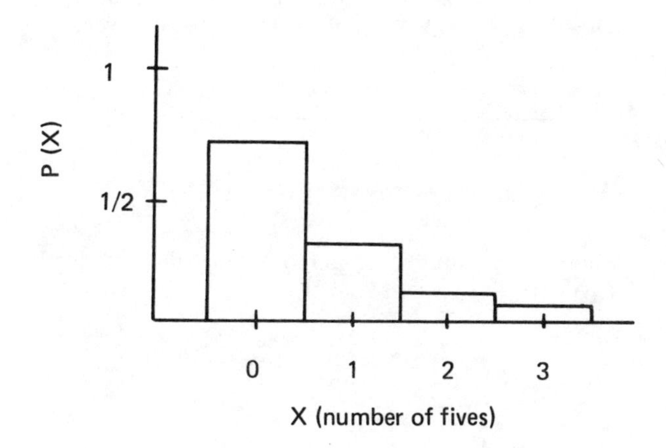

7. (a) Yes, it is a binomial experiment.
(b) }(H,H,H,H),(H,H,H,T),(H,H,T,H),(H,H,T,T),
(H,T,H,H),(H,T,H,T),(H,T,T,H),(H,T,T,T)
(T,H,H,H),(T,H,H,T),(T,H,T,H),(T,H,T,T),
(T,T,H,H),(T,T,H,T),(T,T,T,H),(T,T,T,T)}

(c)

X	0	1	2	3	4
$P(X)$	$\frac{1}{16}$	$\frac{4}{16}$	$\frac{6}{16}$	$\frac{4}{16}$	$\frac{1}{16}$

(d)

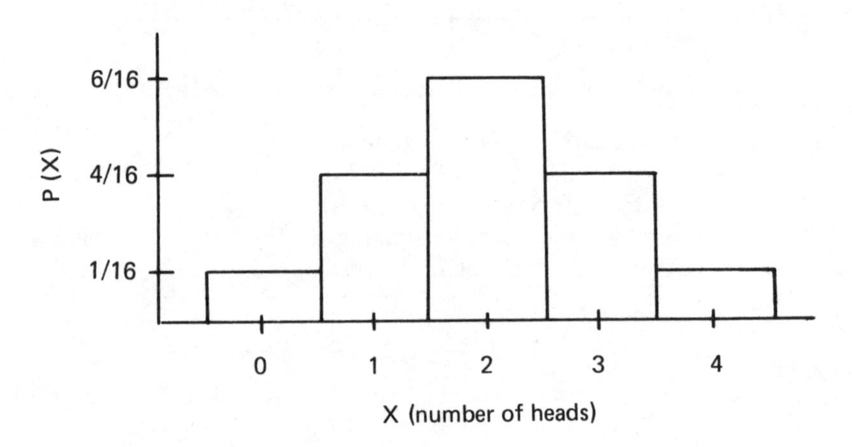

9. (a) Yes, it is a binomial experiment.
(b) {(D,D),(D,N),(N,D),(N,N)}
where D = those who drink Smooth Taste Beer; N = those who do not drink Smooth Taste Beer

(c)

X	0	1	2
$f(X)$	$\frac{4}{25}$	$\frac{12}{25}$	$\frac{9}{25}$

where X = number of persons who do not drink Smooth Taste Beer

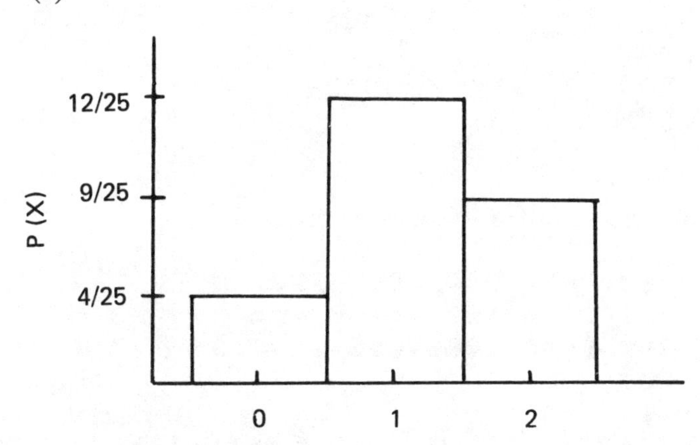

Problem Set 5.2 (using Problem Set 5.1)

1. (a) Yes; (b) $\mu = 198/100$; (c) $\sigma^2 = 198/10{,}000$; (d) $\sigma = \sqrt{0.0198} = 0.1407$
3. (a) Yes; (b) $\mu = 3/2$; (c) $\sigma^2 = 3/4$; (d) $\sigma = \sqrt{0.75} = 0.866$
5. (a) Yes; (b) $\mu = 3/10$; (c) $\sigma^2 = 27/100$; (d) $\sigma = \sqrt{0.27} = 0.5196$
7. (a) Yes; (b) $\mu = 2$; (c) $\sigma^2 = 1$; (d) $\sigma = 1$
9. (a) Yes; (b) $\mu = 6/5$; (c) $\sigma^2 = 12/25$; (d) $\sigma = \sqrt{12/25} = 0.69$

Problem Set 5.3

1. (a) 0.1056
(b) 0.0179
(c) 0.8944
(d) 0.9821
(e) 0.4802
(f) 0.0919
(g) 0.1944
(h) 0.5000
(i) 0.9974
(j) 0.0013
3. (a) 340 (to nearest integer)
(b) 330.6 (331 to nearest integer)
5. (a) 0.1660
(b) 0.4082

Problem Set 5.4

1. 0.002 ($Z = -3.53$)
3. 0.3083 ($Z_1 \approx 0.50$, $Z_2 \approx 3.52$)
5. 0.9997 ($Z \approx 3.50$)

Problem Set 5.5 (using Problem Set 5.4)

```
1. 150 DATA 200,.2,20,20,1
3. 151 DATA 100,.01,2,4,6
5. 152 DATA 1000,.001,4,4.3
   RUN

   STAT5          10:46   06/11/75

1. FIND THE AREA TO THE LEFT OF Z=-3.53553
3. FIND AREA BETWEEN Z= .502519 AND Z= 3.51763
5. FIND THE AREA TO THE LEFT OF Z= 3.50175

   LINE      5:    END OF DATA

   TIME 0.0 SECS.
```

Problem Set 5.6

```
150 DATA 27,64,76,62,60,65,65,74,70,55,61,78,87,51,30,
         64,80,84
151 DATA 73,46,58,37,48,91,70,70,33,65,70,10
152 DATA 21,90,67,85,67,65,75,82,78,84,70,98,74,48,90,
         61,78,95
153 DATA 70,66,100,70,70,1P-0
RUN

            11:00    06/11/75

X                   Z                   N
 64                  2.67082E-02         70.2671
 76                  .813382             78.1338
 62                 -.104404             68.9559
 60                 -.235516             67.6448
 65                  9.22644E-02         70.9226
 65                  9.22644E-02         70.9226
 74                  .68227              76.8227
 70                  .420045             74.2004
```

```
 55          -.563297          64.367
 61          -.16996           68.3004
 78           .944494          79.4449
 87           1.5345           85.345
 51          -.825522          61.7448
 30          -2.2022           47.978
 64           2.67082E-02      70.2671
 80           1.07561          80.7561
 84           1.33783          83.3783
 73           .616714          76.1671
 46          -1.1533           58.467
 58          -.366629          66.3337
 37          -1.74331          52.5669
 48          -1.02219          59.7781
 91           1.79672          87.9672
 70           .420045          74.2004
 70           .420045          74.2004
 33          -2.00553          49.9447
 65           9.22644E-02      70.9226

X             Z                N
 90           1.02272          80.2272
 67          -.760575          62.3942
 85           .635045          76.3504
 67          -.760575          62.3942
 65          -.915644          60.8436
 75          -.140299          68.597
 82           .402442          74.0244
 78           9.23041E-02      70.923
 84           .557511          75.5751
 70          -.527972          64.7203
 98           1.64299          86.4299
 74          -.217834          67.8217
 48          -2.23373          47.6627
 90           1.02272          80.2272
 61          -1.22578          57.7422
 78           9.23041E-02      70.923
 95           1.41039          84.1039
 70          -.527972          64.7203
 66          -.83811           61.6189
 100          1.79806          87.9806
 70          -.527972          64.7203

X             Z                N

LINE    10:   END OF DATA

TIME 0.0 SECS.
```

Problem Set 6.1

1. (a) 1.645
   (b) 3.27
3. Approximately 2.18 to 2.28
5. 8.086 to 39.5136
7. (a) 68.04 to 71.96
   (b) 67.42 to 72.58
   (c) 68.355 to 71.645
   (d) 68.52 to 71.48
   (e) 67.63 to 72.37
   (f) 66.92 to 73.08

Problem Set 6.2 (using Problem Set 6.1)

```
3. 150 DATA 2.23,.0277,1
5. 151 DATA 23.8,.61,2
   RUN

                    11:11   06/11/75


3. THE 95 PERCENT CONFIDENCE INTERVAL
   WITH M= 2.23    AND SD= 2.77000E-02   IS:
    2.17571     TO 2.28429     INCLUSIVE.

5. THE 99 PERCENT CONFIDENCE INTERVAL
   WITH M= 23.8    AND SD= 6.1  IS:
    8.0864      TO 39.5136     INCLUSIVE.


   LINE    10:   END OF DATA

   TIME 0.0 SECS.
```

Problem Set 6.3

```
1. 150 DATA 420,400,100,125
3. 151 DATA 2500,2600,625,28
   RUN
```

```
11:17   Ø6/11/75

1. OBSERVED Z= 2.236Ø7
3. OBSERVED T=-.831385

   LINE     5:   END OF DATA

   TIME Ø.Ø SECS.
```

Conclusions

1. (a) Observed $Z >$ critical Z

$2.236 > 1.645$

Therefore the difference is significant at the 0.05 level of significance. Students in sample perform better than those in general population.

(b) Observed $Z \not>$ critical Z

$2.236 \not> 2.33$

Therefore the difference is not significant at the 0.01 level of significance. Students in sample do not perform better.

3. Observed $t \not<$ critical t

$-0.8463 \not< -1.703$

Retain null hypothesis. There is no significant difference in population mean pounds of garbage collected daily and sample mean.

Problem Set 6.4 (using Problem Set 6.3)

2. $t_{\text{obs}} \not<$ critical $t_{.01}$

$-1 \not< -2.492$

Retain null hypothesis. There is no significant difference in population mean age and sample mean age of such women.

3. $t_{\text{obs}} \not<$ critical $t_{.01}$

$-0.8463 \not< -2.473$

Retain null hypothesis. There is no significant difference in population mean pounds of garbage collected daily and sample mean.

Problem Set 6.5

```
1.     15Ø DATA 1Ø3.67,1Ø2.17,98.89,84.47,6,6
5. (a) 151 DATA 2.Ø3,2.Ø32,.ØØ1,.Ø15,35,4Ø
   (b) 152 DATA 2.Ø38,2.Ø32,.Ø3,.Ø15,35,4Ø
   (c) 153 DATA 2.Ø3,2.Ø32,.ØØ1,.Ø15,35,4Ø
   (d) 154 DATA 2.Ø3,2.Ø38,.ØØ1,.Ø3,35,35
       RUN
```

```
11:27    06/11/75

1.      OBSERVED T= .247699
5. (a) OBSERVED Z=-9.95494E-02
   (b) OBSERVED Z= .170946
   (c) OBSERVED Z=-9.95494E-02
   (d) OBSERVED Z=-.268821

   LINE      5:   END OF DATA

   TIME 0.0 SECS.
```

Conclusions

1. H_m: $\overline{D} > \overline{B}$; H_o: $\overline{D} \not> \overline{B}$

   Observed $t \not>$ critical $t_{.05}$ (where df = 6 +6− 2 = 10)
   0.247699 $\not>$ 1.860

   Therefore, we cannot reject H_o at the 0.05 level of significance.

5. (a) Observed $Z \not<$ critical $Z_{.05}$
   −0.09955 $\not<$ −1.645

   Therefore, we cannot reject H_o at the 0.05 level of significance.

   (b) Observed $Z \not>$ critical $Z_{.05}$
   0.170946 $\not>$ 1.645

   Therefore, we cannot reject H_o at the 0.05 level of significance.

   (c) Critical $Z_{.05}$ < observed Z < critical $Z_{.05}$
   −1.96 < −0.09955 < 1.96

   Therefore, we can reject H_o at the 0.05 level of significance.

   (d) −Critical Z < observed Z < critical Z
   −1.96 < 0.268821 < 1.96

   Therefore, we cannot reject H_o at the 0.05 level of significance.

Problem Set 6.6

1. $n = 400$, $X = 5$, $p = 0.01$
   (a) H_m: $p \leq 0.01$
   H_o: $p > 0.01$

(b) One-tail test
(c) Z-score
(d) Critical $Z = 1.645$
(e) Observed $Z = 0.50$
(f) $0.50 \not> 1.645$. Therefore, we can reject H_o at the 0.05 level of significance.

3. $n = 320$, $X = 235$, $p = 0.75$
(a) H_m: $p = 0.75$
H_o: $p \neq 0.75$
(b) Two-tail test
(c) Z-score
(d) Critical Z-scores are ± 1.96
(e) Observed $Z \approx -0.65$. Therefore, we cannot reject H_o at the 0.05 level of significance.

5. $n = 300$, $X = 25$, $p = 0.1$
(a) H_m: $p < 0.1$
H_o: $p \geq 0.1$
(b) One-tail test
(c) Z-score
(d) Critical $Z = -2.326$
(e) Observed $Z \approx -0.96$. Therefore, we cannot reject H_o at the 0.01 level of significance.

Problem Set 7.1

1. Observed $\chi^2 >$ critical χ^2
$19.8 > 9.21$

Therefore, these differences are significant at the 0.01 level of significance. The residents probably do not have the same church-attendance habits.

3. Observed $\chi^2 >$ critical χ^2
$9.08 > 7.81$

Therefore, the differences are significant at the 0.05 level of significance. The random sampling of housewives probably does not support the company's claim.

5. Observed $\chi^2 >$ critical χ^2
$50 > 3.841$

Therefore, there is a significant difference in favor of tea brand A.

Problem Set 7.2 (using Problem Set 7.1)

```
1. 15Ø DATA 3,Ø,2ØØ,15Ø,6ØØ,3ØØ,3ØØ,2ØØ
3. 151 DATA 4,.333333,5Ø,2Ø,3Ø,8,1ØØ,2Ø,5Ø,1Ø
5. 152 DATA 3,Ø,1ØØ,35,1ØØ,5Ø,1ØØ,65
   RUN

                    11:35    Ø6/11/75

1. OBSERVED CHI-SQUARE VALUE = 19.8718
3. OBSERVED CHI-SQUARE VALUE = 9.Ø6663
5. OBSERVED CHI-SQUARE VALUE = 9

   LINE      5:    END OF DATA

   TIME Ø.Ø SECS.
```

Problem Set 7.3

1. (a) $y_p = 2.9 + 8.9x$
   (b) If $x = 3\frac{1}{2}$, $y_p = 34.05$
   If $x = 8$, $y_p = 74.1$
3. (a) $y_p = 69.8 - 0.52x$
   (b) If $x = 15$, $y_p = 62.05$
   If $x = 25$, $y_p = 56.8$
   If $x = 0$, $y_p = 69.8$
5. (a) $y_p = 250{,}061 - 0.112x$
   (b) If $x = 800{,}000$, $y_p = 160{,}113$
   If $x = 1{,}000{,}000$, $y_p = 137{,}626$
   If $x = 450{,}000$, $y_p = 199{,}465$

Problem Set 7.4 (using Problem Set 7.3)

```
1. 15Ø DATA 7,1,1Ø,2,25,3,22,4,35,5,6Ø,6,55,9,8Ø
3. 151 DATA 7,1Ø,65,2Ø,58,3Ø,5Ø,4Ø,51,5Ø,45,5,7Ø,8,65
5. 152 DATA 5,5ØØØØ,1ØØØØØ,2ØØØØØ,1ØØØØØ,7ØØØØ,3ØØØØ
```

```
153 DATA 500000,-100000,250000,1000000
RUN

STAT11          10:00    11/11/76
```

1. `Y= 2.90789      + 8.88816     X`

3. `Y= 69.8858      -.522702      X`

5. `Y= 250061.      -.112436      X`

```
LINE       5:    END OF DATA

TIME 0.0 SECS.
```

Problem Set 7.5 (using Problem Set 7.3)

1. `150 DATA 7,1,10,2,25,3,22,4,35,5,60,6,55,9,80`
3. `151 DATA 7,10,65,20,58,30,50,40,51,50,45,5,70,8,65`
5. `152 DATA 5,50000,100000,200000,100000,70000,30000`

```
153 DATA 500000,-100000,250000,1000000
RUN

STAT12          10:04    11/11/76
```

1. `COEFF. OF CO RRE LATION R= .963982`
3. `COEFF. OF CORRELATION R=-.967537`
5. `COEFF. OF CORRELATION R=-4.61968E-02`

```
LINE       5:    END OF DATA

TIME 0.0 SECS.
```

Problem Set 7.6

1. (a) Observed $F_{2,12} \not> $ critical $F_{2,12}$

   $0.88199 \not> 3.89$

   Therefore, the differences are not significant at the 0.05 level of significance.

   (b) The differences are not significant at the 0.01 level of significance.

3. Observed $F_{2,15} \not> $ critical $F_{2,15}$
   $.8479 \not> 3.68$

   Therefore, the differences are not significant at the 0.05 level of significance.

5. Observed $F_{3,8} \not> $ critical $F_{3,8}$
   $2.5323 \not> 7.59$

   Therefore, the differences are not significant at the 0.01 level of significance.

Problem Set 7.7 (using Problem Set 7.6)

```
1. 180 DATA 3,5,5,5,2,7,8,3,5,2,1,5,3,3,6,3,2,1,10
3. 181 DATA 3,6,6,6,22,20,30,35,20,25,30,30,21,35,22
   182 DATA 30,25,25,30,30,32,32
5. 183 DATA 4,3,3,3,3,226,218,252,251,237,237,208,
            191,222,247,218,209
   RUN

               12:30    06/11/75

1. OBSERVED F= .881817
   DF. FOR NUMERATOR = 2
   DF. FOR DENOMINATOR= 12
3. OBSERVED F= .847916
   DF. FOR NUMERATOR = 2
   DF. FOR DENOMINATOR= 15
5. OBSERVED F= 2.53282
   DF. FOR NUMERATOR = 3
   DF. FOR DENOMINATOR= 8

   LINE     5:    END OF DATA

   TIME 0.0 SECS.
```

Index